Soil Contamination - Threats and Sustainable Solutions

Edited by Marcelo L. Larramendy and Sonia Soloneski

Published in London, United Kingdom

IntechOpen

Supporting open minds since 2005

Soil Contamination - Threats and Sustainable Solutions
http://dx.doi.org/10.5772/intechopen.87652
Edited by Marcelo L. Larramendy and Sonia Soloneski

Contributors
Amarnath Mishra, Soni Kumari, My Dung Jusselme, Clarisse Balland-Bolou-Bi, Florence Brondeau, Noureddine Bousserrhine, Samir Abbad Andaloussi, Rim Werheni, Abdennaceur Hassen, Yassine Hidri, Govind Kumar, Pankaj Bhatt, Shatrohan Lal, Samuel Tetsopgang, Aurora Santos López, Carmen Dominguez Torre, David Lorenzo Fernández, Lucas Benedet, Guilherme Wilbert Ferreira, Gustavo Brunetto, Arcângelo Loss, Paulo Emílio Lovato, Cledimar Rogério Lourenzi, Sérgio Henrique Godinho Silva, Nilton Curi, Jucinei José Comin, Nikita Bisht, Puneet Singh Chauhan, Sonia Sethi, Payal Gupta, Dotun Nathaniel Arije, Adeyela Ibironke Okunlola, Katherine Olayinka Olajugbagbe, Menaka Salam, Namdol Nilza, Murthy Chavali, Oluwatosin Ogunsola, Odunayo Adeniyi, Victoria Adedokun, Hanuman Singh Jatav, Asik Dutta, Abhik Patra, Surendra Singh Jatav, Sudhanshu Verma, Pavan Singh, Eetela Sathyanarayana, Satish Singh, Sukirtee Chejara, Paras Kamboj, Y.V. Singh, Vikas Tandon, Silpi Sarkar, Manoj Kumar Enamala, G.V.S. Subbaroy Sarma, Mannam Krishna Murthy, Abudukeremu Kadier, Ashokkumar Veeramuthu, K. Chandrasekhar

Notice
Statements and opinions expressed in the chapters are these of the individual contributors and not necessarily those of the editors or publisher. No responsibility is accepted for the accuracy of information contained in the published chapters. The publisher assumes no responsibility for any damage or injury to persons or property arising out of the use of any materials, instructions, methods or ideas contained in the book.

First published in London, United Kingdom, 2021 by IntechOpen
IntechOpen is the global imprint of INTECHOPEN LIMITED, registered in England and Wales, registration number: 11086078, 5 Princes Gate Court, London, SW7 2QJ, United Kingdom
Printed in Croatia

British Library Cataloguing-in-Publication Data
A catalogue record for this book is available from the British Library

Additional hard and PDF copies can be obtained from orders@intechopen.com

Soil Contamination - Threats and Sustainable Solutions
Edited by Marcelo L. Larramendy and Sonia Soloneski
p. cm.
Print ISBN 978-1-83880-753-5
Online ISBN 978-1-83880-754-2
eBook (PDF) ISBN 978-1-83880-758-0

For EU product safety concerns:
IN TECH d.o.o., Prolaz Marije Krucifikse Kozulić 3, 51000 Rijeka, Croatia, info@intechopen.com or visit our website at intechopen.com.

Meet the editors

Marcelo L. Larramendy, Ph.D., serves as Professor of Molecular Cell Biology at the School of Natural Sciences and Museum, National University of La Plata, Argentina. He is appointed as the Senior Researcher of the National Scientific and Technological Research Council of Argentina, and he is a former member of the Executive Committee of the Latin American Association of Environmental Mutagenesis, Teratogenesis and Carcinogenesis. He is the author of more than 450 contributions, including scientific publications, research communications, and conferences worldwide, and he is a recipient of several national and international awards. Professor Larramendy is a regular lecturer at the International A. Hollaender Courses organized by the IAEMS, and he is a former guest scientist at NIH (USA) and at the University of Helsinki (Finland). He is an expert in genetic toxicology and is, or has been, a referee for more than 20 international scientific journals. In 2015, he was a member of the International Panel of Experts at the International Agency for Research on Cancer (IARC, WHO, Lyon, France) for the evaluation of DDT, 2,4-D, and lindane. Presently, Dr. Larramendy is Head of the Laboratory of Molecular Cytogenetics and Genotoxicology at the UNLP.

Sonia Soloneski has a Ph.D. in Natural Sciences and is an Assistant Professor of Molecular Cell Biology at the School of Natural Sciences and Museum of La Plata, National University of La Plata, Argentina. She is a member of the National Scientific and Technological Research Council (CONICET) of Argentina in the Genetic Toxicology field, the Latin American Association of Environmental Mutagenesis, Teratogenesis and Carcinogenesis (ALAMCTA), the Argentinean Society of Toxicology (ATA), the Argentinean Society of Biology (SAB), and the Society of Environmental Toxicology and Chemistry (SETAC). She has authored more than 380 contributions in the field, including scientific publications in peer-reviewed journals and research communications. She has served as a review member for more than 30 scientific international journals. She has been a plenary speaker in scientific conferences and a member of scientific committees. She is a specialist in issues related to genetic toxicology, mutagenesis, and ecotoxicology.

Contents

Preface

Soil is a living, active and nonrenewable reserve and a crucial ecosystem component. Soil presents absorbing and emitting capabilities and is susceptible to contamination by a multiplicity of exogenous and endogenous sources. The study of soil resources and environmental problems is a broad, fascinating field that can take numerous dissimilar directions. Today, many anthropogenic ,as well as geogenic activities, are pressing in soil health, generating a strong ecological concern that requires an interdisciplinary approach involving both regulatory organizations and institutions for promptly resolving this situation. Soil contamination is very complex, and it is also often unsafe and harmful to all living species, including human beings specifically. It most frequently occurs from urban development, agricultural practices, military activities, mine tailings, metal industries, industrial accidents, deposits, or the transport of hazardous chemicals, among several other sources. Like pesticides, chlorinated compounds, and nitrogen, certain trace elements such as arsenic, lead, cadmium, copper, mercury, silver, or nickel pollutants can both be naturally present in the soil and be the consequence of human-made activities that nature cannot—or can only very slowly—decompose or degrade. These interventions can fragment, change, or even destroy soil habitats, significantly modifying the biota that lives in the soil. Whereas soil has a marked self-purification capacity, cleaning up contaminated sites is a long and expensive process. Nonetheless, environmental remediation is an important focus of the green economy, and a wide variety of conventional and emerging technologies must rapidly be employed to remove contaminants from polluted sites in order to restore the soil environment and protect the health of living species, including humans. Finally, ensuring long-term management as a final step is mandatory to evaluate the effectiveness of remedial strategies.

This single volume comprises fifteen high-quality chapters, organized into two sections, describing several issues related to soil contamination. The first section, *Contamination Sources*, comprises seven excellent and detailed chapters, starting with an update in the first chapter about the toxicity of heavy metals in several matrices including soil, water, air, and living organisms, presenting this problem as the main cause of environmental deterioration. This is followed by a second chapter providing information about the advantages and disadvantages of using microbial indices for heavy metal–contaminated and restored soils, highlighting the importance of understanding the mechanism(s) of responding to heavy metal stress and the methods available for the microbial diagnosis of heavy metal–contaminated soils. The third chapter presents an overview of the effect of pentachlorophenol pesticide contamination on microbial diversity, enzymatic activities, microbial biomass, and physicochemical soil characteristics, including a description of a bioremediation process. The fourth chapter summarizes the various sources of cadmium present in the environment and its toxic effects on plants and humans. It also includes a description of some bioremediation approaches to mitigate cadmium pollution in the environment. The fifth chapter provides information about how the prolonged use of high doses of fertilizers of animal origin, such as swine manure, can result in the accumulation of metals and phosphorus in the soil. These accumulations potentially contaminate both the soil and surface water resources, mainly due to losses from runoff and subsurface leaching. The sixth chapter depicts one of the potential sources of soil contamination in agricultural production worldwide: the application of chemical fertilizers employed in excessive and disproportionate quantities. The chapter also summarizes the evidence regarding the employment of biofertilizers as an eco-friendly alternative for improving soil quality. Lastly, the seventh chapter of this section provides information about the application of fine materials from different rock types, such as basalt, trachyte, and volcanic pyroclastic fragments, in addition to sedimentary rocks like limestone and gneiss, as potential fertilizers on tropical soils from Cameroon. The chapter focuses on strategies for soil acidity management and the employment of this type of material as a potential source of phosphorus.

The second section of this book, *Remediation Technologies,* emphasizes integrated remediation approaches for detecting potentially biohazardous contaminants. The eighth chapter evaluates the phytoremediation potential of two ornamental plants, namely *Basella alba* and *Codiaeum variegatum*, in heavy metal–contaminated soils collected from several sites in Nigeria. The study reveals the ability of both plants to remove heavy metals, the heaviest concentration found to be accumulated in the roots rather than the shoots. The ninth chapter reviews the particularities of hexachlorocyclohexane (HCH)-contaminated soils and critically examines the bases and results of the technologies applied, paying special attention to physicochemical remediation processes. The tenth chapter aims to highlight the huge potential of several types of microorganisms being used as an attractive way of removing or remediating pollutants in landfill leachates, emphasizing their cost-effective and environmentally friendly benefits. The eleventh chapter discusses how to take phytoremediation approaches from a proven technology to accepted practice in an urban context. Additionally, it presents an overview of urban soil types following the application of phytoremediation to urban soils, focusing on inorganic and organic pollutants, to provide a frame of reference for the subsequent discussion on the better utilization of phytoremediation. The twelfth chapter aims to highlight the importance of the synergistic association between plants and microbes for the remediation of petroleum hydrocarbons as an effective tool for reclaiming the soil and the environment. The thirteenth chapter offers an interesting overview of the environmental role of nanophytoremediation in the elimination of the bioaccumulation of toxic nanoparticles. This innovative and encouraging technology has gained greater attention due to its current use in research on plants. This chapter describes several plant families that act in the biosynthesis of nanoparticles as well as the physiological process of nanophytoremediation. The fourteenth chapter provides an interesting overview of the role of soil management and conservation in enhancing microbial activity for soil ecological intensification as well as in buffering the soil to neutralize contaminants. The final, fifteenth chapter highlights the employment of green technologies like phytoremediation, biostimulation, and biodegradation for soil sustainable remediation, including information about nanotechnology in the degradation of contaminants.

The editors of ***Soil Contamination - Threats and Sustainable Solutions*** are enormously grateful to all the contributing scientists for sharing their knowledge and insights in this interdisciplinary book project. They have made an extensive effort to arrange the information included in every chapter. The publication of this book is of high importance for researchers, scientists, and engineers with expertise in diverse fields of soil science, health, toxicology, policy making, and other disciplines, who can contribute and share their findings in order to take this area of study forward for future investigations.

Sonia Soloneski Ph.D. and Marcelo L. Larramendy Ph.D.
School of Natural Sciences and Museum,
National University of La Plata,
La Plata, Argentina

Section 1

Contamination Sources

Chapter 1

Heavy Metal Contamination

Soni Kumari and Amarnath Mishra

Abstract

In the era of industrialization, pollution has totally deteriorated the quality and diversity of life. Heavy metal contaminations are the major causes of environment deteriorations. The basic reasons are natural as well as anthropogenic. Chief sources of heavy metal contamination are air pollution, river sediments, sewage sludge, town waste composts, agricultural chemicals like fertilizers and pesticides, and industrial waste like factories releasing chemicals, anthropogenic activities, etc. Agricultural soils in many parts of the world are generally contaminated by heavy metal toxicity such as Cd, Cu, Zn, Ni, Co, Pb, Hg, As, etc. These are due to the long-term use of phosphate fertilizers, sewage sludge, dust from smelters, industrial waste, etc. Heavy metals in soils are detected with some specific instruments like atomic absorption spectroscopy, inductively coupled plasma, inductively coupled plasma-mass spectroscopy, and X-ray fluorescence and spectroscopy. Among all these instruments, atomic absorption spectroscopy (AAS) is the best because it gives the precise quantitative determination. AAS is a method applied for measuring the quantity of the trace elements present in the soil or any other samples.

Keywords: environmental forensics, heavy metals, soil, water, air, public health

1. Introduction

Heavy metals are principally referred as the metals which have high densities and atomic weights or atomic numbers.

Or

In metallurgy, heavy metals are defined on the basis of densities, and in Physics, it is defined on the basis of atomic number while in chemistry on the basis of chemical behaviour. In biochemistry, heavy metals are defined whether it is Lewis's acid (electron pair acceptor) performance of the ions in the aqueous solutions [1].

Some examples of heavy metals are iron (Fe), copper (Cu), tin (Sn), and other elements metals such as silver (Ag), gold (Au), and platinum which are precious; these were the earliest known metals. Other heavy metals are moreover essential nutrient, for example, iron, cobalt, and zinc some comparatively less dangerous such as ruthenium, silver, and indium; but can be toxic in the larger amounts or in their definite form [2]. Then a number of the certain heavy metals which are such as cadmium, mercury, and lead are highly poisonous.

Heavy metals contamination can be observed in soil, water (rivers), air (atmosphere), etc. The composition and physical properties and chemical structure of each of the heavy metals are different. No doubt that the heavy metals are toxic as well as hazardous. They have harmful effects. Heavy metals are needed to be handled with care. Some heavy metals tend to be less reactive, whereas some are very reactive. These are thought to be toxic or highly destructive to the environment. Heavy metals mortify air, water, and soil quality and consequently cause health issues in plants, animals, people, etc. Some toxic elements are also regarded as beneficial, but only in small quantities for human health. These elements are vanadium, manganese, cupper, iron, zinc, strontium, selenium, molybdenum etc. The deficiency of these metals is harmful and may increase susceptibility to heavy metal poisoning. Chronic level ingestion of heavy metals toxic or heavy metals has dangerous effects on human body, and the impacts are observable only after several years of exposure [2].

According to their toxicity to living organisms, the heavy metals are arranged in the following order: Hg > Cu > Zn > Ni > Pb > Cd > Cr > Sn > Fe > Mn > Al [3].

1.1 Properties of heavy metal

- Heavy metals are less reactive than the other lighter metals.
- Heavy metals are relatively scarce in the Earth crust, means concentration of heavy metals are less.
- Heavy metals have usually higher density than the light metals.
- These metals are quite hard whereas soft metals are easy and tend to cut easily.
- Thermal expansively of heavy metals is lower than the light metals.
- Melting point of heavy metals varies from low to high whereas light metals have low melting point.
- Tensile strength of heavy metals is mostly higher.
- Heavy metals are less abundant in earth crust.
- These are extremely insoluble in sulphides.
- These are generally insoluble in hydroxides.
- Heavy metals form coloured solutions in water.
- Complexes of heavy metals are mostly coloured.
- Heavy metals are the micronutrients, required in small concentrations whereas light metals are the macronutrients required in larger concentrations.

1.2 Uses of heavy metals

If heavy metals are toxic, they are useful or beneficial also. Its regular uses depend on the general characteristics of heavy metals such as electrical conductivity and reflectivity or some other general characteristics such as its density, its strength, and its durability. Other uses depend on the quality of the particular element, such as their biological role as nutrients or poisons or some other specific atomic properties. Some examples of the atomic properties include partially filled d or f orbital that enables the formation of coloured complexes. Other uses are in sport, mechanical engineering, military ordnance, and nuclear science, which take the advantage of their relatively high density. In mechanical engineering heavy metals are used as ballast in boats, aeroplanes, and other motor vehicles or in balance weights on wheels and gyroscope and propellers and centrifugal clutches, in situation requiring maximum weight in minimum space. In military ordnance, tungsten and uranium are used in armour plating and armour piercing projectiles, as well as in nuclear weapons to increase the efficiency. Also, denser materials absorbs more radioactive emissions than the lighter ones, so heavy metals are useful for radiation shielding and to focus radiation beams in linear accelerators and radiotherapy applications [3].

2. Heavy metals contamination in soil

Soil is an essential constituent of an ecosystem. Soil has the absorbing and emitting capabilities, which is vulnerable to contamination by a wide variety of sources. The continuous development of the urbanisation leads to constructions, agriculture farming, cemetery, and traffic emission. Soil, where we live, where we do our agriculture, is not safe from these metals. Heavy metals reached into the soils as from the parent materials (lithosphere) and from different human activities. There are some factors which affects the presence and division of the these metals in the earth; they are the composition of the parent rocks, the extent of weathering and the chemical, physical and, biological features of the soil, and it also depends on the climatic conditions. Human activities are the major cause and are gradually increasing day by day which results in the deterioration of environment [4].

2.1 Sources of heavy metal

These metals in soil reach through many sources which include atmospheric deposition, improper disposal of the industrial solid and liquid waste, mining activities, sewage, and using of pesticide and fertilisers. These metals in the environment are released mainly from the gas, dirt, and dust produced by the transportation, energy, metallurgy, and during building. Heavy metal basically reaches into the air in the form of aerosol and finally deposit in the soil through natural sedimentation and rainfall process.

Automobile transport causes some serious heavy metal contamination. Also, heavy metals are brought into soil by irrigative sewage, sanitary sewage, chemical waste water, and industrial mining. Industries release a lot of harmful chemical which are directly released to river or an open area. Also, people nowadays use a lot of fertilisers, pesticides, for agricultural purposes. The long-term extreme application has resulted in the degradation of the soil. Heavy metals are the largely reported pollutant inside the fertilisers. Phosphoric fertilisers generally contain high amount of toxic heavy metals, whereas nitrogen and potash fertilisers contain considerably low amount of heavy metals.

An increased concentration of certain elements is found particularly in vegetables grown in the nearby of industrial plants and busy roads, as well as in crops exposed to municipal and agricultural wastewater [4]. This may result in disrupted balance of these elements, and in turn significantly affect their bioavailability and interaction with other food ingredients as well as other metals. Different branches of industry as well as road traffic have a significant impact on the environmental pollution with the heavy metals such as the following chrome from metallurgical, paint, and tanning industries, nickel mainly from the steel industries and burning of coal and liquid fuel, cadmium from metal smelters, while along reads another sources of contamination with this element may be grease used in motor vehicles and lead from paints industries, metallurgical, and glass industries. The main sources of lead contamination in soils, air, and plants are the exhaust gases releasing from [4].

2.2 Impact of heavy metals contamination of soil

Microbial activity and the enzymatic activity gradually decrease because of heavy metals. Microbial activities are inhibited. Different metals in soil have different effects on it. Low concentration of heavy metals indicates high microbial activity and growth and thus high microbial biomass and high quantity of the metals in the soil indicates very low microbial growth and thus results in low microbial biomass. A significant role in process of organic matter decomposition and nutrient cycling are played by the enzymes. So due to excess heavy metals in soil reduces its enzymatic activity. These degrade the quality and quantity of the soil which are not good for fertilisation. The soil loses its fertility and it is completely degraded. So, the degraded soil is of no use; it will only be considered as barren land.

2.3 Impact of heavy metals contamination of plants

Heavy metals in soil mean heavy metals in plants too. The prime reaction of plants is production of reactive oxygen species (ROS) on the exposure to high level of heavy metals. Through soil it reaches the plants and water. In water, heavy metals reach not only through water but from the environment too. The aerosol particles are present in the atmosphere; these reach the water, soil, from soil to plants and water. These all are related, and all are affected by one another. And high concentration of heavy metals in soil means high concentrations in environment and the surrounding which is a global threat. The heavy metals they are not easy to remove but they get incorporated easily. Exposure to heavy metals is toxic is a serious problem the motor vehicles [5].

The risk of heavy metal presence in vegetables cultivated near old industrial areas is significant. Industrial and municipal sewage is also an important source of heavy metals. Also, wastewater irrigation is an important approach for irrigation in agriculture which is treated biologically. Wastewater irrigation has changed physical and chemical properties of the soil and led to the heavy metals' uptake by plants, mostly vegetables as vegetables are grown mostly. And vegetables grown in such soils has high amount of heavy metals accumulated in them. Also, concentration of heavy metals in the soil also depends on amount of fertilisers and pesticides being used in the soil. The higher we use fertilisers and pesticides in the soil for the plants to grow better and to protect them from the insects and the weeds the higher the chemicals will reach the soil and will be absorbed by the plants or vegetable being grown in that particular soil. So, the amount of fertilisers and pesticides doses also influence the amount of heavy metals in the soil. Also, some other sources of heavy metals are used as plant protectors' products. The dynamics of the heavy metals in the soil and their uptake by plants depend on the soil properties, which play a key role in the bioavailability of these metals. The level and amounts of these heavy metals accumulation in plants depends on among others, soil types, pH, humidity, and micronutrients content as well as the time of crop harvesting. Among these, the acidic pH is considered to be the most important factor influencing the increased in absorption of heavy metals by vegetation. Whereas in alkaline soils, a risk of heavy metal leaching and their bioavailability to plants are lower and the presence of organic matter can inhibit metals uptake from the soil solution. By changing these following soil properties that determine metal solubility in the soil, heavy metals are transferred in its solid phase. Metal mobility and bioavailability may be influenced by addition of organic and inorganic matter [6].

A basic treatment restricting metal mobility is soil de-acidification by liming. Some studies revealed that phosphate stabilises Pb by reducing its availability in the soil. But this effect depends on the soil type. The addition of phosphate increases soil permeability, thus contributing to arsenic migration deeper into soil profile and entering groundwater. Adding materials which are rich in organic compounds such as compost, tree bark, sawdust, or granulated or powdered lignite, are generally recommended to reduce mortality and bioavailability of metals. Some studies also show that regular addition of organic matter in large quantities may inhibit metal uptake from the soil solution. Also, organic matter stabilises trace elements in the soil, and addition of compost may lead to increase in accumulation of metals in cultivated plants. Other studies show that lead and arsenic levels in vegetables were strongly correlated with a total content of these meals in the soil, and not with organic matter content in the soil or a level of compost addition. Some researchers also proved that the soil age also plays a crucial role in modulation of metal bioavailability in plants. Higher content of heavy metals is found in old flat wetlands due to long term discharges of municipal and industrial wastewater. Some analysis shows that bed rock is a source of Fe and Mn, while copper, zinc, cadmium, chromium, nickel, and lead are of anthropogenic origin. After comparing wet ditches and reclaimed wetlands, levels of some heavy metals such as lead, cadmium, and zinc are found in excess or in higher amount in wetlands because the wetlands are richer in mangrove soils. Apart from the soil factors which influence the presence and concentration of heavy metals in soil absorption of metals differ in different types of soils and plants. An important variation in metal concentrations was also seen being depending on the location in plant tissue, species also on the varieties of the same species. Such as concentrations of nickel, chromium, and cadmium were highest in plants roots, while leaves have highest level of lead accumulation. There are differences in the concentrations of heavy metals in edible parts of different plants and vegetables and concentration also varies. Cadmium and zinc are accumulated to a higher extent in lettuce and spinach, and lead was highly accumulated in lettuce and onions. Also, vegetables of the same species accumulated different amount and concentration of the heavy metals depending on the cultivar [6].

Legumes vegetables < melon vegetables < alliums vegetables < root vegetables < solanaceous vegetables < leafy vegetables.

Heavy metals concentration is higher than in potato, carrot, tomato, kohlrabi, and green bean seeds. After some studies, it is concluded that edible parts have highest heavy metal concentration in leafy vegetables, lower in root vegetables, and minimum or lowest in fruit vegetables. Root vegetables in plants are particularly susceptible to cadmium accumulation. Uptake of cadmium is extremely easy, both by the root and the leaves system, generally in proportion to its concentration in the environment, despite of the soil properties. However, acidic soil is considered to be most important factor influencing its increased phyto assimilation. Cadmium is characterised by a high bioaccumulation index up to 10 and its content in plants is commonly directly proportional to its content in the soil [7].

Cultivation of vegetables in regions is characterised by high density of the industrial and factorial, or nearby the busy roads like has traffic 24 hours result in significant contamination of crops with heavy metals. On the other hands, irrigation of agricultural fields with the wastewater results in the considerable contamination with the heavy metals and thereby to the accumulation of metals in vegetables cultured there. It is also important to maintain rational approach to fertilising and the use of plants protection products because they also increase accumulation of these metals in plants. They should be used in a reasonable amount so that could be beneficial for the plants or vegetables as well as the soil. Also if the concentration of heavy metals in the soil and the plants are in limit, they could be taken care off or if they reach into the body of any human being through plants or vegetables; they are in little amount or they can be cured but if once the concentration of heavy metals exceeds its reasonable limit, they then become very difficult to be cure and also causes difficulties in identifying. Leafy vegetables and root vegetables have the metals that have the greatest ability to accumulate the heavy metals absorb from the soil. Therefore, they should not be grown in such types of soil or the soil should be experimented or it should be examined properly before growing of the vegetables. Or they should not be grown at all. Once the plants or animals or human being are contaminated with the heavy metals, it is very difficult to cure them. Some causes damage to the internal organs like damage to the respiratory system, cardiovascular system, excretory system, neuromuscular system, and some other causes damage to the functioning of brain activity, heart, lungs, and kidney activity. In extreme cases, it can lead to paralysis or even death. So, one must take care of themselves. As soil is not the only one factor or way through which heavy metals reach the human body in fact there are many other ways such as through water, through air, and many other ways [7].

2.4 Toxic affects

Soil contaminated with zinc due to fertilisers, urban compost, emission from municipal waste incinerators, and other anthropogenic activities. Also, zinc is as essential nutrient for living organisms while cadmium is non-essential and potentially toxic for higher plants, animals, and also for the humans. Excess concentration of zinc leads to phytotoxicity, limits in its growth of both roots and shoots, senescence, chlorosis in younger leaves, and later extends to older leaves. Copper is a micronutrient for plants and its role is in CO_2 assimilation and ATP synthesis. It reaches to soil through industrial activities and human activities like mining and smelting of Cu containing ores. Excess of Cu in soil leads to growth of plants and its reduction. Mercury reaches the soil mainly in the solid phase through adsorption. It accumulates in the higher aquatic plants. It induces physiological disorder. Chromium is heavy metal and is a serious threat for the environment contamination of soil. Its toxicity is seen in many plants. Its toxicity leads to alterations in the growth of roots, stems, and leaves, deleterious effects on the plants physiological process such as photosynthesis, water relations, and mineral nutrition. Cobalt, this naturally occurs in the earth crust as cobaltite, erythrite, and smaltite. Excess of Ni2+ in soil causes various physiological alterations and diverse toxicity symptoms such as chlorosis and necrosis according to different species, shows impairment of nutrient balance, and resulted in disorder of cell membrane functions. Other symptoms are changes in water balance, etc. Lead accumulations are gradually decreasing in the plants and the nearby highway and roads as when the law came up with using of lead-free petrol. Properties of soil include its texture, capability to absorb water, Ph value, and ion exchange capacity. Lead, mercury, and chromium are usually tightly fixed in soils, and cadmium and zinc are readily absorbed by the plant roots. The heavy metals are absorbed by the plants through soil. These metals are absorbed by plants roots and get accumulated in the softer tissue of plants. These metals are not metabolised by the tissues, so it gets accumulated. From plants, it reaches human body. Human being consumes the plants containing the heavy metals, and the metabolization in humans too is not proper and thus it gets accumulated there. This further becomes toxic for the human and they face the consequences in form of the diseases.

Plants uptake the heavy metals that are present in soluble components in soil solution or those which can be easily solubilised. Some heavy metals are essential for the plants for their growth and upkeep, but in excess amount, these metals become toxic for the plants and even for the humans too. Plants growing in the soils containing heavy metals show reduction in growth, yield, and performance too. Growth reduction because of changes in physiological and biological process in plants growing on heavy metals polluted soil is seen. Continued growth in the size of plants leads to decrease in the yields which eventually lead to food insecurity. Soil properties affect metal availability in many ways. Like pH is the main factor affecting metal availability in soil. Other factors like density and type of charge in soil colloids the degree of complexation with ligands and the soils relative surface area.

Heavy metals reach the soil through the air\atmosphere and absorb by the roots of the plants and get accumulated in the softer tissues of plants. Plants experience oxidative stress on exposure to the heavy metal that finally leads to the cellular damage. Also, plants accumulate metal ions that destroy the cellular ionic homeostasis. To overcome the hazardous effects of heavy metal exposure and their accumulation, plants have developed detoxification method. These mechanisms are mainly based on chelation and sub cellular compartmentalisation [8].

2.5 Impact of heavy metal contamination on humans

Heavy metals reach human body through intake of plants containing it. Heavy metals turn out to be venomous, when not metabolised via the body and thus they accumulate or gather in the softer tissues. Chronic level ingestion of heavy metals toxic or heavy metals has hazardous effects on human body, and the impacts are noticeable only after several years of exposure. Heavy metals enter plants, animals, and human tissue through air (inhalation), diet (humans consuming both plants and animals or animal consuming plants and in turn humans consuming animals). These metals not only cause hazardous effects to humans only but are harmful for soil too. Toxic metals are a threat for both terrestrial and aquatic ecosystem. Heavy metals contaminate water bodies, soils, and sediments, after releasing from both natural or anthropogenic or human activities. Heavy metals are importunate, so they accumulate or deposit in the ground waters. Contamination also occurs when these metals are released into air through volcanic activities and from different industrial emission which ultimately return to the land.

Cadmium toxicity has been identified in the liver, brain, kidney, lungs, placenta, and bones. It causes muscular weakness, vomiting, abdominal cramps, and nausea. In Japan, Itai-Itai diseases brought dangers of cadmium into attention. Zinc is considered nontoxic, however, if taken in excess amount causes vomiting, impairment of growth, diarrhoea, bloody urine, reproductive issues, liver and kidney failure and anaemia. Cupper, exposure to high levels causes hepatic and renal damage, severe gastrointestinal irritation, skin irritation, etc. Heavy metals are detected in soil with some specific instruments like Atomic Absorption Spectroscopy (AAS), Inductively Coupled Plasma (ICP), Inductively Coupled

Plasma-Mass Spectroscopy (ICP-MS), and X-ray Fluorescence and Spectroscopy are used. Among all these instruments, atomic absorption spectroscopy gives the precise quantitative determination. The research basically focuses on the study to find out the presence of heavy metal, and quantity or amount of heavy metals or what the heavy metals are [9].

2.6 Forensic significance

Soil is like the fingerprint because as there are different patterns of fingerprint, the same way soil has its unique characteristics and properties that serve as its identification. Soils can be used as a profitable proof to establish the link between the suspect and the crime scene. Soil is a common blend of chemical, biological, and physical properties. If any heavy metals detected in soil, then it is used to geo-tag any particular location. Atomic Absorption Spectroscopy (AAS) is an apparatus for identification of the metals from the sample (soil, blood, serum, etc.) While investigating a various crimes, soils from different regions like sites of rivers, urban regions, industrial regions, and institutional regions etc. are used. Suppose we found a soil sample from the crime scene, it was sent further to laboratories tor its analysis. Whatever the result is like present of any heavy metals or ay normal or general metals will help to geo tag the location. By tallying both the question soil samples and the reference soil samples conclusions can be drawn [10, 11].

3. Heavy metals contamination of water

Water pollution includes contamination in liquid forms. Water pollution is of both marine pollution and river pollution. Industrial, factorial and consumer waste, fertilisers and pesticides, and acid rain which fall in the soils and finally reache these heavy metals into streams, rivers, lakes, and ground water. Heavy metals tend to bio accumulate and thus they are dangerous. It is well said that because of water, life exists in biosphere. And water is a universal solvent. Various inorganic chemicals and organic chemicals are dissolved in water, and the environment impurities are also dissolved in water. Both aquatic ecosystems be it fresh water or marine all are affected because of pollution. Water contamination is a serious environmental issue and which has direct impact on animals, plants, and human beings. Heavy metals are tremendously lethal to aquatic life even at small concentrations. Histopathological modification in the tissues of aquatic organisms such as fish, insects, and other aquatic life are seen because of the toxicity of the water. As these metals do not decompose, they accumulate in the soft tissues of the aquatic animals as well as in plants too. Contamination by heavy metals from many other various sources is major threat to the water and its organism. One source is effluents which released from the factories and industries [12].

3.1 Sources of heavy metals in water

Environmental pollution from dangerous metals and minerals can appear from natural as well as human activities sources. Natural sources such as leakage from rocks into water, volcanic action, forest fires, etc. With quick industrialisation and changing in lifestyle supply of environment pollution has increased. Chromium through mining and industrial coolants; lead through lead acid batteries, E-wastes, and bangle industry; mercury through fluorescent lamps, hospital waste, and electrical appliances; arsenic through natural process, fuel, and thermal power plants; copper and nickel through mining and electroplating; cadmium is released through waste batteries, e-waste, paints sludge, incinerators, and fuel combustions; zinc through smelting and electroplating [13].

3.2 Impact of heavy metal contamination of water

It releases toxic substances which are harmful for the aquatic life. Heavy metals tend to bio accumulate and thus they are dangerous for all the living organisms. Bio accumulates mean to boost concentration of any chemical in any living organism eventually. Increase in water pollution leads to release of toxic substance, pathogenic germs which requires more oxygen to decompose, for radioactivity, etc. Example: Eutrophication is a condition in which lack of oxygen in a water body is seen which is caused by unnecessary algae growth since it has enrichments of pollutants. It kills all the aquatic plants, small fishes (which in turn feds on the aquatic plants), large fishes (which feeds on the small fishes), and the human beings. And finally degrade the environment. The quality of water everywhere is degrading day by day. It is becoming less suitable for any work, and for drinking purpose, it is completely unfit. But still some people have no choice, they filter the water and then they consume it. And the animals, they have no idea whether is suitable for drinking purpose or not, and after grazing, they come to the river, ponds, or small lakes to drink water. Along with the water, they take the toxic metals in them. And slowly this toxic metal affects their system and makes them ill and finally leads to their death [13].

3.3 Impact of heavy metal contamination on human

Animals graze on the plants which are contaminated with the heavy metals and then gulp the same water. The marine lives that breed in heavy metals contaminated water; both such metals form a mass in their tissue and milk. In short, every single living organism is dependent on each other for their survival. And when humans consume these plants and animals, they eat the toxic substances along with them. Specific metals have specific toxicity. And toxicity also depends on the concentration of the metals consumed. Cadmium is poisonous at a low level [10]. In humans, it leads to renal dysfunction, bone defects, osteoporosis, increases blood pressure, and myocardial dysfunctions. Lead is very harmful and mostly captivated by ingestion with food and water and through inhalation. One of the serious effects is teratogenic effect. Poisoning of lead also causes cessation of the production of haemoglobin, kidney dysfunction, and unceasing damage to central nervous system. Zinc has same effects as lead. Its excessive concentration causes improper growth and reproduction, diarrhoea, bloody urine, vomiting, kidney failure, and anaemia. Mercury is toxic as it causes spontaneous abortion, gastric intestinal disorder, etc. Many health-related issues occur because of protein deficiency including reduction of chondroblastic or osteoblastic activity. In this condition, normal growth and formation of bones and cartilage gets manipulated, which finally results in irreparable body deformities in the growing child and in adults. Fishes are an important and staple food in the coastal regions of many parts of India. Hence, it is very important to know the correct amount of heavy metals present in the coastal amounts to protest both the aquatic and terrestrial life. Terrestrial life suffers damage to liver, kidney, heart, and body joints. The liver and kidney produce metallothioneins causing toxicity which is very hazardous for the human life [14]. In **Figure 1**, there is presentation of bioaccumulation and biomagnification.

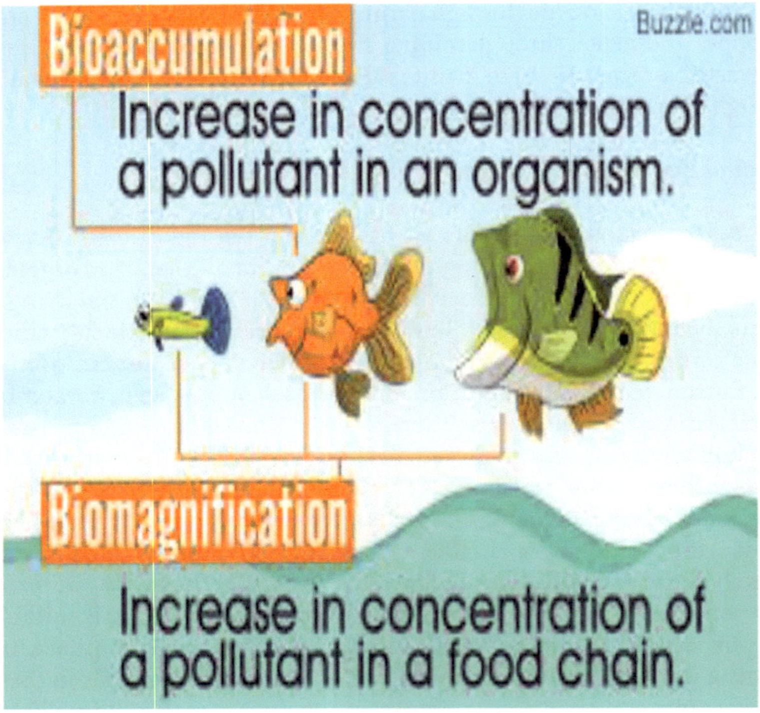

Figure 1.
Bioaccumulation and bio magnification [15].

3.4 Impact of heavy metal contamination on fish

Heavy metals are continuously being released in aquatic environment from natural and anthropogenic activities such as industrial, factories and domestic sewage discharges, mining, farming, electronic waste, anthropic accidents, etc., and metals are easily dissolved in water and are subsequently absorbed by the aquatic organisms such as small and large fishes and invertebrate, including a wide range of biological effects from being important or essential to being lethal. In spite of the fact that some metals are essential at low concentration, at higher concentration, they could be inducing toxic and lethargic effects in organisms, disturbing its growth, metabolism or reproduction [16].

Micronutrients are such as copper, zinc, ferric, manganese, cobalt, molybdenum, chromium and selenium, and macronutrients such as calcium, magnesium, sodium, phosphorous and sulphur. Apart from the essential elements, the non-essential elements such as lead, cadmium, nickel, arsenic, and mercury show its toxic effects at a very low concentration. High levels of metals in the surrounding could be hazardous for the functioning of the natural ecosystem and human health, due to their toxic effects, long persistence, bio accumulative properties, and bio magnifications in the food chain. Due to the possibility that the aquatic biota can accumulate various chemicals, are used to measure the effects of metals on aquatic organism as an essential indicator of water quality [17].

The organic chemical can be metabolised by the organisms easily, but the majority of metals cannot be easily metabolised because of its non-biodegradability. Once they get introduced in the aquatic environment, the metals get redistributed throughout the water column, gets accumulated in sediments, or consumed by biota. Metal residues in contaminated habitats have the ability to bio accumulate in aquatic environment like aquatic flora and fauna, which in turn gets consumed by the human beings or animals as food and enters the human food chain and finally results into health problems. Metal accumulation in sediments occurs through the process of precipitation of certain compounds. Some factors that control the bioavailability of metals are such as size or age, sediments, suspended matter and metal speciation, physical and chemical factors such as temperature, salinity, pH, ionic strength, concentration of the dissolved organic carbon, etc. [18].

The metals are up taken from the permeable epidermis if metals are in dissolved forms or through food ingestion if metals are in particulate forms. The ingestion uptake mainly depends on many factors. The suspended solid particles accumulate the insoluble metal compounds but under certain conditions, the metals reach the interstitial water being dissolved.

4. Discussion

Toxic metals are a threat for both terrestrial and aquatic ecosystem. Heavy metals contaminate water bodies, air, soils, and sediments, after releasing from natural or anthropogenic or human activities. Heavy metals are unrelenting or indestructible, so they both accumulate or deposit in the ground waters. Heavy metals reach water through the industrial, factorial, and consumer waste, through fertilisers and pesticides and also the from acid rain falling down in the earth which releases the heavy metals in the lakes, rivers, streams, and ground water. The release of factories and industrial effluents or waste is the chief source of water pollution as they are released directly into the rivers without being treated, which is a major threat to the aquatic life. Pollution of the rivers and water is a global problem because of its perseverance, bioaccumulation, and biomagnifications in the food chains and toxicity of the heavy metals. Atmosphere pollution is mainly in the gaseous state of the heavy metals and dust emitted by the transport, energy, metallurgy, and during building. Heavy metal principally goes in the atmosphere in the form of atomizer and gets deposited in the soil in the course of the natural sedimentation and rainfall process. Automobile transport causes some serious heavy metal contamination. Also, heavy metal is brought into soil by irrigative sewage, sanitary sewage, chemical wastewater, and industrial mining. Industries release a lot of harmful chemical which are directly released to river or an open area. Also, people nowadays use a lot of fertilisers and pesticides, for agricultural purposes. Factories and industrial outlets in the form of smoke, automobiles emissions and release of variety of compounds and chemical are followed by increase by man's some unwanted pollutants, which are serious threats and problem risking for the environment and for the man himself. Lead, mercury, and cadmium are the common air pollutants.

5. Conclusion

From the description of heavy metals, it can be concluded that these are toxic for all living beings. But the difference is in its concentration. If the concentration is low, the toxicity is also low but if it is high, it could be fatal. Heavy metals are present everywhere. These are present in soil, deposited in water, and present in the atmosphere too. These pollute the environment as well as human's life. Through soil, it is absorbed by plants further by the herbivorous animals which in turn are eaten by the carnivorous animals as well as by the humans. These metals do not go off easily from the plants and animals instead it gets accumulated in the soft tissues of plants and animals and further reaches to humans. The heavy metals which are in the water degrade the quality of water and makes it unfit for drinking for all the living beings and even for other purposes. It is a threat for the aquatic life. Because of this, the amount of dissolve oxygen decreases and thus the life which are under aquatic do not get sufficient oxygen to breathe, which could lead to death of the fishes, aquatic insects, animals, and plants. Heavy metals are present in the atmosphere in the form of aerosols, air, etc. which directly goes into our lungs and then slowly to all our body parts through breathing. These heavy metals reached into the environment through the anthropogenic activities or human activities. By the industrial effluents, discharge of industrial and hospital waste directly into water without being treated, similarly the sewages are also released unto water without being treated, through vehicular emission.

We should control our activities and start thinking about the future. Already a lot of damage has been done to the environment and now it is time to take preventive measures. That includes a forestation, sewage waste should be treated before releasing into rivers, non-biodegradable waste should not be dumped here and there, instead recycle it and then again use it, heavy and radioactive metals should be properly incinerated.

Author details

Soni Kumari and Amarnath Mishra*
Faculty of Science and Technology, Amity Institute of Forensic Sciences, Amity University, Noida, Uttar Pradesh, India

*Address all correspondence to: amishra5@amity.edu; drmishraa1@gmail.com

References

[1] Li C, Zhou K, Qin W, Tian C, Qi M, Yan X, et al. A review on heavy metals contamination in soil: Effects, sources, and remediation techniques. Soil and Sediment Contamination: An International Journal. 2019;**28**(4): 380-394

[2] Zhang MK, Liu ZY, Wang H. Use of single extraction methods to predict bioavailability of heavy metals in polluted soils to rice. Communications in Soil Science and Plant Analysis. 2010;**41**(7):820-831

[3] Su C. A review on heavy metal contamination in the soil worldwide: Situation, impact and remediation techniques. Environmental Skeptics and Critics. 2014;**3**(2):24

[4] Xu J, Liu C, Hsu PC, Zhao J, Wu T, Tang J, et al. Remediation of heavy metal contaminated soil by asymmetrical alternating current electrochemistry. Nature Communications. 2019;**10**(1):1-8

[5] Sharma RK, Agrawal M, Marshall F. Heavy metal contamination of soil and vegetables in suburban areas of Varanasi, India. Ecotoxicology and Environmental Safety. 2007;**66**(2):258-266

[6] Järup L. Hazards of heavy metal contamination. British Medical Bulletin. 2003;**68**(1):167-182

[7] Zheng G, Tian L, Liang Y, Broberg K, Lei L, Guo W, et al. δ-Aminolevulinic acid dehydratase genotype predicts toxic effects of lead on workers' peripheral nervous system. Neurotoxicology. 2011;**32**(4):374-382

[8] Nuris M. Isolasi dan identifikasi bakteri resisten selenate (seo42-) dari sedimen mangrove pantai Banyuglugur Kabupaten Situbondo, Jawa Timur (Doctoral dissertation, Universitas Islam Negeri Maulana Malik Ibrahim)

[9] Wong CS, Li X, Thornton I. Urban environmental geochemistry of trace metals. Environmental Pollution. 2006;**142**(1):1-6

[10] Gupta N, Yadav MR, Nayak MM. Physio-chemical and biological analysis of Gomti River in Lucknow region. 2017;**5**(5):1324-1333

[11] Swaleh SB, Banday UZ, Usmani N. Comparative study of biochemical, histological and molecular biomarkers of heavy metal contamination in Cyprinus carpio collected from warm-monomictic lake and government culture pond. Chemosphere. 2019;**236**:124182

[12] Bhargava DS. Use of water quality index for river classification and zoning of Ganga River. Environmental Pollution. Series B: Chemical and Physical. 1983;**6**(1):51-67

[13] UPPCB. 2013. Available from: http://www.indiawaterportal.org/news/report-Uttar-Pradesh pollution-control board-uppcb-reveals-deteriorating-condition Gomti-river

[14] Available from: https://www.kent.co.in/blog/harmful-effects-of-heavy-metal-contamination-in-drinking-water

[15] Available from: https://sciencectruck.com/difference-between-bioaccumulation-biomagnification-biomagnification

[16] Shinn C, Dauba F, Grenouillet G, Guenard G, Lek S. Temporal variation of heavy metal contamination in fish of the river lot in southern France. Ecotoxicology and Environmental Safety. 2009;**72**(7):1957-1965

[17] Elzwayie A, Afan HA, Allawi MF, El-Shafie A. Heavy metal monitoring, analysis and prediction in lakes and rivers: State of the art. Environmental Science and Pollution Research. 2017;**24**(13):12104-12117

[18] Srivastava SC, Verma P, Tripathi M. Comparative analysis of the microbial load in catfish (Mystus aor) and carp fish (Labio bata) from Gomti River, Lucknow India. Flora & Fauna. 2014;**20**:77-82

Chapter 2

Microbiological Indices for Diagnosis of Heavy Metal Contaminated Soils

Sukirtee Chejara, Paras Kamboj, Y.V. Singh and Vikas Tandon

Abstract

Heavy metal contamination has gained popularity worldwide due to their persistent nature in the environment, on the top of that non-biodegradable nature makes its accumulation easy to toxic levels. Understanding the nature of contamination has become a major concern before heavy metals deteriorate the quality of soil; to diagnose heavy metal pollution suitable indices are required. Microbial indices gaining importance because of their sensitive nature towards change in surrounding, which is the imperative quality required to select microbes as environmental indicators. Albeit enough literature is present related to this topic but the information is scattered so role of this chapter is imperative. The chapter will be helpful for the reader to provide a thorough understanding of merits and demerits of microbiological indices for heavy metal contaminated and restituted soils. The changes in microbiological indices and their mechanism of response towards heavy metal stress are effectively summarized. Research gap and future needs of microbial diagnosis of heavy metal contaminated soils are discussed.

Keywords: heavy metals, soil contamination, microbial indices, soil pollution, soil health

1. Introduction

Heavy metals are very crucial for maintaining the life cycle of living organisms. Albeit they are important but excessive accumulation of heavy metals is harmful for environment as well as for human health. Excessive accumulation of heavy metals in the soil may take different pathways, which may be through industrial residue, chemical manufacturing, pesticides and fertilizers, sewage irrigation, metal plating etc. but all sources are principally result of anthropogenic activities [1]. Heavy metal pollution in soils is present in different parts of the world including Spain, United states, France and India are in worst condition by Cd- contaminated soils [2]. Urban soils of Naples city and Mexico city is contaminated with Cu, Pb and Zn [3, 4]. Due to non- biodegradable nature of heavy metals their persistence in nature is very long which harms soil ecological environment [5]. Earlier studies proved that high heavy metal concentration cause certain harmful effects on microorganism as dysfunction of cell, protein degeneration, and sometimes destruction of plasma membrane [6]. Above that heavy metal contamination affects enzyme activity of microorganism, DNA sequencing as well as abundance is also affected by heavy metal contamination. Thus, it is highly important to choose such indices which are accurate and efficient for the diagnosis and analysis of quality of heavy metal contaminated soils, so that preventive measures chosen beforehand and further deterioration of soil quality can be controlled as well as suitable remediation technique could be employed on time. Soil quality can be diagnose using sensitive microbial indices, which are monitoring of soil microbial changes before and after contamination or some remnant part of land under observation. In a general perspective soil having higher microbial population or activity performs better and can be called as good quality soil. Using microbes for diagnosis have several advantages *i.e.* (1) Microbes are active participants of soil ecosystem [7] highly sensitive for heavy metal contamination than plants and animals growing in the similar conditions [8]; (2) microorganisms are closely related to pollutant degradation and soil fertility conditions [9]; (3) microbial analysis requires a very small amount of sample, quick to perform, simple and cheaper [10]. One should always include some ecologically relevant attributes while diagnosing soil quality so that they give better performance while studying ecosystem quality. Microbiological indicators mainly include study of microbial population, microbial diversity, function and activity. If the indices are correctly selected they will give better information about heavy metal polluted soils. Albeit information about microbial indices are available in literature but that information is scattered. This chapter provides information about merits and demerits of using microbial indices for heavy metal contaminated soils. The changes occurred in different indices and their possible mechanism under heavy metal stress were studied comprehensively and summarized.

2. Diagnosis based on microbial abundance

Abundance of functional gene is a genetic diagnosis method of understanding heavy metal contaminated soils. Presently genes related to nitrogen transformation are gaining popularity in diagnosis of target soil. In the process of nitrification ammonium ($NH_4^+ - N$) is converted to Nitrite (NO_2^-) and ultimately to Nitrate (NO_3^-) [11]. In the nitrification process ammonia oxidation is the rate limiting step in the global N-cycle [12, 13]. Ammonia oxidation is carried out by Ammonia oxidizing archaea (AOA) and ammonia oxidizing bacteria (AOB) [14]. They contains different enzymes to carry forward the process like AMO, HAO and NXR. AMO protein contains alpha, beta and gamma subunits as it is a trimeric membrane-binding protein, units alpha, beta and gamma is encoded by genes amoA, amoB and amoC genes respectively [15]. Nitrite oxidation is carry forward by a group of microbes i.e. nitrite oxidizing bacteria (NOB) [16]. Heavy metal contamination is widely Diagnose using ammonia oxidizing gene as markers mostly amoA gene due to its conservative coding. When abundance of amoA gene is compared for AOB and AOA in a Cu contaminated soil it is found that amoA gene has a negative correlation with Cu concentration [17]. When the sensitivity is compared AOB amoA gene was found more sensitive than AOA amoA gene. AOB and AOA amoA gene abundance is reduced when the soil is contaminated with As and Pb, the sensitivity of AOB was found higher than AOA [18]. Similar results were found in case of sensitivity when studied a Cd contaminated soil [11]. AOA found less sensitive than AOB it may be because of AOA have metal reducing ability and heavy metals are generally less toxic when they are in their reduced state i.e. lower valance state [19] which ultimately is beneficial in metal detoxification. AOA have more rigid cell membrane than AOB.

Just opposite to the above recorded observation, scientist indicated that in a Zn contaminated soil abundance of AOA amoA gene decreased quickly than ABO [20].

In long term Zn tolerance development AOB amoA gene copy and transcript enhanced hence AOB community structure also changed, And AOA failed to respond towards Zn [21, 22]. Albeit the abundance of amoA gene of AOA was dominated in second year but expression from the genes were not detected [20]. Response of AOA community is not that clear till now with the available literature further details are needed to understand whether AOA can adapt to long term contamination. AOA may use other processes to fulfill their energy requirement or they may survive in their dormant state. Despite cultivated AOA clusters are few in numbers, so response of AOA to external environment is so far needed exploration. Remediation of contaminated soil exhibit changes in amount of ammonia oxidation genes. Application of biochar and alfalfa enhanced abundance of amoA gene of AOA and AOB in a heavy metal and fungicide contaminated soil [11]. Abundance of AOB amoA gene increased with application of biochar in a Cu and Pb contaminated soil when the soil is remediate using biochar and compost [23].

However some scientist reported gene copy number is a weak indicator for heavy metal pollution. There was no significant change in gene abundance of AOB or AOA amoA gene when a soil is treated with Hg [24]. This may be because of Hg tolerant ammonia oxidizing community present in soil from before or may be application of Hg may induce tolerance in the community [24] thus from this study it is found that amoA gene did not respond towards heavy metal pollution, but for its confirmation we need further exploration of the nature of gene. Gene transcript number is found a better index than gene abundance when talking about indices of soil quality. In a study it is found that there was a decrease in amoA gene transcript number of AOB and AOA by three and four order of magnitude, while gene copy number remained unchanged in a one week Zn treatment [21]. Hence from the above discussion it can be concluded that heavy metal pollution cannot be predicted accurately on the basis of change in gene abundance of AOA and AOB further research is still needed in this aspect. Furthermore we cannot judge the change in any one of AOA or AOB separately there may be some sort of interaction among both the community while dealing with heavy metal toxicity [19]. So it is recommended to monitor the change in both the community simultaneously other than thinking separately. Sometimes increase in growth of microbial community may be a response of toxic effect [25]. Till now only AMO genes are explored to some extent while HAO gene and NXR gene did not received much attention it may possible they may express well as a diagnosing tool than AMO in heavy metal contaminated soil.

3. Response of denitrification genes

During denitrification nitrate is converted to dinitrogen through several intermediate products $NO_3^- \rightarrow NO_2^- \rightarrow NO \rightarrow N_2O \rightarrow N_2$ [26, 27] different reductase enzymes are involved at different stages of intermediate product for nitrate reduction nitrate reductase (Nar), for nitrite reduction nitrite reductase (Nir), for nitric oxide reduction nitric oxide reductase (Nor) and for nitrous oxide reduction and nitrous oxide reductase (Nos). Nitrate reductases (Nar) encoding is done by operons of nas, nar and nap. Encoding of nitrite reductase (Nir) is done by nirK and nirS, while nitric oxide reductase (nor) consist of norB and norC. Nitrous oxide reductase (nos) contains nosZ, nosR and nosD etc.

Denitrifying enzymes encoding genes are very sensitive towards heavy metal stress, they characterize denitrifier community and helpful in diagnosing soil quality. Different studies have been carried out to study the relation of denitrifying enzymes and heavy metal stress and found that reduction in genetic diversity is the most common inhibitory effect of heavy metal stress [28, 29]. Research conducted on Pb contaminated soils and found that nirK gene community diversity was reduced due to Pb contamination [30]. Enzyme involved in different reduction steps showed significant difference in Cu tolerance in a study conducted it was reported that diversity of nirS, nirK and nosZ genes decreased with the increase in Cu concentration [31]. Increase in Ag concentration lead to decrease in nirK gene copy number but diversity of nirK gene increased [29]. Under Hg stress nirS gene diversity increased. Different denitrification genes respond differently towards same environmental pressure [32, 33] thus, selection of sensitive indicators becomes mandatory for detection of soil pollution.

Abundance of nirS gene changed significantly under Hg stress while no change in nirZ gene was observed under all given treatments, this proves nirS gene more sensitive than nirZ gene [34], while it can be said that nosZ denitrifier is resistant under different pollution condition in soil and shows more stability [21, 27]. Effect of different remediation strategy were observed. Abundance of denitrification genes (narG, nirK, nirS) except nosZ increased with application of alfalfa and biochar in a heavy metal and fungicide contaminated soil [11]. Denitrifying genes shows different patterns while diagnosing heavy metal pollution, hence further research is needed for better information base. However nosZ gene found less sensitive to heavy metal pollution than denitrifying genes, further its resistance need to be study.

4. Microbial biomass

Microbial biomass in soil include living microorganism present in soil i.e. fungi, bacteria, algae and protozoa [35]. Microorganisms contain usually carbon, nitrogen, phosphorus and sulfur but mainly their population is expressed as microbial biomass carbon. Terrestrial ecosystem organic matter dynamics is affected by microbial biomass being an important component of organic matter in soil [36]. Microbial biomass have a direct correlation with soil condition [36], there are sufficient evidences which proves the sensitivity of microbial biomass with increase of heavy metal stress [37–39]. Microbial biomass can be used to predict soil quality. Higher microbial biomass in soil indicates good functional quality and will be able to store more nutrients and regulated nutrient cycles [40]. Heavy metal stress severely inhibits microbial biomass [8, 40]. Increase in Cd concentration leads to decrease in C_{mic} in soil [41]. A negative correlation between soil microbial carbon and heavy metal concentration (Cd, Pb) is indicated [42].

Under heavy metal stress microbes requires more energy for their survival which required more consumption of substrate, resulting less substrate left for other microbes. This limits their growth [5, 39]. Albeit there will be declined microbial biomass but it does not indicate population extinction, more resistant species will fill the gap with their presence, microbial ecosystem will remain enriched [43]. On the other hand remediation strategy helps in increasing microbial biomass, which indicated improved soil condition. Soil replacement found to be helpful in increasing carbon when metal concentration decreased in soil (Cd, Cu, Pb, and Zn) [36]. C_{mic} may not respond sometimes effectively to stimulation of heavy metal, any correlation between microbial biomass and heavy metal was not found [39, 43]. [44] found that there were no significant relation between C_{mic} and soil- soluble Cu. No correlation found between heavy metals (Cr, Cd, Pb, Zn, Cu) and carbon [45]. Microbial biomass Nitrogen (N_{mic}) also serve a good indices for soil quality assessment, it is found associated with heavy metal content in different cases [46]. N_{mic} decreased with increased heavy metal content [43]. While inconsistent and weak downward trend of N_{mic} under metal contaminated sites (Cd, Cu, Pb, Zn) was observed [37]. N_{mic} found less correlated with heavy metal pollution than C_{mic}. Nitrogen at severely polluted areas of metal contamination was 64.4% of non-polluted area while C_{mic} accounted only 31.6% [40]. Albeit individual microbial biomass is highly sensitive towards change in soil condition but it has certain limitation while serving as indices for soil monitoring. One cannot predict change in microbial structure only through microbial biomass observation. Short term response of microbes to heavy metal contamination does not predict soil quality in long run, even if the soil environment is same. At lower metal concentration (Cd/Cu/Zn) C_{mic} changed in long long-term experiment while no change were observed in short laboratory test [47]. Microbial biomass is highly dependent on soil physical, chemical and biological properties, which are helpful to blur the toxicity of heavy metals. Soils with more labile carbon increases C_{org} in soils [48]. Soil particle size also affects toxicity of heavy metals, heavy metal toxicity (Pb, Cd, Zn and Cu) to C_{mic} was more prevalent in coarse fraction of soil than clay fraction [48]. Different biomass related ratios to heavy metals also have been explored. C_{mic}/N_{mic} ratio is helpful in controlling microbiologically operated nutrient cycling and availability [49], this ratio is an important indicator of soil condition. There are sufficient studies available which indicate that heavy metal stress can induce change in C/N ratio [46, 49]. Under heavy metal stress C_{mic}/N_{mic} may increase due to increase in tolerant fungal component. Huge difference between C/N ratio of bacteria and fungi support this increased ratio with increase in fungal population, C/N ratio of bacterial species 3.5:1 while for fungal species this ratio ranges from 10:1 to 15:1 [46].

Soil microbial biomass	Contaminants	Remediation	Results	References
C_{mic}	Cd	Plantation of *Eulaliopsis binata*	Negative correlation	[41, 42, 47]
	Cd, Pb	Plantation of *Sedum plumbizincicola*	Negative correlation, Metal content decreases, C_{mic} increases	[42, 44, 59]
	Cd, Cu, Pb, Zn	Soil replacements in trenches+ planting Eucalyptus in contaminated soil + Natural vegetation	Metal content decreases, C_{mic} increases	[60]
		Soil replacements in trenches+ planting Eucalyptus in contaminated soil along with uncontaminated soil in upper 20 cm soil layer+ *Brachiaria decumbens*	Metal content decreases, C_{mic} increases, C_{mic} was found 100% higher than earlier method	[36]
N_{mic}	Cd, Cr, Cu, Pb, Zn	Cultivation of Eucalyptus binata	No correlation	[45]
	Cd, Pb	Cultivation of Eucalyptus binata	Metal content decreases, N_{mic} increases	[59]
	Cu, Zn, Cd, Pb, Ni, Mn	Cultivation of Eucalyptus binata	Polluted sites, decreased N_{mic}	[40]
	Cd, Cu, Pb, Zn	Cultivation of Eucalyptus binata	Polluted sites, decreased N_{mic}	[37]

Table 1.
Heavy metal pollution with relation to microbial biomass.

Fungal species appear more resistant to heavy metal pollution than actinomycetes and bacterial species [50, 51]. Fungal/bacterial population ratio is considered as a good soil health indicator [48, 52]. Bacteria and fungi play dominant role in nutrient availability and organic matter dynamics being the major population governing soil microbial biomass i.e. about 90% of total microbial biomass [48]. Heavy metal stress cause bacterial mortality which enhances carbon release, this carbon is used by resistant fungal population for their growth [25]. However this index is not generally used for diagnosis of soil pollution. C_{mic}/C_{org} is also a good indicator of soil heavy metal pollution. Different studies indicate under heavy metal pollution C_{mic}/C_{org} ratio decreases [53, 54]. In a study C_{mic}/C_{org} ratio is found negatively correlated with As and Cu contamination [55, 56]. While in a study it was observed that C_{mic}/C_{org} ratio increased with decrease in heavy metal stress (Cu, Zn) [48]. When C_{org} is used by microorganisms for their respiratory metabolism the efficiency of conversion of C_{org} to C_{mic} reduced hence ratio of C_{mic}/C_{org} also declines [55, 57]. Few scientist claim that the ratio of C_{mic}/C_{org} is significant in non-contaminated soils, but for metal contaminated soils this relation even may not exist [58]. Not Any change in C_{mic}/C_{org} nor any obvious trend was present under heavy metal contaminated soil (Zn, Cd, Pb, Cu) [46]. Hence in microbial biomass or in related ratios no consistent and clear change is observed with heavy metal pollution. This ratio does not reveal any change in population structure. Hence none of them is suitable solely as an indicator of soil quality (**Table 1**).

5. Heavy metal contamination diagnosis through change in microbial community structure and diversity

Change in microbial community structure and diversity is a sensitive tool which can be used for diagnosis of heavy metal pollution in soil [47]. Extremely rich microbial diversity in soil [61, 62] can be reduced to 1000 times in a moderately contaminated soil [63] or up to 1% of primitive soils in highly contaminated conditions [64]. Different experiments have been conducted in favor of reduced diversity in metal polluted soils [7, 65] as indicated in the **Table 2**. A reduction in microbial diversity is observed with long term Cr contaminated soil [66]. Microbial community diversity also found decreased with Cu and Zn contamination in long run [68]. Soil remediation techniques show their significance by changing microbial diversity. Use of sepiolite for stabilization of Cr significantly increased community diversity [70]. Iron grit is useful for control of metal contamination (Cd, Cu, Zn) it gives result by improving diversity of microbial communities [71]. Certain findings indicated heavy metal contamination is not always negatively correlated with diversity it may increase diversity [5], while others not found any correlation [62] **Table 2**. Studies indicated that heavy metal contamination directly affects physiology of microbial community thus decreases diversity, Certain communities can withstand this adverse condition while adopting dormant state [62]. Albeit dormancy is an option but it serves the purpose only in short run if exposure is prolong to chronic contamination an obvious adverse effect on functions of community is unavoidable. Communities resistant to contamination may gain their full diversity with time [63]. Soil quality reliably evaluated with Community structure of microbes [42]. Soil microbial community structure significantly changes with heavy metal stress [72, 73]. With long exposure to Cr contamination soil proteobacteria community changed to firmicutes [66]. Pristine soils were dominated with acidobacteria and actinobacteria but population turns into proteobacteria when soil contaminated with Cr, As [62]. Heavy metal contamination may affect one population while not affecting the other one. A study conducted by indicates

Heavy metal	Changes in diversity and structure	Research methodology	References
Cr	Decrease in diversity, community changes	16S r RNA sequencing	[66]
Cu, Zn, Pb	Decrease in diversity, community changes	Pyrosequencing and PFLA techniques	[67]
Cu, Zn	Decrease in diversity, community changes	Metagenomics and functional assays	[68]
Cd	Decrease in diversity, community changes	Metagenomics	[7]
As, Pb	Decrease in diversity, community changes	PCR-DGGE	[69]
Cu	No significant change in diversity, community structure changes	16S r RNA tagcoded pyrosequencing	[63]
V	Diversity first decreases then increases, community structure changes	PCR-DGGE	[5]

Table 2.
Heavy metal pollution with relation to diversity and structure of microbial community in soil.

that Cu contamination changes the community composition for bacteria without affecting fungal community [74]. Heavy metal stress affects bacterial population most than archaea [19]. Archaea shows a positive correlation with Cd while bacterial species exhibit different responses towards Cd like α-proteobacteria shows negative correlation, β-Proteobacteria are positively correlated, γ-proteobacteria and δ-proteobacteria does not show any correlation. Different response of proteobacteria can be explained with complex lifestyle of proteobacteria, it can use different organic matter as a carbon, and energy source [75] this ability enables them withstand in harsh conditions and respond differently to different environments. Different microbial interaction may also help microbes to a better adaption [19]. Consistent conclusion about sensitivity of microbial diversity and structure is not available; one cannot clearly explain which one is more sensitive indicator. Bacterial diversity must be more sensitive than bacterial community structure for heavy metal stress [67]. Soils contaminated with neutral mine effluent and sediments [76] changes bacterial structure significantly than their diversity [77]. It was investigated that both diversity and structure of bacterial population changed under Cd contamination [65]. Increased diversity and structural improvement of microbial community ensures better functioning of soil in heavy metal contaminated soils [74]. In heavy metal contaminated soil sensitive species are replaced with more tolerant species thus it increases species richness [78]. Community dynamics also affected by species evenness [79]. Hence relation between diversity and structure is complex, both need to be use simultaneously in order to evaluate soil quality of a heavy metal contaminated soil. Species richness and evenness may not change simultaneously under stress condition. Mn contamination in soil affects species richness but not evenness to the significant level [80]. In all the previous studies related to heavy metal contamination importance has given to species richness very few literature considered species evenness [80]. Different modern techniques of new era improved our understanding towards cellular constituents like fatty acids, protein, nucleic acid and other compounds related to any specific taxa which proved helpful in recognizing diversity and structure of bacterial community in contaminated soils. Pros and cons of different techniques cannot be avoided; different techniques show certain deviation from other technique **Table 3**. Pyrosequencing does not indicated any significant change in bacterial community structure of a heavy metal Cu, Zn and Pb contaminated soil but using PLFA analysis a significant

Method	Applicability	Advantage	Limitation	References
PLFA	Microbial community	Indicator of living microorganism; act as a biomarker for community structure and physiological state microorganism	Interpretation of PLFA method is difficult; microbial diversity cannot be assessed: Temperature and nutrition can change fatty acid structure; Single acid cannot represent any specific species	[67, 81]
DGGE	Gene cluster; microbial community structure	Sample can be analyze under temporal and spatial variation; easy to operate; multiple samples can be analyzed at a time	It can provide information sequence between primers; if a primer is mismatched it will lead some missing lineages; it only isolates <500 bp fragments effectively; it only detect the microorganism but cannot give any information about species richness.	[82]
ARDRA	Microbial community structure	Identify closely related sequence effectively and inexpensively	Cannot identify polygenetic group; restriction enzyme optimization with this technique is difficult	[83]
High-throughput sequencing	Microbial diversity and community structure	Helpful in tracking biomarker so characteristics of microbial community can be determined	Expansive; data accuracy may get spoiled by some invalid sequence	[84]
T-RFLP	Microbial community	High sensitivity and better resolution	Interpretation needs multiple restriction enzymes; This technique is highly dependent on PCR amplification of 16S/18S r RNA	[81]

PLFA: Phospholipids fatty acids; DGGE: Denaturing gradient gel electrophoresis; T- RFLP: Terminal-restriction fragment length polymorphism; ARDRA: Amplified ribosomal DNA restriction analysis.

Table 3.
Different methods for determination of community structure of microbes.

change is observed [67]. Soil environment also play a significant role in expression of microbial communities in contaminated soils. Soil pH had a significant role in affecting community composition in long term Cu contaminated soil [74]. Soil microbial community structure and diversity not only serves as an indicator of detrition of soil quality but it also predict ways to remediate a deteriorated soil. Metagenomics helps one to understand complicated communities of microorganisms and their working process along with unique ability for identification of new strains and genes [85]. Thermophilic cyanobacterium MTP1 genome is helpful in encoding different resistant system, mainly Cd, Cu, As, Co, Zn, Hg contaminated soils, Which indicates greater potential of this microorganism in remediation of metal contaminated soils [86]. Certain microorganism which are tolerant to contamination for example proteobacteria are tolerant to Cd contamination, possibly can be used to deal with soil Cd contamination [7]. Microbial abundance is less sensitive than microbial community structure and diversity as a indicator for metal contamination [21, 34], but sole dependence on these indicator is not advisable for determination of soil quality. These two indicators do not reflect functioning of system. Different microbial communities may have similar functions which causes superfluity, and in some cases even though microbial diversity is high but activity may be low [48]. However activity of microbial community may recover in long run but it may change its community structure.

6. Diagnosis based on enzyme activity

Sol enzymes, most important component which governs nutrient cycling in soil specially C, N and P cycle [87]. Enzyme system stability and sensitivity makes it an effective indicator of biochemical processes, Hence enzyme system behaves as a biological indicator helpful in diagnosing sol health [87]. High enzyme activity of soil represents good sol health while in presence of pollutant enzyme activity may reduce [88]. Quantitative relation between soil pollution and enzyme activity is not established till today hence only the change in soil enzyme activity after and before contamination is analyzed for determination of soil quality. Sufficient literature is present to support that enzymes are sensitive towards heavy metal pollution [40, 87]. When a contaminated soil is compared with non-contaminated soil dehydrogenase enzyme activity decreased with heavy metal (Cu Cd Zn Pb) contamination [48]. Vanadium (V) concentration shows negative correlation with urease activity [5]. Response of soil enzymes can vary in different ways to heavy metal contamination it may be activation, inhibition and neutral. Most of the studies indicate the depressed enzyme activity, and inhibition may depend on concentration of heavy metal [45]. The mechanism is not certain whether heavy metal direct inhibit enzyme activity or they reduces their release or both the mechanisms are operative simultaneously [89]. Heavy metal seriously inhibit enzyme activity, but with time some recovery was observed [90]. This may be because of sudden exposure to heavy metal contamination but with time microorganism adapt to environment and recovery is seen in enzyme activity. Different soil enzymes react differently to heavy metal stress, it is important to choose the right enzyme which shows maximum response to heavy metal contamination and react as a suitable indicator in determination of soil quality. Enzymes like catalase, urease and dehydrogenase mostly used as bioindicator.

Catalase helps in decomposition of hydrogen peroxide, reduce heavy metal toxicity (Cu, Zn, Pb, As, Cr, and Cd) to microorganisms [87]. Dehydrogenase takes part in oxydative phosphorylation and used in heavy metal contaminated soils [48]. Urease partakes in N cycle and used in V, Zn, Cu, Pb, Ni and Mn contaminated sites [5, 40]. Amylase, phosphatase and protease were also used as biological indicator for metal contaminated sites. Different enzymes have different levels of sensitivity [91] shows that soil contaminated with different heavy metals follow presented order on the basis of their sensitivity; dehydrogenase found highly sensitive followed by urease followed by alkaline phosphatase and lastly acid phosphatases found least sensitive. As and Cd toxicity did not influence dehydrogenase activity [92]. Heavy metal (Cd, Zn and Pb) contaminated soils sensitivity of urease was found higher than other enzymes like invertase, catalase and alkaline phosphatase [93]. Contamination of heavy metals (Zn, Cu, Cd, As, Cr, Ni, Pb) did not affect urease activity significantly [87]. Previously conducted studies and their results indicated that there were many differences during the applicability of experimental results to the actual environment [87]. Synergistic and antagonist relation among different heavy metals also influence their toxicity for enzyme system. In a study conducted by [57] they concluded that combined effect of Cd and Pb was significantly inhibitorier for enzymes (Dehydrogenase, acid phosphatase and urease) than

Cd or Pb alone as a pollutant in the system. Heavy metals (Pb, Cd, Zn) in combination had strong inhibitory action on enzymes (alkaline phosphatase, catalase, invertase and urease) than any single heavy metal [93]. Some researchers found that Cu as a sole heavy metal in a system inhibit enzymes (alkaline phosphatase, acid Phosphatase, dehydrogenases and urease) more than its presence in combination with Cd, Cr, Pb, Ni and Zn. Type of heavy metal and content in a system determines antagonistic or synergistic relationship of heavy metals. Effect of heavy metal on soil enzymes will also be determined by environment (soil grain size, soil organic matter, pH, etc.). Particle size distribution explains the Zn pollution and enzyme resistance to the pollution [94].

Target enzyme	Participation enzyme	Pollutants	Results	References
Catalase	Dehydrogenase, β-glucosidase, urease, alkaline phosphatase, arylsulphatase	As, Cd, Cr, Cu, Hg, Mn, Pb, Zn	Negative correlation	[95]
	Polyphenoloxidase, catalase, amylase, acid phosphatase, urease	As, Cd, Pb, Zn	Positive correlation, polyphenoloxidase was the most sensitive soil enzyme	[87]
	Catalase, alkaline phosphatase, dehydrogenase	Cd, Pb	Negative correlation	[90]
Dehydrogenase	Alkaline phosphatase, dehydrogenases, urease, acid phosphatase	Cd, Cr, Cu, Ni, Pb, Zn	Sensitivity: dehydrogenases > urease > alkaline phosphatase > acid phosphatase	[91]
	Urease, catalase, acid and neutral phosphatase, sucrase	Cu, Zn, Cd, Pb, Ni, Mn	Negative correlation; Sensitivity: dehydrogenase > urease > catalase > neutral phosphatase > sucrase > acid phosphatase	[40]
	Catalase, alkaline phosphatase	Cd, Pb	Negative correlation; Sensitivity: dehydrogenase > catalase, alkaline phosphatase	[90]
	Invertase, urease, arylsulfatase, catalase, alkaline phosphatase	As, Cd	Insignificant	[92]
	Urease	V	Negative correlation; Sensitivity: dehydrogenase > urease	[5]
Urease	Dehydrogenase, catalase, acid and neutral phosphatase, sucrase	Cu, Zn, Cd, Pb, Ni, Mn	Negative correlation;	[40]
Phosphatase	Phosphatase, urease, β-glucosidase, protease	Cd, Ni	Sensitivity: phosphatase > urease > β-glucosidase > protease	[96]
	Catalase, dehydrogenase	Cd, Pb	Negative correlation	[90]

Table 4.
Heavy metal pollution and soil enzymes.

pH also affects enzyme activity in different ways being low and high it controls enzyme activity sites and their dissociation state as well as enzyme stability [87]. Soil organic matter content positively affects soil enzyme activity. There was a quantitative relationship between soil enzymes and organic matter content at Pb concentration of 500 mg/kg, Arylsulfatase activity found higher with organic matter content of more than 1.05%, activity of enzyme decreased gradually with decrease in organic matter content below 1.05% [92]. Dehydrogenase activity was also related to soil organic matter availability [48]. Labile organic carbon not only act as a food source for microorganism but also serve a binding agent for soil particles and in between space of these complexes soil enzymes are being protected [95]. Till now a uniform standard for selection of indicator enzyme is absent, no enzyme serve the purpose of being an universal indicator for soil quality determination. Heavy metals affect different enzymes differently based on their respective environment. All the enzymes used in diagnosis of soilquality can be divided in two classes one oxidoreductase (polyphenoloxidase, catalase etc.) and other one is hydrolases (amylase, urease, phosphatase, etc.). oxidoreductase are bioindicator enzymes, they take part in detoxification of metal contaminated soils hence more sensitive for heavy metal pollution as an indicator [87]. While hydrolases are involved in nutrient cycling hence can be used as auxiliary enzymes. Highly heterogeneous nature of soils demands further verification of this hypothesis over a long time to validate the results. Moreover we need better quantitative relation to understand the nature of heavy metals and enzymes along with their environmental condition (**Table 4**).

7. Conclusion

Different microbiological indices including microbial abundance, diversity structure and function of microbial community have been used to diagnosis of soil health. So far there is not any single method is alone found a suitable indicator of heavy metal pollution. Every indicator has their shortcomings as microbial abundance does not consider population structure change. Community structure does not reflect functions of population. For a better understanding of soil health all the indicators need to be used simultaneously. More study is needed in the direction of heavy metal contamination diagnosis with functional microorganism. Quantitative relationship between physicochemical factors and microbial indicators need to be established in a better way. Harm due to heavy metal on microorganism depends on the speciation and availability of heavy metal not on metal abundance. Heavy metals may change their toxicity after entering the complex soil system [74]. Long term experiments are needed to find the long term effect of heavy metals short term diagnosis of soil quality is unable to reflect long term soil quality changes.

Author details

Sukirtee Chejara[1*], Paras Kamboj[2], Y.V. Singh[3] and Vikas Tandon[1]

1 Department of Soil Science, CCS Haryana Agricultural University, Hisar, India

2 Department of Agronomy, CCS Haryana Agricultural University, Hisar, India

3 Department of Soil Science and Agricultural Chemistry, Institute of Agricultural Sciences, Banaras Hindu University, India

*Address all correspondence to: spsukirtee35@gmail.com

References

[1] Gómez-Sagasti MT, Alkorta I, Becerril JM, Epelde L, Anza M, Garbisu C. Microbial monitoring of the recovery of soil quality during heavy metal phytoremediation. Water, Air, and Soil Pollution. 2012;**223**:3249-3262

[2] Su C, Jiang LQ, Zhang WJ. A review on heavy metal contamination in the soil worldwide: Situation, impact and remediation techniques. Environ Skep Crit. 2014;**3**(2):24-38

[3] Imperato M, Adamo P, Naimo D, Arienzo M, Stanzione D, Violante P. Spatial distribution of heavy metals in urban soils of Naples city (Italy). Environmental Pollution. 2003;**124**:247-256

[4] Morton-Bermea O, Hernández-álvarez E, González-Hernández G, Romero F, Lozano R, Beramendi-Orosco LE. Assessment of heavy metal pollution in urban topsoils from the metropolitan area of Mexico city. Journal of Geochemical Exploration. 2009;**101**:218-224

[5] Xiao XY, Wang MW, Zhu HW, Guo ZH, Han XQ, Zeng P. Response of soil microbial activities and microbial community structure to vanadium stress. Ecotoxicology and Environmental Safety. 2017;**142**:200-206

[6] Leita L, Nobili MD, Muhlbachova G, Mondini C, Marchiol L, Zerbi G. Bioavailability and effects of heavy metals on soil microbial biomass survival during laboratory incubation. Biology and Fertility of Soils. 1995;**19**:103-108

[7] Feng G, Xie T, Wang X, Bai JY, Tang L, Zhao H, et al. Metagenomic analysis of microbial community and function involved in Cd contaminated soil. BMC Microbiology. 2018;**18**:11

[8] Zhang W, Chen L, Zhang R, Lin K. High throughput sequencing analysis of the joint effects of BDE209-Pb on soil bacterial community structure. Journal of Hazardous Materials. 2016c;**301**:1-7

[9] Song B, Zeng G, Gong J, Liang J, Xu P, Liu Z, et al. Evaluation methods for assessing effectiveness of in situ remediation of soil and sediment contaminated with organic pollutants and heavy metals. Environment International. 2017;**105**:43-55

[10] Song B, Zhang C, Zeng G, Gong J, Chang Y, Jiang Y. Antibacterial properties and mechanism of graphene oxide-silver nanocomposites as bactericidal agents for water disinfection. Archives of Biochemistry and Biophysics. 2016;**604**:167-176

[11] Zhang MY, Bai SH, Tang L, Zhang YL, Teng Y, Xu ZH. Linking potential nitrification rates, nitrogen cycling genes and soil properties after remediating the agricultural soil contaminated with heavy metal and fungicide. Chemosphere. 2017;**184**:892-899

[12] Ribbons RR, Levy-Booth DJ, Masse J, Grayston SJ, McDonald MA, Vesterdal L, et al. Linking microbial communities, functional genes and nitrogencycling processes in forest floors under four tree species. Soil Biology and Biochemistry. 2016;**103**:181-191

[13] Leininger S, Urich T, Schloter M, Schwark L, Qi J, Nicol GW, et al. Archaea predominate among ammonia-oxidizing prokaryotes in soils. Nature. 2006;**442**:806-809

[14] Hou J, Song C, Cao X, Zhou Y. Shifts between ammonia-oxidizing bacteria and archaea in relation to nitrification potential across trophic gradients in two large Chinese lakes (Lake Taihu and Lake Chaohu). Water Research. 2013;**47**:2285-2296

[15] Klotz MG, Alzerreca J, Norton JM. A gene encoding a membrane protein exists upstream of the amoA/amoB genes in ammonia oxidizing bacteria: A third member of the amo operon? FEMS Microbiol. Lett. 1997;**150**:65-73

[16] Ge S, Wang S, Yang X, Qiu S, Li B, Peng Y. Detection of nitrifiers and evaluation of partial nitrification for wastewater treatment: A review. Chemosphere. 2015;**140**:85-98

[17] Li XF, Zhu YG, Cavagnaro TR, Chen MM, Sun JW, Chen XP, et al. Do ammonia-oxidizing archaea respond to soil Cu contamination similarly as ammoniaoxidizing bacteria? Plant and Soil. 2009;**324**:209-217

[18] Ollivier J, Wanat N, Austruy A, Hitmi A, Joussein E, Welzl G, et al. Abundance and diversity of ammonia-oxidizing prokaryotes in the root-rhizosphere complex of miscanthus × giganteus grown in heavy metalcontaminated soils. Microbial Ecology. 2012;**64**:1038-1046

[19] Li X, Meng D, Li J, Yin H, Liu H, Liu X, et al. Response of soil microbial communities and microbial interactions to long-term heavy metal contamination. Environmental Pollution. 2017;**231**:908-917

[20] Mertens J, Broos K, Wakelin SA, Kowalchuk GA, Springael D, Smolders E. Bacteria, not archaea, restore nitrification in a zinc-contaminated soil. The ISME Journal. 2009;**3**:916-923

[21] Ruyters S, Mertens J, T'Seyen I, Springael D, Smolders E. Dynamics of the nitrous oxide reducing community during adaptation to Zn stress in soil. Soil Biology and Biochemistry. 2010;**42**:1581-1587

[22] Ruyters S, Nicol GW, Prosser JI, Lievens B, Smolders E. Activity of the ammonia oxidising bacteria is responsible for zinc tolerance development of the ammonia oxidising community in soil: A stable isotope probing study. Soil Biology and Biochemistry. 2013;**58**:244-247

[23] Li M, Ren L, Zhang J, Luo L, Qin P, Zhou Y, et al. Population characteristics and influential factors of nitrogen cycling functional genes in heavy metal contaminated soil remediated by biochar and compost. Sci. Total Environ. 2019;**651**:2166-2174

[24] Liu YR, Zheng YM, Shen JP, Zhang LM, He JZ. Effects of mercury on the activity and community composition of soil ammonia oxidizers. Environmental Science and Pollution Research. 2010;**17**:1237-1244

[25] Fernández-Calviño D, Bååth E. Interaction between pH and Cu toxicity on fungal and bacterial performance in soil. Soil Biology and Biochemistry. 2016;**96**:20-29

[26] Ward BB, Devol AH, Rich JJ, Chang BX, Bulow SE, Naik H, et al. Denitrification as the dominant nitrogen loss process in the Arabian Sea. Nature. 2009;**461**:78-81

[27] Zhang MY, Xu ZH, Teng Y, Christie P, Wang J, Ren WJ, et al. Non-target effects of repeated chlorothalonil application on soil nitrogen cycling: The key functional gene study. Sci Total Environ. 2016b;**543**:636-643

[28] Moffett BF, Nicholson FA, Uwakwe NC, Chambers BJ, Harris JA, Hill TC. Zinc contamination decreases the bacterial diversity of agricultural soil. FEMS Microbiology Ecology. 2003;**43**:13-19

[29] Throbäck IN, Johansson M, Rosenquist M, Pell M, Hansson M, Hallin S. Silver (Ag+) reduces denitrification and induces enrichment of novel nirK genotypes in soil. FEMS Microbiology Letters. 2007;**270**:189-194

[30] Sobolev D, Begonia M. Effects of heavy metal contamination upon soil microbes: Lead-induced changes in general and denitrifying microbial communities as evidenced by molecular markers. International Journal of Environmental Research and Public Health. 2008;**5**:450-456

[31] Magalhães CM, Machado A, Matos P, Bordalo AA. Impact of copper on the diversity, abundance and transcription of nitrite and nitrous oxide reductase genes in an urban European estuary. FEMS Microbiology Ecology. 2011;**77**:274-284

[32] Hai B, Diallo NH, Sall S, Haesler F, Schauss K, Bonzi M, et al. Quantification of key genes steering the microbial nitrogen cycle in the rhizosphere of sorghum cultivars in tropical agroecosystems. Applied and Environmental Microbiology. 2009;**75**:4993-5000

[33] Yoshida M, Ishii S, Otsuka S, Senoo K. nirK-harboring denitrifiers are more responsive to denitrification-inducing conditions in rice paddy soil than nirS-harboring bacteria. Microb. Environ. 2010;**25**:45-48

[34] Zhou Z, Zheng Y, Shen J, Zhang L, Liu Y, He J. Responses of activities, abundances and community structures of soil denitrifiers to short-term mercury stress. Journal of Environmental Sciences. 2012;**24**:369-375

[35] Charlton A, Sakrabani R, Tyrrel S, Casado MR, McGrath SP, Crooks B, et al. Long-term impact of sewage sludge application on soil microbial biomass: An evaluation using meta-analysis. Environmental Pollution. 2016;**219**:1021-1035

[36] Dos Santos JV, Varón-López M, Soares CRFS, Leal PL, Siqueira JO, de Souza Moreira FM. Biological attributes of rehabilitated soils contaminated with heavy metals. Environmental Science and Pollution Research. 2016;**23**:6735-6748

[37] Liu YZ, Zong T, Crowley D, Li LQ, Liu D, Zheng JW, et al. Decline in topsoil microbial quotient, fungal abundance and c utilization efficiency of rice paddies under heavy metal pollution across South China. PLoS One. 2012;7(6):e38858

[38] Zhang X, Wang H, He L, Lu K, Sarmah A, Li J, et al. Using biochar for remediation of soils contaminated with heavy metals and organic pollutants. Environmental Science and Pollution Research. 2013;**20**:8472-8483

[39] Zhang C, Nie S, Liang J, Zeng G, Wu H, Hua S, et al. Effects of heavy metals and soil physicochemical properties on wetland soil microbial biomass and bacterial community structure. Sci. Total Environ. 2016a;**557**:785-790

[40] Hu XF, Jiang Y, Shu Y, Hu X, Liu L, Luo F. Effects of mining wastewater discharges on heavy metal pollution and soil enzyme activity of the paddy fields. Journal of Geochemical Exploration. 2014;**147**:139-150

[41] Zhang Y, Zhang HW, Su ZC, Zhang CG. Soil microbial characteristics under long-term heavy metal stress: A case study in Zhangshi wastewater irrigation area. Shenyang. Pedosphere. 2008;**18**:1-10

[42] Khan S, Hesham AB, Qiao M, Rehman S, He JZ. Effects of Cd and Pb on soil microbial community structure and activities. Environmental Science and Pollution Research. 2010;**17**:288-296

[43] Zhang FP, Li CF, Tong LG, Yue LX, Li P, Ciren YJ, et al. Response of microbial characteristics to heavy metal pollution of mining soils in Central Tibet, China. Applied Soil Ecology. 2010;**45**:144-151

[44] Wang Y, Shi J, Wang H, Lin Q, Chen X, Chen Y. The influence of soil heavy metals pollution on soil microbial biomass, enzyme activity, and community composition near a copper smelter. Ecotoxicology and Environmental Safety. 2007a;**67**:75-81

[45] Deng L, Zeng G, Fan C, Lu L, Chen X, Chen M, et al. Response of rhizosphere microbial community structure and diversity to heavy metal co-pollution in arable soil. Applied Microbiology and Biotechnology. 2015;**99**:8259-8269

[46] Dai J, Becquer T, Rouiller JH, Reversat G, Bernhard-Reversat F, Lavelle P. Influence of heavy metals on C and N mineralisation and microbial biomass in Zn-, Pb-, Cu-, and Cd-contaminated soils. Applied Soil Ecology. 2004;**25**:99-109

[47] Song J, Shen Q, Wang L, Qiu G, Shi J, Xu J, et al. Effects of Cd, Cu, Zn and their combined action on microbial biomass and bacterial community structure. Environmental Pollution. 2018;**243**:510-518

[48] Chen J, He F, Zhang X, Sun X, Zheng J, Zheng J. Heavy metal pollution decreases microbial abundance, diversity and activity within particle-size fractions of a paddy soil. FEMS Microbiology Ecology. 2014;**87**:164-181

[49] Muhammad A, Xu J, Li Z, Wang H, Yao H. Effects of lead and cadmium nitrate on biomass and substrate utilization pattern of soil microbial communities. Chemosphere. 2005;**60**:508 514

[50] Gao Y, Zhou P, Mao L, Zhi YE, Shi WJ. Assessment of effects of heavy metals combined pollution on soil enzyme activities and microbial community structure: Modified ecological dose-response model and PCR-RAPD. Environment and Earth Science. 2010;**60**:603-612

[51] Pan J, Yu L. Effects of Cd or/and Pb on soil enzyme activities and microbial community structure. Ecological Engineering. 2011;**37**:1889-1894

[52] Rajapaksha RMCP, Tobor-Kapłon MA, Bååth E. Metal toxicity affects fungal and bacterial activities in soil differently. Applied Microbiology and Biotechnology. 2004;**70**:2966-2973

[53] Bastida F, Zsolnay A, Hernández T, García C. Past, present and future of soil quality indices: A biological perspective. Geoderma. 2008;**147**:159-171

[54] Bhattacharyya P, Tripathy S, Chakrabarti K, Chakraborty A, Banik P. Fractionation and bioavailability of metals and their impacts on microbial properties in sewage irrigated soil. Chemosphere. 2008;**72**:543-550

[55] Shukurov N, Pen-Mouratov S, Steinberger Y. The influence of soil pollution on soil microbial biomass and nematode community structure in Navoiy Industrial Park, Uzbekistan. Environ. Int. 2006;**32**:1-11

[56] Wang YP, Shi JY, Lin Q, Chen XC, Chen YX. Heavy metal availability and impact on activity of soil microorganisms along a Cu/Zn contamination gradient. Journal of Environmental Sciences. 2007b;**19**:848-853

[57] Liao M, Xie XM. Effect of heavy metals on substrate utilization pattern, biomass, and activity of microbial communities in a reclaimed mining wasteland of red soil area. Ecotoxicology and Environmental Safety. 2007;**66**:217-223

[58] Barajas-Aceves M. Comparison of different microbial biomass and activity measurement methods in metal-contaminated soils. Bioresource Technology. 2005;**96**:1405-1414

[59] Yu H, Xiang Y, Zou D. The effect of Eulaliopsis binata on the physi-chemical

properties, microbial biomass, and enzymatic activities in Cd-Pb polluted soil. Environmental Science and Pollution Research. 2016;**23**:19212-19218

[60] Jiang J, Wu L, Li N, Luo Y, Liu L, Zhao Q, et al. Effects of multiple heavy metal contamination and repeated phytoextraction by sedum plumbizincicola on soil microbial properties. European Journal of Soil Biology. 2010;**46**:18-26

[61] Griffiths BS, Philippot L. Insights into the resistance and resilience of the soil microbial community. FEMS Microbiology Reviews. 2013;**37**:112-129

[62] Sheik CS, Mitchell TW, Rizvi FZ, Rehman Y, Faisal M, Hasnain S, et al. Exposure of soil microbial communities to chromium and arsenic alters their diversity and structure. PLoS One. 2012;**7**:e40059

[63] Berg J, Brandt KK, Al-Soud WA, Holm PE, Hansen LH, Sørensen SJ, et al. Selection for Cu-tolerant bacterial communities with altered composition, but unaltered richness, via long-term Cu exposure. Applied and Environmental Microbiology. 2012;**78**:7438-7446

[64] Gołębiewski M, Deja-Sikora E, Cichosz M, Tretyn A, Wróbel B. 16S rDNA pyrosequencing analysis of bacterial community in heavy metals polluted soils. Microbial Ecology. 2014;**67**:635-647

[65] Wu B, Hou SY, Peng DH, Wang Y, Wang C, Xu F, et al. Response of soil micro-ecology to different levels of cadmium in alkaline soil. Ecotoxicology and Environmental Safety. 2018;**166**:116-122

[66] Desai C, Parikh RY, Vaishnav T, Shouche YS, Madamwar D. Tracking the influence of long-term chromium pollution on soil bacterial community structures by comparative analyses of 16S rRNA gene phylotypes. Research in Microbiology. 2009;**160**:1-9

[67] Chodak M, Gołębiewski M, Morawska-Płoskonka J, Kuduk K, Niklińska M. Diversity of microorganisms from forest soils differently polluted with heavy metals. Applied Soil Ecology. 2013;**64**:7-14

[68] Singh BK, Quince C, Macdonald CA, Khachane A, Thomas N, Al-Soud WA, et al. Loss of microbial diversity in soils is coincident with reductions in some specialized functions. Environmental Microbiology. 2014;**16**:2408-2420

[69] Shen TL, Liu L, Li YC, Wang Q, Dai JL, Wang RQ. Long-term effects of untreated wastewater on soil bacterial communities. Sci. Total Environ. 2019;**646**:940-950

[70] Sun Y, Sun G, Xu Y, Wang L, Liang X, Lin D, et al. Assessment of natural sepiolite on cadmium stabilization, microbial communities, and enzyme activities in acidic soil. Environmental Science and Pollution Research. 2013;**20**:3290-3299

[71] Kaplan H, Ratering S, Hanauer T, Felix-Henningsen P, Schnell S. Impact of trace metal contamination and in situ remediation on microbial diversity and respiratory activity of heavily polluted Kastanozems. Biology and Fertility of Soils. 2014;**50**:735-744

[72] Fernández-Calviño D, Arias-Estévez M, Díaz-Raviña M, Bååth E. Bacterial pollution induced community tolerance (PICT) to Cu and interactions with pH in long-term polluted vineyard soils. Soil Biology and Biochemistry. 2011;**43**:2324-2331

[73] Yang J, Huang JH, Lazzaro A, Tang Y, Zeyer J. Response of soil enzyme activity and microbial community in vanadium-loaded soil. Water, Air, and Soil Pollution. 2014;**225**(1-10):2012

[74] Cavani L, Manici LM, Caputo F, Peruzzi E, Ciavatta C. Ecological

restoration of a copper polluted vineyard: Long-term impact of farmland abandonment on soil biochemical properties and microbial communities. Journal of Environmental Management. 2016;**182**:37-47

[75] Bouskill NJ, Barker-Finkel J, Galloway TS, Handy RD, Ford TE. Temporal bacterial diversity associated with metal-contaminated river sediments. Ecotoxicology. 2010;**19**:317-328

[76] Yin HQ, Niu JJ, Ren YH, Cong J, Zhang XX, Fan FL, et al. An integrated insight into the response of sedimentary microbial communities to heavy metal contamination. Scientific Reports. 2015;**5**:14266

[77] Pereira LB, Vicentini R, Ottoboni LM. Changes in the bacterial community of soil from a neutral mine drainage channel. PLoS One. 2014;**9**(5):e96605

[78] Tipayno SC, Truu J, Samaddar S, Truu M, Preem J, Oopkaup K, et al. The bacterial community structure and functional profile in the heavy metal contaminated paddy soils, surrounding a nonferrous smelter in South Korea. Ecology and Evolution. 2018;**8**:6157-6168

[79] Wittebolle L, Marzorati M, Clement L, Balloi A, Daffonchio D, Heylen K, et al. Initial community evenness favours functionality under selective stress. Nature. 2009;**458**(7238):623-626

[80] Zhang J, Wang LH, Yang JC, Liu H, Dai JL. Health risk to residents and stimulation to inherent bacteria of various heavy metals in soil. Sci. Total Environ. 2015;**508**:29-36

[81] Malik S, Beer M, Megharaj M, Naidu R. The use of molecular techniques to characterize the microbial communities in contaminated soil and water. Environment International. 2008;**34**:265-276

[82] Huang Z, Ke X, Lv X, Liu Z, Ni L. Unique sequence characteristics account for good DGGE separation of almost full-length 18s rDNAs. World Journal of Microbiology and Biotechnology. 2016;**32**:48

[83] Das S, Sen M, Saha C, Chakraborty D, Das A, Banerjee M. Isolation and expression analysis of partial sequences of heavy metal transporters from Brassica juncea by coupling high throughput cloning with a molecular fingerprinting technique. Planta. 2011;**234**:139-156

[84] Gomez-Alvarez V, Teal TK, Schmidt TM. Systematic artifacts in metagenomes from complex microbial communities. The ISME Journal. 2009;**3**(11):1314-1317

[85] Zhou J, He Z, Yang Y, Deng Y, Tringe SG, Alvarez-Cohen L. High-throughput metagenomic technologies for complex microbial community analysis: Open and closed formats. MBio. 2015;**6**(1)

[86] Hallenbeck PC, Grogger M, Mraz M, Veverka D. Draft genome sequence of a thermophilic cyanobacterium from the family Oscillatoriales (strain MTP1) from the Chalk River, Colorado. Genome Announcements. 2016;**4**(1)

[87] Yang JS, Yang FL, Yang Y, Xing GL, Deng CP, Shen YT, et al. A proposal of "core enzyme" bioindicator in long-term Pb-Zn ore pollution areas based on topsoil property analysis. Environmental Pollution. 2016;**213**:760-769

[88] Lebrun JD, Trinsoutrot-Gattin I, Vinceslas-Akpa M, Bailleul C, Brault A, Mougin C, et al. Assessing impacts of copper on soil enzyme activities in regard to their natural spatiotemporal variation under long-term different land uses. Soil Biology and Biochemistry. 2012;**49**:150-156

[89] Mench M, Renella G, Gelsomino A, Landi L, Nannipieri P. Biochemical parameters and bacterial species richness in soils contaminated by sludge-borne metals and remediated with inorganic soil amendments. Environmental Pollution. 2006;**144**:24-31

[90] Khan S, Cao Q, Hesham AB, Xia Y, He J. Soil enzymatic activities and microbial community structure with different application rates of Cd and Pb. Journal of Environmental Sciences. 2007;**19**:834-840

[91] Wyszkowska J, Kucharski J, Lajszner W. The effects of copper on soil biochemical properties and its interaction with other heavy metals. Polish Journal of Environmental Studies. 2006;**15**:927-934

[92] Xian Y, Wang M, Chen W. Quantitative assessment on soil enzyme activities of heavy metal contaminated soils with various soil properties. Chemosphere. 2015;**139**:604-608

[93] Yang ZX, Liu SQ, Zhang DW, Feng SD. Effects of cadium, zinc and lead on soil enzyme activities. Journal of Environmental Sciences. 2006;**18**:1135-1141

[94] Kucharski J, Wyszkowska J, Kucharski M, Borowik A. Resistance of dehydrogenases, catalase, urease and plants to soil contamination with zinc. Journal of Elementology. 2012;**19**:929-946

[95] Liang Q, Gao R, Xi B, Zhang Y, Zhang H. Long-term effects of irrigation using water from the river receiving treated industrial wastewater on soil organic carbon fractions and enzyme activities. Agricultural Water Management. 2014;**135**:100-108

[96] Moreno JL, Garcia C, Hernandez T. Toxic effect of cadmium and nickel on soil enzymes and the influence of adding sewage sludge. European Journal of Soil Science. 2003;**54**:377-386

Chapter 3

Effect of PCP Pesticide Contamination on Soil Quality

Rim Werheni Ammeri, Yassine Hidri and Hassen Abdenaceur

Abstract

In recent years, soil contamination with pesticides has become a crucial news issue with serious short- and long-term effects on human health and its environment. Pesticides play a significant role in the success of modern farming and food production. These compounds have potential for toxicity and adverse effects on human health and ecological soil systems. Pentachlorophenol (PCP) is one of the most recalcitrant chemicals polluting the environment for its stable aromatic ring system and chloride content. Nowadays, many sites are contaminated with this substance. In these areas, concentrations may stay high for a long time because of slow degradation in the soil due to the negative effects that PCP has on soil microbial populations. Bioremediation of PCP contaminated sites can be realized introducing directly, into a contaminated system, microorganisms able to consume selectively the target compound (bioaugmentation) or increasing the microbial indigenous population by addiction of nutrients in form of organic and/or inorganic fertilizers and biosolids (biostimulation). In the present chapter, we present an overview of the effect of PCP pesticide contamination on soil microbial populations (density and diversity), enzymatic activity and physicochemical parameters. Additionally, the bioremediation process will be detailed.

Keywords: soil contamination, pentachlorophenol, bioremediation, microorganisms, ecological system

1. Introduction

Soil is an active, dynamic, and nonreplaceable reserve, and its situations impact its construction, environmental efficacy, and total stability [1, 2]. The charge of soil reposes on in portion on its ordinary structure and on the variations affected by human use and administration [3]. Soil, as the main interface with other environmental compartments, plays an important role in the fate of organic pollutants. During the treatment of crops, most of the quantities of pesticides applied reach the soil, either because pesticides are directly applied to it, or because the rain has washed the foliage of treated plants (crops and/or weeds). The ground therefore occupies a position central in regulating the fate of pesticides in the environment, and it will have a dual role of storage and purification [4].

Pollutants come into contact with soil primarily through deliberate application, dispersion, and atmospheric deposition. The soil therefore represents a storage reservoir for these substances. These compounds can also be lost from the soil or remain at high concentrations. Consequently, the fate and behavior of organic contaminants in soils have been the subject of intense research, with particular interest in the bioavailability of these contaminants [5]. The greatest difficulties in studying and estimating the retention and degradation of pollutants in soil are the diversity of chemical structures and reactivity of these compounds on the one hand and the high level of diversity of soils in terms of structure and composition, soil, and climatic conditions, especially soil temperature and humidity on the other [6]. Soil pollutants are very diverse, and they are also often harmful and toxic to all living forms and more specifically humans (Roger and Jacq, 2000). The occurrence of pollution most often results from industrial accidents, deposits, or the transport of dangerous materials [7]. Like pesticides, chlorinated solvents, nitrogen, and certain trace elements such as copper, mercury, or silver, pollutants can both be naturally present in the soil or be the result of human activities. This generalized use of PCP has led to the contamination of water and soil systems, with PCP currently considered to be a product of priority for decontamination studies [8]. These compounds are, in fact, the source of many concerns for companies operating in the wood preservation sector. Chlorine compounds are harmful, and they are found in the effluents of many industries, such as the chemical and petrochemical industries, those of resins and coking plants, pesticides, textiles, paper, and even in the pharmaceutical industry and many others [9]. This is why these chlorinated compounds appear as the most frequently encountered pollutants in various natural environments such as forests, rivers, marine waters, industrial discharges, urban effluents and even in the groundwater. The soil has a marked self-purification capacity. It is in fact capable of degrading the polluting compounds or immobilizing them inside it so that the volatilization and leaching processes are drastically reduced to the benefit, also, of the other environmental sectors. However, an excess of pollutant exceeded the soil storage capacity or a change in environmental conditions [10]. Soil, as the main interface with other environmental compartments, plays an important role in the fate of organic pollutants. Pollutants come into contact with soil primarily through deliberate application, dispersion, and atmospheric deposition. The soil therefore represents a storage reservoir for these substances. These compounds can also be lost from the soil or remain at high concentrations. Consequently, the fate and behavior of organic contaminants in soils have been the subject of intense research, with particular interest in the bioavailability of these contaminants [5]. The greatest difficulties in studying and estimating the retention and degradation of pollutants in soil are the diversity of chemical structures and reactivity of these compounds on the one hand and the high level of diversity of soils in terms of structure and composition, soil, and climatic conditions, especially soil temperature and humidity, on the other [6].

2. Soil and pesticide pollution

Vigorous soil is a necessity for a healthy existence. Vigor, value, and sustainability of soils are contingent on their physical, chemical, and biological variety. Hence, soil biodiversity that really tops midair biodiversity is vital for ecosystem permanence and service area. A detailed association occurs among soil biodiversity and agricultural soil organization [11]. Pesticides or biocides are chemicals, organic or inorganic, intended for the fight against undesirable organisms such as bacteria, fungi, insects, and weeds. The use of pesticides appears beneficial, or in the absence of treatments, the yields of agricultural production and quality (essentially the development of crops in the agricultural sector) or industrial (such as wood treatments and railways, textiles and food), would be reduced and/or weakened. Over 500 different pesticide formulations are authorized worldwide to control different types of pests in the agricultural sector [12]. Pesticides are classified into different

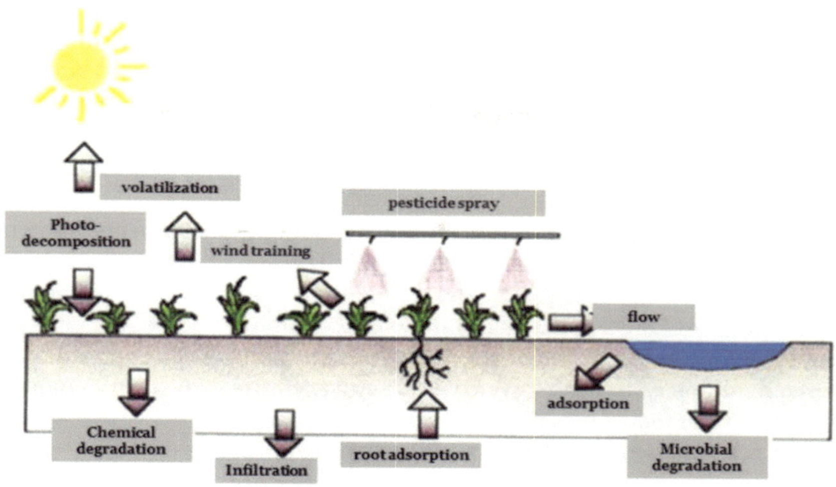

Figure 1.
Behavior of pesticide molecules in the natural environment [17].

categories according to their target, their mode of action, their time of action or their chemical nature, and recently, in response to pressure social stressing the danger of pesticides mainly for humans and the environment [12]. In this study, we will focus more specifically on the study of a pesticide very well used in the wood treatment industries, namely PCP. Soil pollution is often thought because of chemical contamination. The use of poor-quality water and application of excessive amounts of pesticides and fertilizers can result in soil contamination. To some degree, most of the soils are capable of adsorbing and detoxifying many pollutants to harmless levels through chemical and biochemical processes. Polluted water and soil pose a serious threat to plants, affecting the yield [13]. Furthermore, soil has the ability to adsorb pesticides in the humus and clay contents [14]. However, soil plays an important role in pesticide degradation [15].

The dispersion of pesticides in the various compartments of the ecosystem (air, soil, and water) is very rapid [4, 16]. When an organic compound is applied and penetrates into the soil, in substance, it may be subject to relocation or alteration of its chemical structure. These mechanisms can be on the one hand abiotic, of a physical nature (volatilization, adsorption by the soil, leaching, etc.) or of a chemical nature (hydrolysis, photodegradation, etc.) and on the other hand biotic when it occurs. Acts of absorption and metabolism by the various microorganisms living in the medium (**Figure 1**). All of these processes are strictly influenced by the physicochemical properties of the soil and pesticides as well as environmental factors [18].

3. Pentachlorophenol and soil contamination

In 1936, the American company Dow and Monsanto Chemical introduced PCP [19]. Due to its high availability and very favorable price, it has been increasingly used in different functions in several countries around the world. Its main use was accentuated in the wood industry as a preservative (80%). PCP (C_6Cl_5OH) is a highly substituted aromatic compound, prepared by reacting chlorine with phenol in the presence of a catalyst at high temperature and does not have isomers. It was last manufactured in Canada in 1983. Petroleum oils used as a carrier for PCP are generally sourced from

Canadian sources. PCP is solid at room temperature. It is a stable organic compound, slightly soluble in water and highly soluble in organic solvents. Also, PCP a proven carcinogen, immunotoxic, produces an oxidative stress and metabolic disorders [20]. PCP is a highly recalcitrant compound with toxic and carcinogenic properties. PCP is a respiratory poison with both noncarcinogenic and carcinogenic health effects. The PCP molecule can be an endocrine disruptor and inflict high toxicity on all types of organisms [21].

Due to its toxicity and carcinogenicity, as well as the large number of known sites contaminated with PCP around the world, it has been placed on the Priority Pollutants Worldwide List. The products treated with PCP are mainly telephone and power line poles and railway ties. It has therefore become the preferred impregnation product for a wide variety of other special purpose products, such as guardrail posts, signposts, retaining walls [22]. It is also used as an antimicrobial agent in industrial cooling systems, in food packaging, as the main active ingredient in exterior stains and paints. It is found in dental care products [23], in antibacterial soaps, in dermatological medical products [24] and as agricultural biocides and fungicides. PCP is toxic to humans as well as to animals. PCP toxicity is due to the fact that it decouples oxidative phosphorylation making the cell membranes permeable to protons and thus dissipating the gradient transmembrane of H + ions and electric potential [25]. It is therefore responsible for alterations in the functionality of the membranes [26]. PCP can be absorbed by mammals through the skin from the ground; it is corrosive to the skin and can cause burns and blisters. In mammals, acute exposure can increase body temperature, causes breathing difficulties, increase blood pressure, causing hypoglycemia and cardiovascular stress [24]. Chronic exposures to PCP can have serious adverse health effects. PCP is a carcinogenic, teratogenic suspect and is highly embryotoxic in addition, potential chronic effects can include kidney, liver, lung and system damage central nervous [24].

Environmental pollution from PCP can occur due to release into the environment during the production, storage, transport or use as a preservative of the wood in place. Also, the production of its sodium salt and the secondary use as fungicide, bactericide, algaecide, herbicide, etc. can cause environmental pollution.

The PCP then enters the surface and deep waters of factories, wood treatment plants, and sites for the accumulation of hazardous waste or for spillage, disposal of hazardous waste and for its use as a pesticide. In soils, due to the stability of its structure and high degree of chlorination, PCP is persistent in the environment and is one of the most common soil contaminants. The dispersion of pesticides in the various compartments of the ecosystem (air, soil, water) is very rapid [16]. When an organic compound is applied and penetrates the soil, in substance, it may be subject to relocation or to a change in its chemical structure. These mechanisms can be on the one hand abiotic, physical in nature (volatilization, adsorption by soil, leaching, etc.) or chemical in nature (hydrolysis, photodegradation, etc.), and on the other hand biotic when it acts of absorption and metabolism by the various microorganisms living in the environment. All these processes are strictly influenced by the physicochemical properties of the soil and pesticides as well as environmental factors [18].

4. Pesticide PCP effect in bacterial ecological system

Microbial communities in soils are among the most diverse on Earth [27]. In doing so, soil microorganisms mainly perform several soil functions such as the nutrient cycle and the detoxification of terrestrial ecosystems [28]. By affecting this diversity, contamination of natural environments constitutes a significant risk that can reduce the ability of ecosystems to resist and recover from the various disturbances they must undergo. The diversity of natural ecosystems is therefore an asset to be preserved. Indeed, it has been shown that the most diverse ecosystems are the most resistant and resilient to natural and anthropogenic disturbances. Since the start of the industrial era, the diversity of natural ecosystems has been in constant decline due to, among other things, contamination of soil, air, and waterways. In order to predict the effect of a substance on a biological community and thus control or limit its use, it is necessary to produce toxicological information on a wide range of organisms. Over 1.75 million different species have been listed for the eukaryotic domain alone [29]. The total number of eukaryotic species has been estimated by several authors and is generally between 5 and 10 million [30]. With regard to prokaryotes (archaea and bacteria), 10,000 species have been described, but this could constitute only around 0.1% of the total diversity of these two domains [31], for an approximate total of around 10 million. It is important to mention that the concept of species in biology and microbiology is different [32]. Each gram of soil can contain more than 1000 species of single-celled fungi [33] and 6000 species of bacteria [34]. This genetic (and therefore metabolic) diversity allows microbial communities to be involved in a multitude of processes that allow ecosystems to function well. Soil microorganisms are important contributors to the different biogeochemical cycles of carbon, nitrogen, and phosphorus in soils [28]. It has been estimated that this community could withstand between 80 to 90% of the biochemical reactions occurring in the soil [35]. Communities of soil microorganisms, via their diverse metabolic capacities, also show a response to soil pollution and thus participate in the detoxification of natural environments [35].

The communities established in polluted soils are very different from those present in unpolluted soils, whether from the point of view of total abundance or specific diversity [36]. It follows from these disturbances of communities of microorganisms a modification of the enzymatic activities carried out by the microorganisms [37]. The specific diversity and the total abundance of microorganisms can be influenced by pollution. This reaction would depend both on the nature of the pollutants and their abundance [37].

In the environment, PCP is a topic to a diversity of biological and physicochemical procedures, counting biodegradation, photodegradation, evaporation, and sorption, and leaching [38]. These procedures happen in all kinds of natural ecosystems with variable efficacy and have a direct influence on the last rate of this chemical. The main way to eliminate PCP from the environment is through biodegradation by microorganisms [39]. Studies with experimental ecosystems have designated that ecological properties may occur at PCP levels as low as those causing chronic toxicity in sensitive species in single-species tests [40, 41]. The final rate that produced adverse effects in these studies was 15.8 $\mu g/L^{-1}$, which caused a reduction in numbers of individuals and species in a marine benthic community [42]. The diversity and activity of microorganisms in the soil effect the working of ecosystems and thus plant development and health, including the quality and quantity of the crop yield [43]. However, the variety and movement of microbes are actually prone to various stresses counting chemical pollution [44]. The attendance of soil bacteria can improve the extent of pesticide degradation [15] as well as degradation of other organic pollutants [45].

5. PCP degradation

In the soil, pesticides are affected by diverse physical, chemical, and biological procedures, which will condition their degradation, their transmission to other compartments of the environment (water, plant, and atmosphere) and thus their potential influence on exposed living beings [4]. The behavior of pesticides will be more particularly controlled by the phenomena of retention on soil constituents (organic matter, clays) and degradation [46].

The remediation of a PCP contaminated site can take place through abiotic processes such as volatilization, photodecomposition, and immobilization in the soil. There biotic degradation can occur through absorption by plants or animals and through microbial degradation. Three processes are responsible for PCP biodegradation: hydroxylation, oxygenation and dechlorination. The most common formulation for PCP is that of sodium salt which, being not volatile, causes that the contribution of volatilization to the entire abiotic degradation is normally negligible [24]. The biological degradation of pollutants in the soil, or biodegradation, is carried out by living organisms and / or by the associated enzyme kit. During the biodegradation process one or more organisms metabolize the contaminant in an inorganic compound (such as CO_2, H_2O, NH_3), the autotrophs derive the necessary resources for their growth and development [47]. This catabolic activity of which microorganisms are capable, and which allows them to degrade the contaminants present in the soil, is fundamental for the fertility and health of soils. Several researchers have developed methods to treat and degrade PCP, among these techniques the use of Fenton reagent [48], photocatalytic degradation using TiO_2 [49], the combination of the two methods, namely the Fenton reaction and photocatalytic degradation [50], the ultrasonic method recommended by Francony and Pétrier [51] and also by ozonation. Another method, using the purifying capacities naturally present in certain organisms, is bioremediation. In this case, it is the microorganisms present in contaminated environments that are used to degrade the pollutants (**Figure 2**). Bacteria play an important part in this natural decontamination, due to their ability to evolve very quickly in the presence of selection pressure. Indeed, thanks to point mutations, endogenous rearrangements and horizontal transfers, they can adapt to the presence of pollutants by developing the enzymes making it possible to degrade and/or use this pollutant for their survival and their

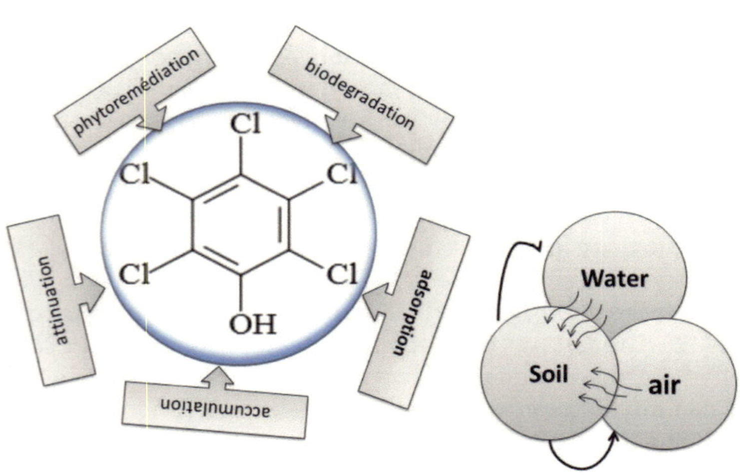

Figure 2.
PCP interactions in environment.

development. Thus, sometime after the appearance of xenobiotic molecules having no equivalent in nature, we can witness the appearance of new metabolic pathways allowing the degradation of this compound [52]. Thus, microorganisms can adapt to resistance to a broad spectrum of diverse pollutants [21]. They therefore constitute an interesting path in the development of natural techniques for removing pollutants.

Bioremediation can be carried out in different forms: natural attenuation, biostimulation and bioaugmentation. Natural attenuation is a process that uses the capacities of microorganisms present in polluted ecosystems or soils. Even if this decontamination technique does not theoretically require human intervention, it is nevertheless necessary to eliminate or neutralize the source of pollution and to constantly monitor the site until the end of treatment [53].

This type of bioremediation is very inexpensive, since it does not require a lot of resources, but it does require long periods of treatment. Biostimulation is the stimulation of the native microflora by adding nutritive molecules, specific or not, to promote bio-pollution (ex-situ or in-situ). Bioaugmentation consists of the addition or inoculation of specific bacterial cultures to stimulate the biodegradation used in bioreactors and ex situ systems.

Numerous works have monitored PCP removal by bacteria and fungi and the usage of plants for its biological elimination [54–57]. Organic objects such as wood chips, sawdust, straw of wheat have been revealed to motivate microorganisms in the removal of PCP in soil [58, 59].

6. Some enzymes responsible for the degradation of PCP

For the bacterium *Sphingomonas chlorophenolica*, the first step is catalyzed by the enzyme PCP-4 monooxygenase encoded by the pcpB gene. According to Orser et al. [60], the pcpB gene would probably be present on the chromosome and not on a mobile element (plasmid or operon). The degradation of PCP by *Sphingomonas chlorophenolica* sp. nov is carried out via four structural genes pcpA, pcpB, pcpC, and pcpD as well as by the regulatory gene pcpR [61]. Transcription of the pcpB gene is induced by the presence of PCP in the bacteria Flavobacterium sp. ATCC 39723 [60]. The presence of PCP is also necessary in the bacterium *Rhodococcus chlorophenolicus*, during the degradation of several chlorophenols [62]. For the pcpA gene, its location relative to that of pcpB is not yet known. This enzyme is responsible for the conversion of di-p-Hydroquinone to chloromaleylacetate. The pcpB enzyme is responsible for the conversion of PCP to tetrahydroquinone via the elimination of chloride ions and by hydroxylation at the para position [63]. Subsequently the enzyme tetrachloro dehalogenase reductive (pcpC) converts tetrachlorohydroquinone into trichloro-chlorohydroquinone and itself responsible for the conversion into trichlorohydroquinone (TeCHQ) into dichloro-p-hydroquinone (2,6-DCHQ) or the compound TeCHQ is sequentially dehalogenated. The latter compound is converted into chloromaleylacetate by the enzyme 1,2-dioxygenase (pcpA) which will subsequently be converted into 2-maleylacetate by the enzyme pcpE; this degradation of 2,6-DCHQ occurs by cleavage of the cycle, leading to the formation of 2-chloromaleylacetate which is more degraded via the Krebs cycle [63]. The presence of an electron donor and acceptor is essential. Biodegradation would not occur if one of the two is missing. The degradation rates of organic compounds depend on their chemical structure. The more a molecule is substituted, the more difficult it is to degrade. The position of substituent also plays a role since the ortho and meta positions increase the resistance of the molecules, as well as the substitutions on alpha carbon, compared to that in omega. Under aerobic conditions, substituted chlorine inhibits the activity of monooxygenase and dioxygenase, which are the main aerobic cleavage enzymes of the benzene nucleus. PCP is resistant to aerobic degradation. Reductive dechlorination has been suggested to be the first step in the biodegradation mechanism of PCP [64]. Low-substituted chlorophenols are more sensitive and labile to aerobic degradation [65]. The optimal condition is that these reductive dechlorination metabolites are subsequently degraded via the aerobic process. This can reduce the inhibition produced by the dechlorinated intermediates. The progressive mineralization of the component generates a series of microbial communities and enzymatic activities that enables an efficient dissipation of pesticides in soil and avoids metabolite accumulation [45].

6.1 PCP effect in enzymatic soil activity

It must also take into reason that soil is a specific active micro-habitat, everywhere organic and inorganic constituents, microbes, enzymes, nutrients, and environmental influences collaborate with each other and alteration with period and place. Evidently, these communications can control spatial variety of soil bacterial communities and enzyme activities and affect their appearance and association levels, in turn depending on diverse soil properties [66]. Consequently, difficulties in the approximation of the total bacteriological community and its dynamic portion, characterized by enzyme actions, can raise level if progressive methods have been utilized in their control. Many studies recommended that soil enzyme activities as appropriate and reliable indicators of soil quality by Gianfreda and Bollag [67] and Drijber et al. [68]. The study of Siczek et al. [40, 41] improved that the soil biological parameters can increase the activities of the enzymes involved in the N and P cycle (protease and acid phosphomonoesterase) and total activity (dehydrogenase). Some biological analysis confirmed that the addition of PCP had a significant impact on the metabolic potential of soil bacteria. Several studies have described changes in the enzymatic activities of soil contaminated with PCP [69]. PCP degradation is a process that can be completed through three ways: oxygenolysis, hydroxylation, or reductive dehalogenation [70] (**Figure 3**). Since, soil microorganisms can produce various extracellular compounds like oxidoreductases, such as peroxidases, laccases, and tyrosinases. The laccase is known as the benzene-oxygen oxidoreductase; EC 1.0.3.2. has been subjected to intensive research in the last decades. This enzyme oxidizes a great variety of aromatic compounds with a concomitant reduction of oxygen to water [71, 72]. Thus, this kind of enzyme is involved in the oxidative coupling processes of chlorophenols [73]. The residual products of enzymatic reactions, laccase, and peroxidase are usually less toxic than the parent components according to Gianfreda and Bollag [74]. PCP removal from soil can occur either by abiotic [58] or enzymatic oxidative processes [75]. According to Liang et al. [76], the incorporation of some organic compounds to soil allowed effectively stimulation of the dehydrogenase activity since the added organic material may contain some intra- and extracellular enzymes allowing stimulation of the microbial activity in the soil. Also, bioaugmentation is known as a bioremediation choice allowed by increasing the natural in-situ microbial population in the polluted environment [77].

PCP also troubled the activities of intracellular enzymes, which are measured to be an indicator of the active microbial biomass, since they are active within the living cells of microorganisms [40]. Zhang et al. [78] originate that phenol contaminants (including PCP) significantly reduced dehydrogenase, respiration, and urease activity in comparison with soil, which had not been contaminated. As dehydrogenase contributes to the biological oxidation of soil organic matter by hydrogen

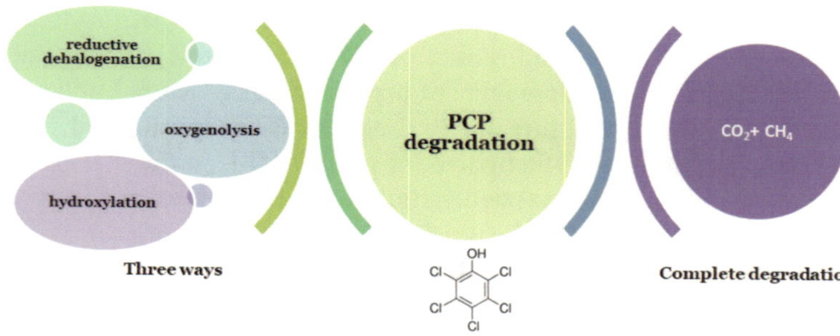

Figure 3.
PCP degradation ways.

relocation from the organic substrate to inorganic acceptors, the lower activity of this enzyme could designate an inferior rate of decomposition of soil organic matter after PCP treatment. A similar conclusion may be drawn from a respiration analysis; this activity was also reduced by PCP [40, 41].

On the other hand, PCP increased the amount of phosphorus transformation, as showed by an acid phosphomonoesterase analysis showing an important coefficient correlation (r = 0.850) with PCP. Wang et al. [79] showed the opposite effect of PCP on acid phosphomonoesterase was create in our study (it increases activity), which could be the result of different soil properties and different experimental conditions.

In a micro-environment study, the destructive effect of PCP on manganese peroxidase activity was controlled during the first 14 days, though, after that period the movement augmented [58]. Additionally, laccase movement decrease to PCP. A laboratory research presented that the influence of PCP on enzymes was reliant on its rate [80]. A study by Urrutia et al. [57] achieved with rhizotrons showed that here was no impact of growing PCP rate in soil from 50 to 250 mg kg − 1 on the microbial biomass in the ryegrass rhizosphere. However, PCP negatively affected soil activity through reducing the dehydrogenase as well as β-glucosidase activities as the PCP rate augmented. A considerable rate of literature has been published in relation to the influence of organic contaminants including PCP on soil enzyme and microorganism activity [41, 81]. However, some studies mainly concerned laboratory experiments, proved that PCP significantly reduced dehydrogenase, respiration, protease, urease, and β-glucosidase activity. This shows that PCP was a substantial factor in decreasing microbe activity in soils [80].

6.2 PCP effect in physicochemical parameters

PCP absorbs to organic matter causing removal of PCP from water into sediment depending on the chemical structure and environmental conditions[82, 83]. Bio-elimination of chemicals arises through the actions of logically arising microorganisms and biomass population. Soil influences, such as moisture content, pH, and temperature, also show a significant character. The removal is improved in the soil pH range of 5.5–8.0, with an optimal value of about 7 [84], and tends to rise with temperature [85]. The result of soil moisture satisfied on the biodegradation of pesticides, though, is not completely assumed. It is acknowledged that the accessibility of soil moisture is obligatory for improved biomass movement. The amount of pesticide removal under saturated soil situations is also acknowledged to be very slow [86]. With upper soil moisture content and soil temperature in the summer months, the pesticide may destroy quickly, thus dropping the hazard of water pollution. It can, though, be renowned that the moisture content in the soil profile is not preserved at the similar level during sub-irrigation; it is close to saturation near the water table and reductions with distance overhead the water table. Yet, when the soils had a low level of organic matter (>10%) will be take a great affinity for organic pollutants due to the presence of humic acid, fulvic acid, and reactive clay such as Al and Fe hydroxide groups [58]. In the literatures, there are plenty of studies indicated that denitrification can be disturbed by several environmental pollutants, such as heavy metal and synthetic organic compounds [87]. For example, the Zinc oxide nanoparticles have been observed to inhibit the denitrifying reductase, which further led to more nitrate accumulation. Zheng et al. [88] found that, it is essential to explore the effect of PCP on the metabolism and function of denitrifying bacteria. The contact of PCP to P. *denitrificans* bacteria induced the reduced the key enzymes activities connected to glycolysis process, caused the trouble of the metabolism of glucose utilization and the cell growth, and subsequently disturbed the generation of electron donor (NADH) for denitrification via NAD+ decrease. Denitrification procedure was significantly inhibited by the PCP at upper amount of PCP, which would further disturb the nitrogen cycle in soil [89].

This may indicate that less nitrogen was available for the plants, and that the plants contaminated with PCP may suffer from nitrogen deficiency, which confirmed our analysis of plant N content. It is worth noting that relatively speaking the most harmful effect of PCP was noted for enzymes related to the nitrogen cycle, e.g. protease and urease [57]. In satisfactory situations of development e.g. pH, temperature and moisture and adequate supply of nutrients like vitamins, magnesium, manganese, copper, sulfur, potassium, phosphorus and nitrogen, microbes can biodegrade/biotransform the complex hazardous organic chemicals into simpler and harmful ones. After the usage of "super bug" in elimination of oil spills, there has been numerous efficacious stories of microbial method in clean-up of polluted lands and soils [90]. The Microbiological Resource Centers (MIRCENS) at Cairo, Egypt is examining the use of microbes in degrading persistent pesticides pollutants (UNEP Reports, 1996–2006).

6.3 PCP bioremediation in soil

Soils are open, porous, multi-compound of biogeochemical systems containing solids, liquids, and gases [91]. At the same time, they are a preferred sink for dangerous pollutants like hydrophobic organic compounds and multiple other compounds that are increasingly finding their place in the environment [92, 93]. The main difficulties encountered in biological treatment methods are the lack of knowledge concerning the bacterial population degrading PCP under unfavorable environmental conditions [94, 95]. Biodegradation is a biological degradation carried out by living beings (bacteria, fungi, plants, etc.). It is due to the abundance and variety of organisms in the environment considered [96]. For example, the attack of a chemical molecule by microorganisms often results in its mineralization and the production of low molecular weight metabolites (**Table 1**). Two types of biodegradation are most often cited and distinguished:

Primary biodegradation: It corresponds to metabolism and co-metabolism. These can be done by substitution or rearrangement of the structure of the compound, by redox or isomerization, or by addition and loss of substituent. This is a partial attack on the molecule. In some cases, it can lead to the appearance of persistent metabolites, more bioavailable and/or more toxic than the initial molecule.
Ultimate biodegradation: It is a complete degradation leading to the formation of carbon dioxide, methane, water, and mineral elements. This biodegradation, if it

Bacterial strains	Degraded chlorophenols	References
Pseudomonas sp. UG25 et UG30	PCP	[97]
Pseudomonas sp. RA2	PCP	[98]
Pseudomonas sp. strain SR3	PCP	[99]
Pseudomonas sp. strain IST103 PCP	PCP	[100]
Pseudomonas mendocina NSYSU	PCP	[101]
Mycobacterium sp. strain CG-2	PCP	[102]
Mycobacterium chlorophenolicum PCP-1	PCP	[103]
Sphingomonas sp. strain P5	PCP	[104]
Sphingomonas chlorophenolica RA2	PCP	Nohynek et al., 1995; Ederer et al., 1997; [103]
Novosphingobium lentum MT1	PCP	[105]
Sphingomonas chlorophenolica	PCP	[106]
Rhodococcus sp. CP-2 et CG1	PCP	[102, 62, 107]
Strain KC-3	PCP	[108]
Flavobacterium sp. ATCC39723	PCP	[109, 60]
Flavobacterium sp.	PCP	Gonzalez and Hu, 1991
Flavobacterium strains	PCP	[110, 111]
Arthrobacter strain NC	PCP	[112]
Corynebactrium	PCP	[108]
Burkholderia	PCP and CP	[113]

Table 1.
Examples of some bacterial strains competent to degrade PCP.

occurs quickly, leads to the total elimination of the pollutant from the environment. A substance that undergoes ultimate biodegradation is one that poses less risk to the environment than a substance that undergoes primary biodegradation.

Bacteria can feed on all kinds of compounds. These are what we call electron donors. In addition, they can breathe with different compounds. These are the electron acceptors. In the case of stimulated biodegradation, the electron donor or electron acceptor is contamination. In this context, several researchers have focused their attention on studying microbial biodegradation which has been reported as a main mechanism of the dissipation of pesticides in the soil environment ([114]; Pieuchot al., 1996). As an electron acceptor or as an electron donor, the degradation of these molecules is an integral part of metabolism and directly serves the production of energy for microorganisms. The substance appears to be metabolized by the body. A compound is said to be biodegradable if it is completely transformed by living organisms into CO_2, H_2O, and cellular biomass. Mineralization corresponds to the bioconversion of organic matter into mineral products (CO_2, CH_4, H_2O, NH_3, HCI, etc.). It is the reverse biological process of the synthesis of organic matter (mainly photosynthesis and methanogenesis). Some molecules are resistant to any degradation action over very long periods. The stability of these molecules is linked to their chemical structure, their concentration, and the characteristics of the surrounding environment. Generally, the more a molecule is substituted, the more it is resistant to biodegradation. The position of the substituents also plays a role [64]. Replacing carbon with other atoms such as O, N, S, such as multiple branching on the same carbon atom, changes the resistance to biotransformation of organic products. The presence of the substrate in too high concentration may result in the inhibition or inactivation of one or more enzymes involved in microbial metabolism. Many species of soil bacteria have been isolated from samples of soil contaminated with PCP (**Table 1**).

Whereas microbial remediation (bioremediation) is a fixed technology for the removal of organic soil contaminants, the use of microorganisms to transform organic contaminants similar PCP is still being explored. Bioremediation of soils includes numerous technologies, counting bioaugmentation and also biostimulation, to augment the elimination of PAHs. Bio-augmentation, it is the addition of microorganisms that biodegrade (toxic organic compounds) a specific contaminant. Microbial remediation depends upon the appearance of suitable microorganisms in the correct amounts and in mixtures and in appropriate environmental conditions. Biostimulation and bio-augmentation are two indispensable factors inducing bioremediation by microbes. In the bio-Stimulation procedure, the adding of the amendments serves to rise the number or activity or both, of naturally happening micro-organisms available for bioremediation. The in-situ bacteriological remediation approaches might necessity to combination with phytoremediation process with suitable hyper-fixator plants that can successfully acceptance the pollutant (made bioavailable by the microorganisms) from soil and bioaccumulate them in their roots and shoots, thus stopping their reprocessing in soil. Bioremediation is the procedure by which active organisms destroy or transform hazardous organic contaminants to inorganic components, such as CO_2, H_2O, and NO_3- [115], which are also formed during the elimination of organic matter in soil. A numeral of procedures upstream of the biocatalysis, e.g., dispersal in solid matrixes, bioavailability, weathering, and abiotic catalysis of pollutants, and downstream, stress, predation, and competition, are acknowledged to oblige the procedure [116]. PCP degraders are ubiquitous at contaminated sites with widespread PCP contamination, but their degradation amounts are relatively low in soil due to low solubility/bioavailability of PCP, poor nutrient level and inappropriate soil redox conditions [69]. The variation of some enzymatic activities and mainly of those partially involved in the contaminant transformation will occur. On the other side, many studies have shown that the addition of supplemental nutrients known as biostimulation procedures, like carbon, nitrogen (C:N) [117]; phosphorous (as phosphates) should mainly increase the rate of xenobiotic compounds degradation such PCP [118]. However, the relationship between nutrient supplementation and microbial degradation of organic contaminants does not appear to be completely straight forward [117].

6.4 The mechanism of microbial remediation of toxic pesticide

Researches on microbial elimination of pesticide residues created in 1940s, and as people reimbursement more consideration to the environment, the research on the elimination procedure and degradation mechanism of organic contaminants has been intensely considered [119]. Bacteria in normal conditions could destroy the pesticide residues, with little cost and environmentally friendly and it would not cause secondary pollution [120]. But the efficacy was moderately slow, and the natural environment was complex and variable, which may disturb the viability and productivity of microbial degradation of pesticides. Consequently, researchers have showed fine studies of bacteria and had a clear considerate of the degradation mechanism of organic pesticides. Numerous microorganisms have been known in nature, which can disturbance depressed the dangerous organic substances in the environment (soil and water) comprising the xenobiotic composites such as pesticides, polycyclic aromatic hydrocarbons (PAHs) and the chlorinated substances approaching polychlorinated biphenyls (PCBs) in due course of time. General of the organochlorines looks to be bio-change, create conjugates with the soil humic matter. Bacterial mineralization of toxic organics logically happening aerobic bacteria decompose both natural and the synthetic hazardous organic materials to harmless CO_2 and water (**Figure 4**). However, mechanism of microbial act in removal of toxic composites is attained by biodegradation and biotransformation of compound toxic chemicals into inoffensive simpler biochemical produces [121].

Microorganisms acclimatize to eliminate "novel artificial compounds" either by using catabolic enzymes they previously possess or by obtaining novel metabolic pathways. Microorganisms break down the complex hydrocarbons in the dangerous waste by via the three general mechanisms-aerobic and anaerobic respiration and fermentation. Aerobic procedure needs satisfactory supply of oxygen, the biodegradation procedure is fast and more complete, and there are no problematic products similar methane and hydrogen sulfide. In anaerobic degradation, for example, there is a sequential, biologically destructive process in which the complex "hydrocarbons" of hazardous wastes are converted into simpler molecules of "carbon dioxide" and "methane." PCP readily degrades in the environment by chemical, microbiological and photochemical procedures. Degradation in soil is affected by numerous chemical, physical, and biological factors. PCP degrades more quickly in flooded or anaerobic soil than in aerobic moist soil. Numerous pathways of degradation have been studied.

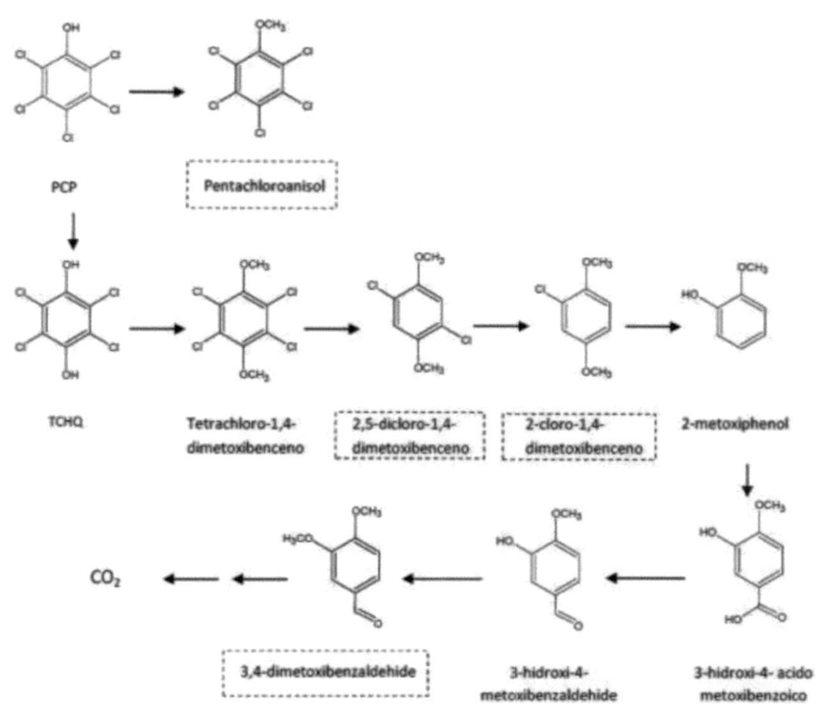

Figure 4.
Biological aspects involved in the degradation of organic pollutants [121].

7. Conclusions

This chapter investigated the effects of PCP soil contamination on microbial diversity, enzymatic activities, microbial biomass, and physicochemical soil characters. In general, the results verified the damaging consequence of PCP on soil activity and variations in soil microbe and genetic variety. PCP negatively affected the intracellular actions of soil microbes and the amount of nitrogen alteration. This may result in the deterioration of soil role and procedures connected to nutrient availability to plants and soil organic matter decomposition and, so, unfavorably affect plant development and health. Moreover, the results presented that the soil fungal community is more sensitive to PCP pollution than the bacterial community. However, enzyme activity can be inhibited at PCP contaminated soil. The denitrification process was significantly reduced by the PCP at a higher rate of PCP, which would further interrupt the nitrogen cycle in soil. Finally, it is necessary to study more details about the effect of PCP accumulation in long-term in soil.

Author details

Rim Werheni Ammeri[1,2,3*], Yassine Hidri[4] and Hassen Abdenaceur[1,2,3]

1 Water Treatment and Recycling Laboratory, Soliman, Tunisie

2 Water Research and Technology Center (CERTE), Technopole Borj-Cédria, Soliman, Tunisie

3 Faculty of Sciences of Tunis (FST), University of Mathematical, Physical and Natural Sciences of Tunis El Manar, Tunisie

4 Laboratory of Integrated Olive Production (LR 16IO3), Tunis, Tunisie

*Address all correspondence to: rimwerheni@gmail.com

References

[1] Dorana JW, Zeiss MR. Soil health and sustainability: Managing the biotic component of soil quality. Applied Soil Ecology. 2000;**15**(1):3-11

[2] Fatichi S, Or D, Walko R, Vereecken H, Young MH, Ghezzehei TA, et al. Soil structure is an important omission in earth system models. Nature Communications. 2020;**11**:522

[3] Pierce FJ, Larson WE, Dowdy RH, Graham WAP. Productivity of soils: Assessing long-term changes due to erosion. Journal of Soil and Water Conservation. 1983;**38**:39-44

[4] Barriuso E, Calvet R, Schiavon M, Soulas G. Les pesticides et les polluants organiques des sols: Transformations et dissipation. In: Etudes et Gestion des Sols, numéro spécial: Le sol, un patrimoine menacé? Vol. 3. Boca Raton, Florida: Lewis Publishers; 1996. pp. 279-295

[5] Semple SJ. Illustrations of damnation in late Anglo-Saxon manuscripts. Anglo-Saxon England. 2003;**32**:31-45

[6] Reid BJ, Stokes JD, Jones KC, Semple KT. Nonexhaustive cyclodextrin-based extraction technique for the evaluation of PAH bioavailability. Environmental Science & Technology. 2000;**34**:3174-3179

[7] Riser-Roberts E. Remediation of Petroleum Contaminated Soils: Biological, Physical, and Chemical Processes. Boca Raton: CRC Press; 1998

[8] Xiao P, Kondo R. Biodegradation and biotransformation of pentachlorophenol by wood-decaying white rot fungus *Phlebia acanthocystis* TMIC34875. Journal of Wood Science. 2020;**66**:2

[9] Chanama S, Crawford RL. Mutational analysis of pcpA and its role in pentachlorophenol degradation by *Sphingomonas* (Flavobacterium) chlorophenolica ATCC 39723. Applied and Environmental Microbiology. 1997;**63**(12):4833-4838

[10] Werheni R, Mokni TS, Mehri I, Badi S, Hassen A. Pentachlorophenol biodegradation by *Citrobacter freundii* isolated from forest contaminated soil. Water, Air, and Soil Pollution. 2016;**227**:367

[11] Thiele-Bruhn S, Bloem J, de Vries FT, Kalbitz K, Wagg C. Linking soil biodiversity and agricultural soil management. Current Opinion in Environmental Sustainability. 2012;**4**(5):523-528

[12] Arias-Estevez M, Lopez-Periago E, Martinez-Carballo E, Simal-Gandara J, Mejuto JC, Garcia-Rio L. The mobility and degradation of pesticides in soils and the pollution of groundwater resources. Agriculture, Ecosystems and Environment. 2008;**123**:247-260

[13] Suresh KR, Nagesh MA. Experimental studies on effect of water and soil quality on crop yield. In: International Conference on Water Resources, Coastal and Ocean Engineering (ICWRCOE). Aquatic Procedia. Vol. 4. 2015. pp. 1235-1242

[14] Castillo MDP, Torstensson L. Effect of biobed composition, moisture, and temperature on the degradation of pesticides. Journal of Agricultural and Food Chemistry. 2007;**55**(14):5725-5733

[15] Yadav M, Srivastva N, Singh RS, Upadhyay SN, Dubey SK. Biodegradation of chlorpyrifos by *Pseudomonas* sp. in a continuous packed bed bioreactor. Bioresource Technology. 2014;**165**:265-269. DOI: 10.1016/j.biortech.2014.01.098

[16] Louis M, Strandholm K, Kevin Steensma H, Mark Weaver K. The moderating effect of national

culture on the relationship between entrepreneurial orientation and strategic alliance portfolio extensiveness. Entrepreneurship Theory and Practice. 2002;**26**:145-160

[17] Siampiringue M. Phototransformation de polluants organiques à la surface de sol (études cinétique et analytique sur supports modèles et sur sol réel). Clermont-Ferrand: Laboratoire de Photochimie Moléculaire et Macromoléculaire, Université Blaise Pascal, École doctorale des sciences fondamentales; 2011

[18] Testa RJ, Sciacca LM, Wang F, Hendricks ML, Goldblum P, Bradford J, et al. Effects of violence on transgender people. Professional Psychology: Research and Practice. 2012;**43**(5):452-459

[19] Gunther FA, Buzzetti F, Westlake WE. Residue behavior of polynuclear hydrocarbons on and in oranges. Residue Reviews. 1967;**17**:81-104

[20] Maheshwari P, Du H, Sheen J, Assmann SM, Albert R. Model-driven discovery of calcium-related protein-phosphatase inhibition in plant guard cell signaling. PLoS Computational Biology. 2019;**15**(10):29-74

[21] Santosh P, Ramavadh R, Shanthakumar SP. Isolation and characterization of phenol degrading bacteria *Stenotrophomonas* sp. SKC_BP54 Kasbekar. Research Journal of Biotechnology. 2018;**53**(1):39-45

[22] Domtar Corporation. Politics and Business Magazines. 395 de Maisonneuve Boulevard West Montreal, Quebec H3A 1L6 Canada; 1996

[23] Jones MJ. Conservation systems and crop production. In: Paper Presented to Conference on Agricultural Development in Drought-Prone Africa. London: Royal Commonwealth Society; 1985

[24] Crosby DG, Beynon KI, Greve PA, Korte K, Still GG, Vonk JW. Environmental chemistry of pentachlorophenol; 1981

[25] Steiert JG, Thoma WJ, Ugurbil K, Crawford RL. 31P nuclear magnetic resonance studies of effects of some chlorophenols on *Escherichia coli* and a pentachlorophenol-degrading bacterium. Journal of Bacteriology. 1988;**170**(10):4954-4957

[26] Smejtek P, Barstad AW, Wang S. Pentachlorophenol-induced charge of c-potential and gel-to-fluid transition temperature in model lecithin membranes. Chemico-Biological Interactions. 1989;**71**:37-61

[27] Roesch LFW, Fulthorpe RR, Riva A, Casella G, Hadwin AKM, Kent AD, et al. Pyrosequencing enumerates and contrasts soil microbial diversity. The International Society for Microbial Ecology Journal. 2007;**1**:283-290

[28] van der Heijden MGA, Bardgett RD, van Straalen NM. The unseen majority: Soil microbes as drivers of plant diversity and productivity in terrestrial ecosystems. Ecology Letters. 2008;**11**:296-310

[29] Groombridge M, Jenkins D, editors. World Atlas of Biodiversity: Earth's Living Resources in the 21st Century. Berkeley, CA: University of California Press; 2002. pp. 195-223

[30] Baillie JEM, Upham K. Species diversity within and among ecosystems. In: Leemans R, editor. Ecological Systems. 2013. pp. 257-271

[31] Tamames J, Rosselló-Móra R. On the fitness of microbial taxonomy. Trends in Microbiology. 2012;**20**:514-516

[32] Ereshefsky M. Darwin's solution to the species problem. Synthese. 2009;**175**(3):405-425

[33] Buée M, Reich M, Murat C, Morin E, Nilsson RH, Uroz S, et al. 454 pyrosequencing analyses of forest soils reveal an unexpectedly high fungal diversity. New Phytologist. 2009;**184**:449-456

[34] Pires C, Franco AR, Pereira SIA, Henriques I, Correia A, Magan N, et al. Metal(loid)-contaminated soils as a source of culturable heterotrophic aerobic bacteria for remediation applications. Geomicrobiology Journal. 2017;**34**(9):760-768

[35] Nannipieri P, Ascher J, Ceccherini MT, Landi L, Pietramellara G, Renella G. Microbial diversity and soil functions. European Journal of Soil Science. 2003;**54**:655-670

[36] Giller K, Witter E, Corbeels M, Tittonell PA. Conservation agriculture and smallholder farming in Africa: The Heretics' view. Field Crops Research. 2009;**114**(1):23-34

[37] Brandt A, Papagiannouli F, Wagner N, Wilsch-Bräuninger M, Braun M, Furlong EE, et al. Developmental control of nuclear size and shape by Kugelkern and Kurzkern. Current Biology. 2006;**16**(6):543-552

[38] Verbrugge SAJ, Schonfelder M, Becker L, Yaghoob Nezhad F, Hrabe de Angelis M, Wackerhage H. Genes whose gain or loss-of-function increases skeletal muscle mass in mice: A systematic literature review. Frontiers in Physiology. 2018;**9**:553

[39] Lopez-Echartea E, Macek T, Demnerova K, Uhlik O. Bacterial biotransformation of pentachlorophenol and micropollutants formed during its production process. International Journal of Environmental Research and Public Health. 2016;**13**:1146

[40] Siczek A, Frąc M, Gryta A, Kalembasa S, Kalembasa D. Variation in soil microbial population and enzyme activities under faba bean as affected by pentachlorophenol. Applied Soil Ecology. 2020a;**150**:103466

[41] Siczek A, Wielbo J, Lipiec J, Kalembasa S, Kalembasa D, Kidaj D, et al. Nod factors improve the nitrogen content and rhizobial diversity of faba bean and alter soil dehydrogenase, protease, and acid phosphomonoesterase activities. International Agrophysics. 2020b;**34**(1):9-15

[42] Tagatz ME, Tobia M. Effects of barite ($BaSO_4$) on development of estuarine communities. Estuarine and Coastal Marine Science. 1978;**7**:401-407

[43] Jacoby R, Peukert M, Succurro A, Koprivova A, Kopriva S. The role of soil microorganisms in plant mineral nutrition - Current knowledge and future directions. Frontiers in Plant Science. 2017;**8**:16-17

[44] Gałązka A, Gawryjołek K, Gajda A, Furtak K, Księżniak A, Jończyk K. Assessment of the glomalins content in the soil under winter wheat in different crop production systems. Plant, Soil and Environment. 2018;**64**(1):32-37

[45] Castillo L, Martínez AI, Gelis S, Ruiz-Herrera J, Valentín E, Sentandreu R. Genomic response programs of *Saccharomyces cerevisiae* following protoplasting and regeneration. Fungal Genetics and Biology. 2008;**45**(3):253-265

[46] Bosso N, Gugliotta A, Zampieri N. Strategies to simulate wheel-rail adhesion in degraded conditions using a roller-rig. Vehicle System Dynamics. 2015;**53**(5):619-634

[47] Chaplain V, Défossez P, Richard G, Tessier D, Roger-Estrade J. Contrasted effects of no-till on bulk density of soil and mechanical resistance. Soil and Tillage Research. 2011;**111**(2):105-114

[48] Watts RL, Glatt SL, Koller WC. Objective measures of motor disability. In: Koller WC, Paulson G, editors. Therapy of Parkinson's Disease. New York: Marcel Dekker; 1990. pp. 31-61

[49] Pecchi G, Reyes P, Sanhueza P, Villaseñor J. Photocatalytic degradation of pentachlorophenol on TiO_2 sol–gel catalysts. Chemosphere. 2001;**43**(2):141-146

[50] Fukushima M, Tatsumi K. Degradation pathways of pentachlorophenol by photo-Fenton systems in the presence of iron(III), humic acid and hydrogen peroxide. Environmental Science and Technology. 2001;**35**:1771-1778

[51] Francony A, Pétrier C. Sonochemical degradation of carbon tetrachloride in aqueous solution at two frequencies: 20 kHz and 500 kHz. Ultrasonics Sonochemistry. 1996;**3**:77-82

[52] Boubakri H. Le Maghreb et les migrations de transit: le piège. In revue. Migrations et sociétés. Septembre-Octobre 2006;**107**:20

[53] Mulligan CN, Yong RN, Gibbs BF. Remediation technologies for metal-contaminated soils and groundwater: An evaluation. Engineering Geology. 2001;**60**(1-4):193-207

[54] Hechmi N, Aissa NB, Abdenaceur H, Jedidi N. Evaluating the phytoremediation potential of *Phragmites australis* grown in pentachlorophenol and cadmium contaminated soils. Environmental Science and Pollution Research. 2014;**21**:1304

[55] Kumari R, Kumar R, Lynn A. g_mmpbsa—A GROMACS tool for high-throughput MM-PBSA calculations. Journal of Chemical Information and Modeling. 2014;**54**(7):1951-1962

[56] Marichal R, Grimaldi M, Feijoo MA, Oszwald J, Praxedes C, Ruiz Cobo DH, et al. Soil macroinvertebrate communities and ecosystem services in deforested landscapes of Amazonia. Applied Soil Ecology. 2014;**83**:177-185

[57] Urrutia P, Aguirre P, Esparza A, Tapia V, Mena NP, Arredondo M, et al. Inflammation alters the expression of DMT1, FPN1 and hepcidin, and it causes iron accumulation in central nervous system cells. Journal of Neurochemistry. 2013;**126**:541-549

[58] Cea M, Jorquera M, Rubilar O, Langer H, Tortella G, Diez MC. Bioremediation of soil contaminated with pentachlorophenol by *Anthracophyllum* discolor and its effect on soil microbial communities. Journal of Hazardous Materials. 2010;**181**:315-323

[59] Lin X, Wang F, Li Y, Zhai C, Wang G, Zhang X, et al. The SCF ubiquitin ligase slimb controls Nerfin-1 turnover in *Drosophila*. Biochemical and Biophysical Research Communications. 2018;**495**(1):629-633

[60] Orser CS, Lange CC. Molecular analysis of pentachlorophenol degradation. Biodegradation. 1994;**5**:277-288

[61] Copley J. Ecology goes underground. Nature. 2000;**406**:452-454

[62] Apajalahti JHA, Salkinoja-Salonen MS. Degradation of polychlorinated phenols by *Rhodococcus chlorophenolicus*. Applied Microbiology and Biotechnology. 1986;**25**:62-67

[63] Ohtsubo N, Mitsuhara I, Koga M, Seo S, Ohashi Y. Ethylene promotes the necrotic lesion formation and basic PR gene expression in TMV-infected tobacco. Plant & Cell Physiology. 1999;**40**:808-817

[64] Wang X, Mann CJ, Bai Y, Ni L, Weiner H. Molecular cloning, characterization, and potential roles of

cytosolic and mitochondrial aldehyde dehydrogenases in ethanol metabolism in *Saccharomyces cerevisiae*. Journal of Bacteriology. 1998;**180**(4):822-830

[65] Reddy GVB, Gold MH. Degradation of pentachlorophenol by *Phanerochaete chrysosporium*: Intermediates and reactions involved. Microbiology. 2000;**146**:405-413

[66] Kirk JL, Beaudette LA, Hart M, Moutoglis P, Klironomos JN, Lee H, et al. Methods of studying soil microbial diversity. Journal of Microbiological Methods. 2004;**58**:169-188

[67] Gianfreda L, Bollag J-M. Influence of natural and anthropogenic factors on enzyme activity in soil. In: Stotzky G, Bollag J-M, editors. Soil Biochemistry. Vol. 9. New York: Marcel Dekker; 1996. pp. 123-194

[68] Drijber RA, Doran JW, Parkhurst AM, Lyon DJ. Changes in soil microbial community structure with tillage under long-term wheat-fallow management. Soil Biology and Biochemistry. 2000;**32**:1419-1430

[69] Scelza R, Rao MA, Gianfreda L. Response of an agricultural soil to phenanthrene and pentachlorophenol pollution and different bioremediation strategies. Soil Biology and Biochemistry. 2008;**40**:2162-2169

[70] Field JA, Sierra-Alvarez R. Microbial degradation of chlorinated dioxins. Chemosphere. 2008;**71**:1005-1018

[71] Couto SR, Herrera JI. Laccases in pollution control. Terrestrial and Aquatic Environmental Toxicology. 2006;**2**(1):34-45

[72] Couto SR, Toca-Herrera JL. Laccase production at reactor scale by filamentous fungi. Biotechnology Advances. 2007;**25**(6):558-569

[73] Gianfreda L, Bollag JM. Isolated enzyme for the transformation and detoxification of organic pollutants. In: Burns R, Richard D, editors. Enzyme in the Environment. Activity, Ecology and Applications. New York: Marcel Dekker; 2002

[74] Gianfreda L, Bollag JM. Effect of soils on the behavior of immobilized enzymes. Soil Science Society of America Journal. 1994;**58**(16):681-721

[75] Bollag JM. Enzymes catalyzing oxidative coupling reactions of pollutants. Metal Ions in Biological Systems. 1992;**28**:205-217

[76] Liang Y, Nikolic M, Peng Y, Chen W, Jiang Y. Organic manure stimulates biological activity and barley growth in soil subject to secondary salinization. Soil Biology and Biochemistry. 2005;**37**:1185-1195

[77] Innemanova P, Velebova R, Filipova A, et al. Anaerobic in situ biodegradation of TNT using whey as an electron donor: A case study. New Biotechnology. 2015;**32**(6):701-709

[78] Zhang M, Hanna M, Li J, Butcher S, Dai H, Xiao W. Creation of a hyperpermeable yeast strain to genotoxic agents through combined inactivation of PDR and CWP genes. Toxicological Sciences. 2010;**113**(2):401-411

[79] Wang G, Post WM, Mayes MA. Development of microbial-enzyme-mediated decomposition model parameters through steady-state and dynamic analyses. Ecological Applications. 2013;**23**(1):255-272

[80] Diez JMC, Gallardo FA, Saavedra GM, Cea L, Gianfreda L, Alvear ZM. Effect of pentachlorophenol on selected soil enzyme activities in a Chilean Andisoil. Soil Nutrition Vegetal. 2006;**6**(3):40-51

[81] Jia Q, Liu S, Wen D, Cheng Y, Bendena WG, Wang J, et al. Juvenile hormone and 20-hydroxyecdysone coordinately control the developmental timing of matrix metalloproteinase-induced fat body cell dissociation. The Journal of Biological Chemistry. 2017;**292**(52):21504-21516

[82] Coats JR, Karr LL, Drewes CD. Toxicity and neurotoxic effects of monoterpenoids: In insects and earthworms. Entomology at Iowa State University digital repository. In: Hedin P, editor. Natural Occuring Pest Bioregulators, ACS Symposium Series. Washington, D.C.; 1991a. pp. 1-9

[83] Coats JR, Karr LL, Drewes CD. Toxicity and neurotoxic effects of monoterpenoids: In insects and earthworms. In: Entomology at Iowa State University Digital Repository. 1991b. pp. 1-9

[84] Sparks DL, Page AL, Helmke PA, Loeppert RH, Soltanpour PN, Tabatabai MA, et al. Methods of Soil Analysis. Part 3: Chemical Methods. 3rd ed. Madison, WI: SSSA and ASA; 1996. pp. 46-64

[85] Savage DC. Microbial ecology of the gastrointestinal tract. Annual Review of Microbiology. 1977;**31**:107-133

[86] Goswami KP, Green RE. Microbial degradation of the herbicide atrazine and its 2-hydroxy analog in submerged soils. Environmental Science & Technology. 1971;**5**:426-429

[87] Behera PC, Tripathy DP, Parija SC. Shatavari: Potentials for galactogogue in dairy cows. Indian Journal of Traditional Knowledge. 2013;**12**(1):9-17

[88] Zheng X, Su Y, Chen Y, et al. Zinc oxide nanoparticles cause inhibition of microbial denitrification by affecting transcriptional regulation and enzyme activity. Environmental Science & Technology. 2014;**48**:800-807

[89] Chen Y, Yu S, Tang S, Li Y, Liu H, Zhang X, et al. Site-specific water quality criteria for aquatic ecosystems: A case study of pentachlorophenol for Tai Lake, China. Science of the Total Environment. 2016;**541**:65-73

[90] US Geology Survey (USGS). Coastal Wetlands and Global Change: Overview. USGS FS-089-97; 1997

[91] McBride MB. Environmental Chemistry of Soils. New York: Oxford University Press, Inc.; 1994

[92] Cordova-Rosa SM, Dams RI, Cordova-Rosa EV, Radetski MR, Corrêa AXR, Radetski CM. Remediation of phenol-contaminated soil by a bacterial consortium and *Acinetobacter calcoaceticus* isolated from an industrial wastewater. Journal of Hazardous Materials. 2009;**164**:61-66

[93] Cornelissen G, Gustafsson Ö, Bucheli TD, Jonker MTO, Koelmans AA, van Noort PCM. Extensive sorption of organic compounds to black carbon, coal, and kerogen in sediments and soils: Mechanisms and consequences for distribution, bioaccumulation, and biodegradation. Environmental Science & Technology. 2005;**39**(18):6881-6895

[94] Kaoa CM, Chaib CT, Liub JK, Yehc TY, Chena KF, Chend SC. Evaluation of natural and enhanced PCP biodegradation at a former pesticide manufacturing plant. Water Research. 2004;**38**:663-672

[95] Thakur A, Chitoor B, Goswami AV, Pareek G, Atreya HS, D'Silva P. Structure and mechanistic insights into novel iron-mediated moonlighting functions of human J-protein cochaperone, Dph4. The Journal of Biological Chemistry. 2012;**287**(16):194-205

[96] Errampalli D, Trevorns JT, Lee H, Leung K, Cassidy M, Knoke K.

Bioremediation: A perspective. Journal of Soil Contamination. 1997;**6**:207-218

[97] Leung W, Malkova A, Haber JE. Gene targeting by linear duplex DNA frequently occurs by assimilation of a single strand that is subject to preferential mismatch correction. Proceedings of the National Academy of Sciences of the United States of America. 1997;**94**(13):51-68

[98] Radehaus PM, Schmidt SK. Characterization of a novel *Pseudomonas* sp. that mineralizes high concentrations of pentachlorophenol. Applied and Environmental Microbiology. 1992;**58**(9):79-85

[99] Resnick SM, Chapman PJ. Physiological properties and substrate specificity of a pentachlorophenol-degrading *Pseudomonas* species. Biodegradation. 1994;**5**:47-54

[100] Thakur V, Guptan RC, Kazim SN, Malhotra V, Sarin SK. Profile, spectrum and significance of HBV genotypes in chronic liver disease patients in the Indian subcontinent. Journal of Gastroenterology and Hepatology. 2002;**17**:165-170

[101] Kao MH, Doupe AJ, Doupe AJ, Brainard MS. Contributions of an avian basal ganglia-forebrain circuit to real-time modulation of song. Nature. 2005;**433**(7026):38-43

[102] Häggblom MM, Nohynek LJ, Salkinoja-Salonen MS. Degradation and O-methylation of chlorinated phenolic compounds by *Rhodococcus* and *Mycobacterium* strains. Applied and Environmental Microbiology. 1988;**54**:3043-3052

[103] Wittmann T, Hyman T. Chapter 7 recombinant p50/Dynamitin as a tool to examine the role of Dynactin in intracellular processes. Methods in Cell Biology. 1998;**61**:137-143

[104] Rutgers M, Breure AM, van Andel JG, Duetz WA. Growth yield coefficients of *Sphingomonas* sp. strain P5 on various chlorophenols in chemostat culture. Applied Microbiology and Biotechnology. 1997;**48**:656-661

[105] Tiirola MA, Busse HJ, Kampfer P, Mannisto M. *Novosphingobium lentum* sp. nov., a psychrotolerant bacterium from a polychlorophenol bioremediation process. International Journal of Systematic and Evolutionary Microbiology. 2005;**55**:583-588

[106] Yang Z, Khoury C, Jean-Baptiste G, Greenwood MT. Identification of mouse sphingomyelin synthase 1 as a suppressor of Bax-mediated cell death in yeast. FEMS Yeast Research. 2006;**6**(5):751-762

[107] Crawford RW, Giangrande P, Murray D. Fibrin sealant reduces blood loss in total hip arthroplasty. HIP International. 1999;**9**(3):127-132

[108] Chu JP, Kirsch EJ. Metabolism of pentachlorophenol by an axenic bacterial culture. Applied Microbiology. 1972;**23**(5):1033-1035

[109] Hu ZC, Korus RA, Levinson WE, Crawford RL. Adsorption and biodegradation of pentachlorophenol by polyurethane-immobilized *Flavobacterium*. Environmental Science & Technology. 1994;**28**(3):491-496

[110] Saber DL, Crawford RL. Isolation and characterization of *Flavobacterium* strains that degrade pentachlorophenol. Applied and Environmental Microbiology. 1985;**50**:1512-1518

[111] Steiert JG, Crawford RL. Catabolism of pentachlorophenol by a *Flavobacterium* sp. Biochemical and Biophysical Research Communications. 1986;**141**(2):825-830

[112] Stanlake GJ, Finn RK. Isolation and characterization of a pentachlorophenol-degrading bacterium. Applied and Environmental Microbiology. 1982:1421-1427

[113] Cobos I, Long JE, Thwin MT, Rubenstein JL. Cellular patterns of transcription factor expression in developing cortical interneurons. Cerebral Cortex. 2006;**16**(suppl 1):82-88

[114] Cox RT, Kirkpatrick C, Peifer M. Armadillo is required for adherens junction assembly, cell polarity, and morphogenesis during *Drosophila* embryogenesis. The Journal of Cell Biology. 1996;**134**(1):133-148

[115] Paul R, Augustyn A, Klin A, Volkmar FR. Perception and production of prosody by speakers with autism spectrum disorders. Journal of Autism and Developmental Disorders. 2005;**35**(2):205-220

[116] Di Lorenzo E, Schneider N, Cobb KM, Franks PJS, Chhak K, Miller AJ, et al. North Pacific gyre oscillation links ocean climate and ecosystem change. Geophysical Research Letters. 2008;**35**:8607

[117] Providenti MA, Lee H, Trevors JT. Selected factors limiting the microbial degradation of recalcitrant compounds. Journal of Industrial Microbiology. 1993;**12**:379-395

[118] Harmsen EW, Converse JC, Anderson MP, Hoopes JA. A model for evaluating the three-dimensional groundwater dividing pathline between a contaminant source and a partially penetrating water-supply well. Journal of Contaminant Hydrology. 1991;**8**(1):71-90

[119] Audus LJ. The biological detoxication of 2:4-dichlorophenoxy-acetic acid in soil. Plant and Soil. 1949;**2**:31-36

[120] Akbar S, Sultan S. Soil bacteria showing a potential of chlorpyrifos degradation and plant growth enhancement. Brazilian Journal of Microbiology. 2016;**47**:563-570

[121] Diez MC. Biological aspects involved in the degradation of organic pollutants. Journal of Soil Science and Plant Nutrition. 2010;**10**(3):244-267

Chapter 4

Toxicity of Cadmium in Soil-Plant-Human Continuum and Its Bioremediation Techniques

Asik Dutta, Abhik Patra,
Hanuman Singh Jatav, Surendra Singh Jatav,
Satish Kumar Singh, Eetela Sathyanarayana,
Sudhanshu Verma and Pavan Singh

Abstract

Cadmium (Cd) toxicity is highly detrimental for the human and largely originated from faulty industrial and agricultural practices. Cadmium toxicity can be observed in minute concentration and highly mobile in the soil–plant system and availability in soil is mainly governed by various physio-chemical properties of the soil. Cereals and vegetables cultivated in peri-urban areas, former mining and industrial areas accumulate Cd in toxic limit as they receive Cd from multiple ways. In general, when the total cadmium (Cd) concentration in soil exceeds 8 mg kg^{-1}, or the bioavailable Cd concentration becomes >0.001 mg kg^{-1}, or the Cd concentration in plant tissue reaches 3–30 mg kg^{-1} most plants exhibit visible Cd toxicity symptoms. The impacts of Cd toxicity are seed germination, growth, photosynthesis, stomata conductance, enzyme activities and alteration in mineral nutrition. The major source of Cd in human is food chain cycle and causes disorders like "*itai-itai*" disease, cancer, and nephrotoxicity. Cadmium harms kidney, liver, bone and reproductive body parts and may be fatal in serious condition. WHO recommended the tolerable monthly Cd intake are 25 μg kg^{-1} body weights and in drinking water Cd concentration should not exceed 3 μg L^{-1}. It is hard to remove these potent and hazardous metals from the environment as they have long mean residence time but, can be converted into less toxic form through bioremediation. This chapter focuses on the effect of Cd toxicity in soil–plant-human continuum and its bioremediation techniques to mitigate the Cd- toxicity.

Keywords: bioremediation, cadmium, carcinogen, food safety, soil contamination

1. Introduction

Cadmium (Cd) is an element which is extremely toxic to humans and can cause adverse effects even in small doses. Cadmium is a non-essential trace metal, which plays no recognized role in human, plant and animal development and growth. Various Environmental Protection Agency classified Cd as one of the pollutant element and include it in the list of 126 priority pollutants [1]. Lithosphere, hydrosphere and atmosphere take part in the exchange of Cd in its bio-geo-chemical

IntechOpen

cycle [2]. The aggregate industrial emission of Cd is vast and significantly contributed to bio-geo-chemical cycles, resulting Cd deposition in many ecosystems and hastening buildup of Cd both in nature and human food chain. Therefore, a variety of detrimental health effects of Cd have been identified in various parts of the world and these symptoms are increases progressively [3]. Cadmium (Cd), a hazardous heavy metal, falls into Group IIB of the periodic table and, its amounts ranging from 0.1 to 1 mg kg^{-1} in environment [4]. According to recent data collected in 2011, 7500, 2500 and 2000 t of Cd was emitted by China, Republic of Korea and Japan whereas globally it was 21,500 t $yr.^{-1}$. After the industrial revolution, man-made activities have greatly intensified the CD level in environment. The produce and use of Cd containing batteries, dyes, electroplating, combustion of crude oil, paints (Cd use as stabilizer), phosphate fertilizer processing and waste water applications have added 3–10 folds higher Cd than natural methods to the ecology. The release of Cd into to the soil environment is responsible for some natural disasters, such as volcanic eruption, sea salt spray, wild fires, weathering of Cd containing minerals and rock, transportation and accumulation of Cd-polluted soil by water and wind [5]. Cadmium, resulting from occupational and non-occupational contact, has detrimental impact on human health through build-up of Cd in human body. Occupational contamination is primarily observed by the extraction and smelting of non-ferrous metals, the manufacturing and handling of composite-containing CDs, and e-waste recycling activities. Non-occupational Cd contamination is mainly done by smoking, feeding behavior and atmospheric Cd particles [5]. Cadmium is ingested into multiple organs within the human body

Country	Adults N 19	Children	Adolescent 14–18 years	References
MAL/RDA[a]	5.0E−02[b]	—	—	[11]
RfD (oral reference dose)	1.0E−03	1.0E−03	—	[12]
Netherland	2.01E−02	4.10E−02	1.60E−02	[13]
USA	1.08E−05	2.21E−05	8.63E−06	[14]
Bangladesh	5.17E−05	1.06E−04	4.13E−05	[15]
Italy	1.54E−05 to 5.48E−05	3.16E−05 to 1.12E−04	1.23E−05 4.38E−05	[16]
Ethiopia	1.16E−04	2.37E−04	9.24E−05	[17]
Zimbabwe	8.87E−04	1.81E−03	7.09E−04	[18]
China	2.05E−04 to 2.805E−03	4.18E−04 to 5.72E−03	1.63E−04 2.23E−03	[19]
Sweden	6.95E−05	1.42E−04	5.55E−05	[14]
Uganda	8.22E−05	1.68E−04	6.56E−05	[11]
India	8.03E−04 to 4.92E−03	1.64E−03 to 1.00E−02	6.41E−04 3.93E−03	[13]
Pakistan	3.67E−05 to 8.10E−04	7.49E−05 to 1.66E−03	2.93E−05to 6.47E−04	[17]
France	5.78E−03	1.18E−02	4.62E−03	[12]

[a]*MAL/RDA maximum allowable limit/recommended dietary allowance.*
[b]*E−02 represents 1 × 10^{-2}.*

Table 1.
Daily dietary intake of Cd (mg kg^{-1} day^{-1}) through consumption of Cd contaminated vegetables.

i.e., kidney, liver, lungs, thymus testes, heart, epididymis, prostate, and salivary glands, leading to malfunctioning of multi-organ and ultimately death [6, 7]. The Itai-Itai epidemic with 184 patients and 388 possible victims was a well-known environmental hazard associated with Cd infection. Faulty farming practices and the use of hazardous plant agro-chemicals allow Cd to invade the food chain of humans. Commonly, trace elements level is typically higher in the roots, however in certain leafy vegetables (*e.g.*, lettuce and spinach), Cd is accumulated in plant leaves owing to its fast absorption and mobility within the plant system [8]. The estimation of quantities of Cd content in food materials indicates that vegetables and grains are the key factor of Cd in the food material, even though they are often present in animal products with a low quality. It is estimated that the everyday Cd ingestion by food material is 10.0–30.0 μg for adults in various countries [9, 10] (**Table 1**). Satarug et al. [20] reported that Cd level in vegetables varied from 0.001 to 0.124 mg kg^{-1} and intake of vegetables accounts >70–90% Cd susceptibility to humans. Remediation measures like washing the matrix, excavation and burial, and filed mechanization techniques have been followed in both limited and commercial scale but, not economically viable. An alternative strategy to mitigate the harmful effects Cd on soil–plant could be the use of bioremediation using suitable plants and microbes. So, in this chapter in brief the importance of Cd as a toxic element, its dynamics in the soil and plant and environment friendly measures to eliminate Cd pollution is discussed.

2. Cadmium contamination in soil and water

Cadmium (Cd) is a hazardous trace element disseminated extensively in the environment and causes implacable impact on human health even in very minute content [21]. Cadmium in lithosphere, sedimentary rocks and soil content 0.2, 0.3 and 0.53 mg kg^{-1} however in soil water and groundwater 5.0 and 1 μg L^{-1}, respectively [22, 23]. Cadmium contamination in soils and groundwater arises due to both natural and anthropogenic activities and cause harmful impact as its goes into human body through drinking water and foods [24]. Cadmium is mostly geogenic by origin whereas, majority comes from natural weathering and other sources are mining, casting and smelting, irrigation with sewage water, factories and vehicular discharges, and agrochemicals are major man-made causes of Cd pollution [25, 26]. Moreover, unmonitored and unsafe garbage dumping activities have intensely raised Cd levels in soil and water bodies. At end of 1980's it was reported that geogenic and anthropogenic sources mobilizes Cd to the biosphere 24,000 and 4.5 t $yr.^{-1}$, respectively which depicted the supremacy of man-made activity [27].

Among the natural sources windblown soil particles are the main reason for atmospheric Cd contamination followed by wildfires, sea spray, volcanic emissions, and meteoric dust. In California, Burke et al. [28] estimated that forest fire enhanced the average Cd level in water bodies by 2 folds. Pacyna and Pacyna [29] and Richardson et al. [30] reported that the Global average annual emission of natural Cd is about 1400 t however, from anthropogenic sources it was 2983 t. In nature, Cd is present ubiquitously in all areas and interestingly it's presence can be seen in remote places like ice peak of the Himalaya and North and South poles [31]. In southern Germany mainly relies on agricultural activities has Cd concentration in soil deposition was upto 0.25 g $(ha^*a)^{-1}$ however, in industrial western Germany the Cd deposition was quite high upto 1.4 g $(ha^*a)^{-1}$ [32]. Thus, indicates that anthropogenic activities have greater potential in Cd pollution.

Cadmium content in the soil is positively correlated with the weathering of parent material but, unscientific practices have worsen the input, output balance *i.e.,* input through atmospheric precipitation, factory or agricultural operations, minus its output through leaching, erosion and uptake by the crops [33]. The average Cd concentration in unpolluted soils in worldwide is 3.6%, while amounts which might be differ across continents, countries and type of soils. Cadmium in soil >30% is critically consider as Cd pollution limit, however, it was found that Cd level in soil reduces proportionately as the distance between manufacturing units and urban areas increases [34, 35]. In soil, the predominant source of Cd contamination is through weathering of various rocks and minerals present in the soil [25]. Maximum quantity of Cd was found in sedimentary rocks (0.1 to 26%) as compared to metamorphic and igneous rocks which contains Cd in the range of 1.1–10% and 0.7–2.5%, respectively [36, 37]. Similarly, Liu et al. [36] reported that in mudstone and siltstone has higher Cd content (46%) whereas, carbonate rocks has only 17% Cd content. He et al. [38] documented that soils generated from metamorphic rock like shales are highly prone to Cd toxicity. The **Table 2** illustrated the various Cd containing rocks and minerals that may be recognize important for the incidence of Cd in the soil and water. Zinc (Zn) from sphalerite (ZnS) or smithsonite ($ZnCO_3$), and iron (Fe) from pyrite (FeS_2) and hydrous oxides of iron can be easily substituted by Cd [39]. Due to similarity in ionic radius Cd can able to replace several divalent cations (*i.e.,* Ca, Fe, Zn, Pb, and Co) from their rocks [37]. Gnandi and Tobschall [40] stated that Ca in apatite mineral can be substituted by Cd therefore

Rock type	Average Cd content (%)	Mineral	Composition	Average Cd content (%)
Carbonate stone	0.1	Apatite	$Ca_5(F,Cl)(PO_4)_3$	1.4–1.5
Ultramafic rocks	0.2	Sphalerite	(Zn,Cd)S	2
Schists	0.2	Smithonite	$ZnCO_3$	< 2.35
Sandstone	0.3	Magnetite	Fe_3O4	< 3.1
Red shales	0.3	Silicates	—	0.3–58
Gneisses	0.4	Arsenopyrite	FeAsS	< 50
Mafic rocks	1.1	Scorodite	$FeAsO_4.\ 2H_2O$	< 10–58
Granitic rocks	1.2	Otavite	$CdCO_3$	65.2
Basalt	2.2	Greenockite	CdS	77.8
Obsidian	2.5	Pyromorphite	$Pb_5Cl(PO_4)_3$	< 10–80
Organic sediment	5.0	Calcite	$CaCO_3$	< 10–230
Red clay	5.6	Marcasite	FeS_2	< 500
Bituminous shale	8.0	Chalcopyrite	$CuFeS_2$	< 1100
Limestone	10	Bindheimite	$Pb_2Sb_2O_6(O,OH)$	1000–10,000
Shale and claystone	10	Tetrahedrite	$(Cu,Fe,Zn,Ag)_{12}SbAs_4S_{13}$	800–20,000
Bentonite	14	Anglesite	$PbSO_4$	1200 to >10,000
Marlstone	26	Mn-oxides	$MnO.\ nH_2O$	< 10,000
Oceanic manganese oxides	80	Limonite	$FeO(OH).\ nH_2O$	< 10,000
Phosphorites	250	Galena	PbS	< 30,000

Table 2.
Cadmium contents in different rocks and minerals.

Source	Type of pollution	Country/Area	Maximum Cd level	Reference
Mining				
Pb mining and refinery	Atmospheric deposition	Příbram, Czech Republic	Soil: 48 mg kg^{-1}	[45]
Cu mining	Waste water	Canchaque, Peru	Soil: 499 mg kg^{-1}	[28]
Pb–Zn mining/ refinery	Waste water	Coeur d'Alene basin, Idaho, USA	Groundwater: 77 μg L^{-1}	[46]
Fe–Ni–Co mining	Waste material	Several sites in Albania	Soil: 14 mg kg^{-1}	[47]
Au–Ag–Pb–Zn mining	Waste water	Chloride, Arizona USA	Groundwater: 19 μg L^{-1}	[48]
As refinery	Waste material	Reppel, Belgium	Soil: 79 mg kg^{-1}	[49]
Phosphorite mining	Mining waste, transport	Kpogamé, Hahotoé, Togo	Soil: 43 mg kg^{-1}	[50]
Zn smelter	Atmospheric deposition	Hezhang County, China	Soil: 74 mg kg^{-1}	[51]
Zn smelter	Waste material	Celje, Slovenia	Soil: 344 mg kg^{-1}	[52]
Pb–Zn mining/ refinery	Atmospheric deposition and waste water	Jinding, China	Soil: 531 mg kg^{-1}	[53]
Mining activities	Waste water	BacKan province, North Vietnam	Soil: 4.26 mg kg^{-1} Irrigation water: 2.51 μg L^{-1}	[54]
Au–Cu mining	Waste water	Bolnisi, Georgia	Soil: 121.5 mg kg^{-1}	[55]
Coal mining	Mining waste and deposition	Anhui province, eastern China	Soil: 0.05–0.87 mg kg^{-1}	[56]
Cu, Mo and Ni mining	Mining waste and deposition	Yangjiazhangzh and Dexing, China	Soil: 22.8 mg kg^{-1} Sediment: 66.1 mg kg^{-1}	[57]
Coal mines	Atmospheric deposition and waste water	Singrauli, India	Groundwater: 108 ppb	[58]
Industries				
Cement factory	Atmospheric deposition	Qadissiya, Jordan	Soil: 13 mg kg^{-1}	[59]
Various (*e.g.*, textile, electroplating)	Waste water	Coimbatore, India	Soil: 12.8 mg kg^{-1}	[42]
Ceramic industry	Sewage sludge	Castellon, Spain	Soil: 72 mg kg^{-1}	[60]
Pigment manufacture	Atmospheric deposition	Staffordshire, UK	Soil: 16 mg kg^{-1}	[61]
Textile industry	Waste water	Haridwar, India	Soil: 83.6 mg kg^{-1} Groundwater: 40 μg L^{-1}	[62]
Metal industry	Atmospheric deposition	Unnao, India	Groundwater: 74 μg L^{-1}	[63]
Ceramic industry	Atmospheric deposition	Yixing, China	Soil: 5.9 mg kg^{-1}	[64]
Paper mill	Waste water	Morigaon, India	Soil: 31.01 mg kg^{-1}	[65]

Source	Type of pollution	Country/Area	Maximum Cd level	Reference
Power industry and industrial plants	Atmospheric deposition and waste water	Malopolska province, southern Poland	Soil: 16.9 mg kg^{-1}	[66]
Zinc-smelter plant	Irrigation through industrial effluents	Rajasthan, India	Soil: 96.8 mg kg^{-1}	[67]
Atlas Cycle factory	Irrigation through industrial effluents	Haryana, India	Soil: 9.81 mg kg^{-1}	[67]
Waste management				
Disposal facilities	Leachate	Great lakes region, USA	Soil: 32 mg kg^{-1}	[40]
Household wastes	Waste water	Ikare, Nigeria	Groundwater: 580 μg L^{-1}	[6]
Landfill	Leachate	Taoyuan, Taiwan Alexandria, Egypt	Soil: 378 mg kg^{-1} Groundwater: 51 μg L^{-1}	[68]
Sewage and waste disposal	Waste water	Sekondi-Takoradi Metropolis, Ghana	Groundwater: 90 μg L^{-1}	[69]
Sewage disposal	Waste water and physical mixing	Sundarban, India	Soil: 1.70 mg kg^{-1}	[70]
Brownfield	Waste water	Xiangjiang River, China	Groundwater: 474 μg L^{-1}	[71]
Oil spill accident	Waste deposition and physical mixing	Sundarban, Bangladesh	Sediment: 0.82 mg kg^{-1}	[38]
Electronical waste recycling	Waste water	Krishna Vihar, India	Soil: 47.7 mg kg^{-1} Groundwater: 280 μg L^{-1}	[72]
Agriculture				
Sewage sludge application	Irrigation	Several sites in Spain	Soil: 90 mg kg^{-1}	[73]
P fertilizer production	Atmospheric deposition	Rio Grande, Brazil	Soil: 9.3 mg kg^{-1} Groundwater: 3 μg L^{-1}	[32]
P fertilizer application	Infiltration	Cauvery River basin, India	Groundwater: 60 μg L^{-1}	[74]
Urban agriculture	Atmospheric pollution and soil contamination	Belo Horizonte, Brazil	Soil: 0.20 mg kg^{-1}	[75]
Sewage sludge application	Soil application	Jiangsu Province, China	Leachate: 0.14 mg kg^{-1}	[76]
Urban areas				
Sewerage	Leakage	Rastatt, Germany	Groundwater: 5 μg L^{-1}	[1]
Road traffic	Infiltration	Celle, Germany	Groundwater: 2.34 μg L^{-1}	[9]
Over populated, E-wastes and industrialized	Infiltration and physical mixing	Western Uttar Pradesh, India	Groundwater: 0.07 mg L^{-1}	[77]

Table 3.
Various types of cadmium contamination in soil and waterbodies.

Cd may be a natural adulteration in phosphate (P) minerals and phosphorite rocks that are essential for the manufacture of phosphate fertilizers. Unlike Eastern Europe, there is considerably higher Cd in agricultural fields of Western European and one of the reasons for this is use of P fertilizer from distinct source [41]. The Cd bioavailability is governed by several factors such as: pH, moisture content, soil texture, clay content and type, cation exchange capacity, quantity and type of organic matter (OM), hydrous oxides, etc. [38]. Cadmium is easily mobilize in the soil due to its weaker bonding between soil exchange sites (*i.e.,* OM, carbonate, and hydrous oxide) [42] and that is the key factor to increase bio-availability of Cd to plants, ground water as well as plant products.

Geogenic sources input only 10 percent Cd in the environment however, man-made emission input 90 percent Cd in the environment. Among the various man-made sources major contribution is from manufacturing and application of P fertilizers, petroleum oil burning, smelting and casting industries, effluents from cement factories, vehicular emission, sewage sludge, landfills, municipality solid wastes, and mining activities [43, 44]. The **Table 3** explained various anthropogenic activities and their impact on Cd build-up in soil and groundwater. Cadmium is mainly used in stabilization of plastics, pigments manufacturing, solar panels, nickel-cadmium batteries, and rust resistant steel production, agri-chemicals, solders, engine oil, and rubber and fabric industries [78, 79]. Brown et al. [80] reported that in 2015, globally Cd manufacture was ~24,900 metric tons and it was increases in the coming years. Among the anthropogenic sources mining and metal industries are the main reason for environmental Cd pollution followed by textiles industries, nonmetallic mineral products, fertilizers and agro-chemicals production, and leathers industries [81]. Landfills and municipal solid waste deposition are the major causes of soil pollution with Cd and in European countries municipal solid waste contain Cd level up to 3 to 12% [62]. Leachates from various sources are the main cause of Cd pollution in groundwater and Belon et al. [35] estimated that leachate form FYM, atmospheric deposition, inorganic fertilizers and municipal solid waste ranges from 10 to 25, 15–50, 30–55 and 2–5%, respectively. Another important source of Cd pollution in soil through the use of P fertilizers and P fertilizer used in various countries like Eastern Mediterranean countries, European countries and Germany the Cd content is as high as 770, 360 and 600%, respectively [37, 82]. Cadmium discharge and emitted from multiple sources gradually enters into the soil and then eventually bio-accumulates in food grains which ultimately leads to human health hazard.

3. Mechanism of Cd accumulation in plants and consequences

Cadmium (Cd) is a potent pestilential metal which enters primarily via plant roots, get distributed and accumulated in plant parts in different proportions and concentrations, hampering crop yield and deteriorating the quality of produce. It ultimately makes it way to enter food chain thereby possessing serious threat to human and animal health. Cadmium ranks 7 among the top 20 toxins and it enter to arable land through various industrial processes and farming practices [83].

3.1 Accumulation of Cd in plants

Accumulation of Cd in plant is facilitated by its mobilization, uptake and transport/distribution in various plant parts. Unscientific agricultural practices and industrial effluents are the major contributor of Cd in soil [84]. Phosphaic fertilizer and sewage-sludge contribute to Cd pollution in agricultural soil. Concentration of Cd in plants is also an indicative of its concentration in soil; however various other factors including soil pH, organic matter content, interaction with other ions and plant species govern its availability in plants [85–87]. Meta data analysis of 162 wheat and 215 barley grain samples by Adams and associates, [88] showed grain Cd concentration is positively correlated with soil total cd content and soil reaction (pH). They also highlighted the fact that higher microbial activity, nitrification and application of sewage sludge increased the chance of Cd toxicity but, reclaiming the soil with liming may abate the chance of toxicity. Sauvé et al. [89] found that organic matter had almost 30 times more sorption affinity for Cd when compared with mineral soil in Canada which indicates the importance of quality of organic matter in binding and accumulating Cd. It is assumed that lowering of pH will facilitate Cd availability to plants, but it might not hold true for soils with lower pH and high organic matter.

Before apprehending the mechanism of Cd accumulation in plants, one has to understand uptake and translocation of Cd inside plants. Ability of plants to take up Cd depends upon numerous factors like total Cd content in soil solution, soil reaction (pH), redox potential (Eh) and moisture content, soil organic carbon content, soil temperature, and last but not the least interaction among different elements. Primarily Cd enters plant through roots. Once in roots, Cd can get stored or exported to shoots through xylem. Cadmium is both xylem and phloem mobile [54, 74]. There are two possible mechanisms of Cd translocation into the plants and subsequently to the grains. These are: (i) Xylem mediated translocation to the sink i.e. grains (ii) Active transportation to various plant parts culm, rachis, flag leaves, external parts of the panicles and followed by phloem mediated mobilization to grains [90] and Schematic representation of Cd uptake and subsequent translocation in rice was shown in **Figure 1**. Root cell membrane located transporters take key role in Cd uptake in plants [91].

Cadmium uptake and accumulation in plants must undoubtedly be under control of multiple genes which contribute quantitatively in stage-specific, tissue-specific, environment-specific to Cd transport, accumulation and sequestration in plants [92]. In a study conducted by Hédiji et al. [72] on long term exposure of

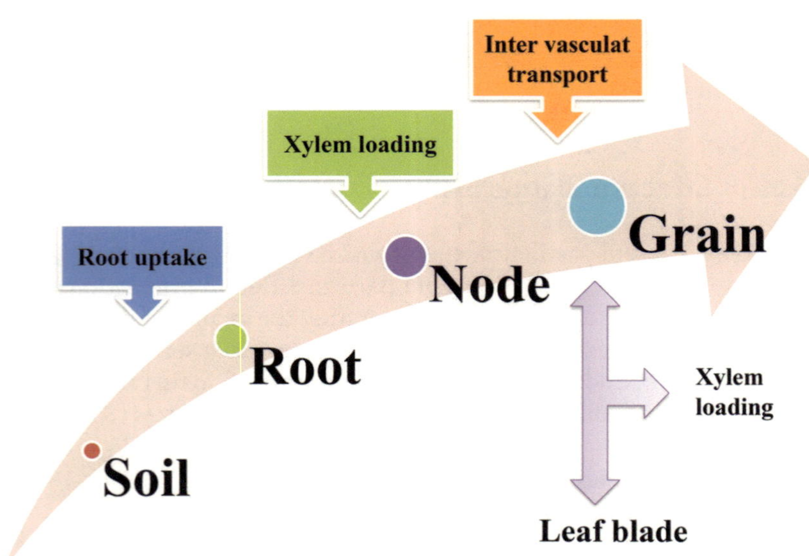

Figure 1.
Schematic model of Cd uptake process from soil to grains in rice.

Cd on tomato (*Solanumlycopersicum* L.) concluded that, impact of Cd toxicity is highly dose specific and significantly correlated with soil nutrient status. Whereas, in higher dose severely affecting the plant growth and metabolism by altering the nutrient partitioning. Several genes are responsible to carry out these processes.

3.2 Consequences to plant health

The impact of Cd toxicity in plants is still a closed book thing but, recent advances in plant physiological studies helped the researchers to answer the questions. Clemens [54] reported that the major influence on Cd toxicity in plants is nutrient imbalance by regulating the normal work of transporters peculiarly in fruit plants. For instance, the concentration of K, Zn, and Fe in developing fruits falls off drastically at the expense of Ca and Mg. The antagonistic relationship between Cd and K is well documented like sub-optimal K concentration in the pericarp which disrupts the normal bio-chemical cycles like bio-synthesis of protein, enzymatic activity and membrane bound activities such as sustaining cellular turgidity [54].

4. Consequences to human health

According to International Agency for Research on Cancer, Cd is highly inimical and labeled as class-I carcinogenic compound to mammalian health. Cadmium may not be toxic to the plants that accumulate it, yet are toxic to animals and humans feeding upon it. Cadmium makes it entry to human body either from food, water or breath and a little amount enters through skin. Majority of Cd entering to human body is either breathed out or excreted in feces, whereas only one-quarter of it gets into human body through breath and one-twentieth from food. People working in industries that release Cd are more prone to get affected by Cd toxicity because they might breath, eat or drink Cd in air, food or water. Cadmium with biological half-life of 10–30 years, generally gets accumulated in kidneys and liver and slowly leaves human body through urine or feces [93, 94]. Researches around the world indicate that daily cadmium intake from all sources is very low in case of general population which range between 10 and 25 $\mu gday^{-1}$, however the tolerable daily intake established by WHO is 60 and 70 10–25 $\mu g\ day^{-1}$ respectively, for adult women and men.

5. Cadmium toxicity in humans

Human health due to Cd is an emerging issue and needs urgent attentions [52]. During the process, 10–50% of the cadmium dust is consumed according to the particle size. Digestion is higher for people that have an iron, calcium or zinc deficiency. The main source of human cadmium toxicity is considered to be tobacco smoking other than industrial exposures and food habit [95–98]. Cd toxicity is developing gradually in the human body and eventually causes different negative health effects, particularly bone loss and nephron toxicity.

5.1 Absorption and distribution

Cd is passed across the body after assimilation, usually linked with a bunch of sulfhydryl containing protein such as metalllothionine. Typically 30% stores in liver and kidney; the remaining spread across the body, with an independence half-life about a quarter of a century [99]. Blood, hair and urine Cd levels are indicator of

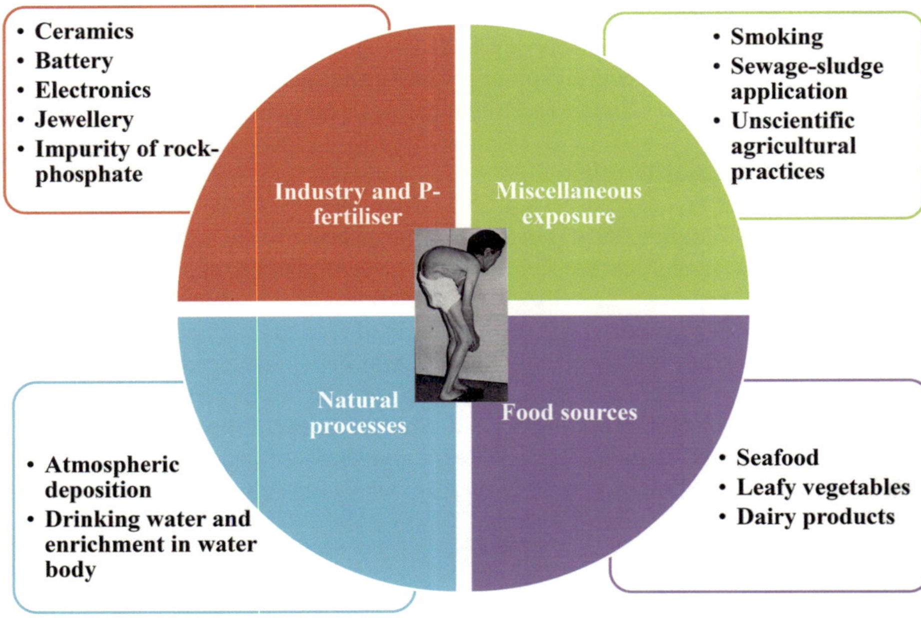

Figure 2.
Mechanisms of cadmium toxicity in humans.

potential toxicity but, to get the actual toxicity level urine stimulation test with the subjects body weight is highly important [100].

5.2 Mechanisms of toxicity

As previously mentioned, Cd induced epigenetic changes in DNA articulation by oxidative pressure, impediments or guidance for transport pathways particularly in the kidney [98] (**Figure 2**). Extreme impedance to the physiological function of Zn or Mg is introduced by other pathological mechanisms [99]. Restriction of the heme and the weakening of mitochondrial work which is likely to cause apoptosis [47]. Glutathione explosion has been found alongside the auxiliary protein contortion attributable to the official Cd in sulfhydryl bunches [100]. Cooperation with other hazardous metals, such as lead (Pb) and arsenis (As) hastens these impacts [101, 102].

6. Clinical toxicity

The major site of Cd toxicity is kidney where a fragment S1 of the proximal tubule is a majorly targeted and disruption in mitochondrial protein synthesis due reabsorption of glucose, bicarbonate and phosphate clinically known as Fanconi disorder [76, 103]. Cadmium can also inhibit the digestion of vitamin D in the kidneys with progressively rises of issues like osteomalacia, osteoporosis, renal-around broking and calcium malabsorption [103–105]. Cadmium has multiple deleterious effects on the cardiovascular framework like adverse impact on vascular endothelium consistency [95, 106]. Cd links to sudden coronary death marginal blood vessel dysfunction, increased intima media thickness and scattered myocardial necrosis [64, 107]. In comparison, low-recurrence listening was substantially decreased by people with elevated urinary Cd levels [108]. In comparison, high-urinary Cd rates have decreased cognitive power. Cadmium is assumed to be the carcinogenic agent Class B1 by the United States Environmental Protection Agency [46]. Conflicting research links Cd adoption and denies bosom malignant development [88, 94, 109]. Cd was associated to pancreas and lymphoma cell disturbance [88]. Vegetables developed in Cd-defiled soils can possibly cause toxicological issues in people particularly in developing women [110]. A few different components like low admission of Ca, vitamin D, and minor components, for example, Cu and Zn can build this sum. Thus, daily entry of Cd by Cd is exceptional due to the fusion of Cd in diets and the human dietary propensities. The mean daily use of Cd (DICd) uses the following formula as a general basis:

$$DI_{Cd} = (C_{Cd} \text{ x } C_{factor} \text{ x } D_{\text{food intake})}/BW_{\text{average weight}}$$

DI_{Cd} symbolizes daily intake of Cd, C_{Cd}Cofactor, intake of D_{food} and $W_{average}$ weight are Cd fixations in vegetables, transition factor (new weight to dry weight), and human consumption of vegetables every day and regular body weight respectively. **Table 2** describes the DI_{Cd} figures given in different countries by the use of Cd-sullied vegetables. The number of inhabitants in the Netherlands unmistakably ingests the most notable Cd from the available information through defiled vegetables, followed by France and USA. The introduced data shows that the use of Cd contaminated nourishments is a significant implementation course. In these lines, in order to avoid harmful health consequences, the intake of infected vegetables should be reduced to the fullest degree possible. Different remediation steps can also be introduced in infected soil to carry the Cd concentration to a reasonable amount. In contrast, $DI_{Cd's}$ principles are based on a few experiments worldwide. To describe incidents and potential dangers more thoroughly, further studies are needed. Furthermore, day-to-day vegetable intake, eating patterns, general status and the overall body weight of a person should be taken into account. Cadmium (Cd) is a toxicity ia result of long term exposure and "*itai-itai*" infection in Japan during 1950's is an eye opening instance. Arrangement of rules and rules has been created in numerous nations and worldwide associations to manage the examination on wellbeing impact of Cd contamination [111].

7. Bioremediation of cadmium

According to EPA, bioremediation can be defined as "technique which uses naturally occurring microorganisms to break down hazardous substances into less toxic or non-toxic substances [111]."

7.1 Techniques of bioremediation

a. ***In-situ* Bioremediation:** This technique follows on-site remediation of polluted soil using sustainable technologies [112, 113].

b. ***Ex-situ* Bioremediation:** This method based on cleaning contaminated site elsewhere i.e. not in the site of pollution.

7.2 Types

a. **Phytoremediation:** Phytoremediation is an eco-friendly option for rejuvenating contaminated site using plants and microbes. Plants suitable for phytoremediation techniques must have important characters like high above ground biomass with vigorous growth, proliferated root system and metal accumulating characters [114].

b. **Phytoextraction:** Phytoextraction can be described as a metal extracting character by plant roots and subsequently plants are subjected to burial in some other place or incineration. Taxonomically plants species which are excellent metal extractor's belongs to families like *Scrophulariaceae*, *Lamiaceae*, *Asteraceae*, *Euphorbiaceae*, and *Brassicaceae.* However, plant species like *Celosia argentea* L. [115], *Salix mucronata* L. [61], *Cassia alata* L. [116], *Solanummelonaena* L., *Momordicacharantia* L. [117], *Kummerowiastriata* L. [118], and *Swieteniamacrophylla* L. [65], may be used as potential plant choices to increase the process of Cd phytoextraction. Moreover, a sub-division of phytoextraction, known as chelate-assisted phytoextraction, is also used as a possible solution for metals that have no hyperaccumulator species using EDTA or citric acid [66, 119].

7.3 Microorganisms for bioremediation

Microbe's works in both active and passive mode and microbial species like bacteria, fungi and alage can be used as a potential option for eco-friendly remediation techniques [93]. Bacteria's are very effective for cleaning contaminated site due to its unique metabolic characters and tolerance to harsh conditions [120]. Several heavy metals have been tested using bacteria species like *Flavobacterium*, *Pseudomonas*, *Enterobacter*, *Bacillus*, and *Micrococcus* sp. Their great bio-sorption ability is due to high surface-to-volume ratios and the potential active chemosorption sites (teichoic acid) on the cell wall [121]. Abioye and his coworkers [122] reported successful use of bacterial species like *Bacillus subtilis* L., *B. megaterium* L., *Aspergillusniger* L., and *Penicillium* sp. for revive soils contaminated with lead (Pb) and cadmium (Cd). Fungal species like *Coprinopsisatramentariais* L. can bio accumulate more than 75% of Cd of the contaminated site by 1 mg L^{-1} [123]. Goher and his co-authors [68] reported cleaning of Cd- contaminated site using dead algal cells of *Chlorellavulgaris* L.

8. Conclusions

This present chapter summarizes the various sources of Cd in environment and its toxic effects on plant and human being as well as suggested some approaches of bioremediation to mitigate the Cd pollution from environment. Anthropogenic activities are the key pathway to contaminate the environment with Cd which ultimately accumulated in various leafy vegetables and food grains. Consumption of this high Cd containing food causes several toxic symptoms in human being and leads to malfunctioning of multiple human organs. To reduce the Cd accumulation in food grain various amelioration strategies has been adopted among them use of microbes to decrease Cd uptake by plants seems to have great prospective. Moreover, some microbes may increase amounts of Cd due to their biochemical processes, and their implementation may also worsen problems with soil pollution. Use It is also suggested to characterize the microbes and tested them in laboratory and field condition prior to their use in agricultural soils, thus maintaining soil quality and food safety.

Author details

Asik Dutta[1], Abhik Patra[2], Hanuman Singh Jatav[2,3]*, Surendra Singh Jatav[2], Satish Kumar Singh[2], Eetela Sathyanarayana[4], Sudhanshu Verma[5] and Pavan Singh[2]

1 Crop Production Division, ICAR-Indian Institute of Pulse Research, Kanpur, India

2 Department of Soil Science and Agricultural Chemistry, Institute of Agricultural Sciences, Banaras Hindu University, Varanasi, Uttar Pradesh, India

3 Sri Karan Narendra Agriculture University, Jobner, Jaipur, Rajasthan, India

4 Professor Jayashankar Telangana State Agricultural University, Hyderabad, Telangana, India

5 Department of Agriculture, Integral Institute of Agricultural Science and Technology, Integral University, Lucknow, Uttar Pradesh, India

*Address all correspondence to: hanumaniasbhu@gmail.com; hsjatav.soils@sknau.ac.in

References

[1] Jin T, Lu J, Nordberg M. Toxicokinetics and biochemistry of cadmium with special emphasis on the role of metallothionein. Neurotoxicology. 1998;**19**: 529-36

[2] Luoma SN, Rainbow PS. Metal Contamination in Aquatic Environments. Science and Lateral Management. Cambridge University Press, New York, 2008, pp. 573

[3] Weis JS, Bergey L, Reichmuth J, Candelmo A. Living in a contaminated estuary: behavioral changes and ecological consequences for five species. BioScience. 2011;**5**:375-85.

[4] Stoeppler M. Metals and Their Compounds in the Environment. Occurrence, Analysis, and Biological Relevance. FW Jr (Ed.), VCH, Weinheim. 1991:803

[5] Schwartz GG, Reis IM. Is cadmium a cause of human pancreatic cancer?. Cancer Epidemiology and Prevention Biomarkers. 2000;**9**:139-45

[6] Godt J, Scheidig F, Grosse-Siestrup C, Esche V, Brandenburg P, Reich A, Groneberg DA.The toxicity of cadmium and resulting hazards for human health. Journal of occupational medicine and toxicology. 2006;**1**:1-6.

[7] Takamure Y, Shimada H, Kiyozumi M, Yasutake A, Imamura Y. A possible mechanism of resistance to cadmium toxicity in male Long–Evans rats.Environmental toxicology and pharmacology. 2006;**26**:231-4

[8] Peijnenburg WJ, Baerselman R, De Groot A, Jager T, Leenders D, Posthuma L, Van Veen R. Quantification of metal bioavailability for lettuce (*Lactuca sativa* L.) in field soils. Archives of Environmental Contamination and Toxicology. 2000;**39**:420-30

[9] Joint FA, WHO Expert Committee on Food Additives, World Health Organization. Evaluation of certain food additives and contaminants: fifty-fifth report of the Joint FAO/WHO Expert Committee on Food Additives. World Health Organization; 2001

[10] Sahmoun AE, Case LD, Jackson SA, Schwartz GG. Cadmium and prostate cancer: a critical epidemiologic analysis. Cancer investigation. 2005;**23**:256-63

[11] Affum AO, Osae SD, Nyarko BJB, Afful S, Fianko JR, Akiti TT, Adomako D, Acquaah SO, Dorleku M, Antoh E, Barnes F, Affum EA. Total coliforms, arsenic and cadmium exposure through drinking water in the Western Region of Ghana: application of multivariate statistical technique to groundwater quality. Environmental Monitoring and Assessment. 2015;**187**:1

[12] Avkopashvili G, Avkopashvili M, Gongadze A, Gakhokidze R. Eco-Monitoring of Georgia's Contaminated Soil and Water with Heavy Metals. Carpathian Journal of Earth and Environmental Sciences. 2017;**12**:595-604

[13] Vangronsveld J, Herzig R, Weyens N, Boulet J, Adriaensen K, Ruttens A, Thewys T, Vassilev A, Meers E, Nehnevajova E, van der Lelie D, Mench M. Phytoremediation of contaminated soils and groundwater: lessons from the field. Environmental Science and Pollution Research. 2009;**16**:765-794

[14] Kauppinen T, Pukkala E, A. Saalo A, and A. J. Sasco AJ. "Exposure to chemical carcinogens and risk of cancer among Finnish laboratory workers," American Journal of Industrial Medicine, 2003; **44**:343-350

[15] Rahman MM, Asaduzzaman M, Naidu R. Consumption of arsenic

and other elements from vegetables and drinking water from an arsenic-contaminated area of Bangladesh. Journal of Hazardous Material.2014; **262**:1056-1063

[16] Beccaloni E, Vanni F, Beccaloni M, Carere M. Concentrations of arsenic, cadmium, lead and zinc in homegrown vegetables and fruits: Estimated intake by population in an industrialized area of Sardinia, Italy. Microchemical Journal. 2013;**107**:190-5

[17] Khan K, Lu Y, Khan H, Ishtiaq M, Khan S, Waqas M, Wei L, Wang T. Heavy metals in agricultural soils and crops and their health risks in Swat District, northern Pakistan. Food and chemical toxicology. 2013;**58**:449-58

[18] Mapanda, F., Mangwayana, E.N., Nyamangara, J., Giller, K.E. Uptake of heavymetals by vegetables irrigated using wastewater and the subsequent risks in Harare, Zimbabwe. Physics and Chemistry of Earth, Parts A/B/C. 2007; **32**, 1399-1405

[19] Li N, Kang Y, Pan W, Zeng L, Zhang Q, Luo J. Concentration and transportation of heavy metals in vegetables and risk assessment of human exposure to bioaccessible heavy metals in soil near a waste-incinerator site, South China. Science of the total environment. 2015 ;**521**:144-51

[20] Satarug S, Haswell-Elkins MR, Moore MR. Safe levels of cadmium intake to prevent renal toxicity in human subjects.British Journal of Nutrition. 2000:**84**:791-802

[21] Khan A, Khan S, Khan MA, Qamar Z, Waqas M. The uptake and bioaccumulation of heavy metals by food plants, their effects on plants nutrients, and associated health risk: a review. Environmental Science and Pollution Research. 2015;**22**:13772-13799

[22] Kabata-Pendias A, Pendias H. Trace Element in Soil and Plants. 3rd ed. CRC Press, Boca Raton, USA. 2001

[23] Naseem S, Hamza S, Nawaz-ul-Huda S, Bashir E. Geochemistry of Cd in groundwater of Winder, Balochistan and suspected health problems. Environmental Earth Sciences. 2014;**71**:1683-1690

[24] Pan LB, Ma J, Wang XL, Hou H. Heavy metals in soils from a typical county in Shanxi Province, China: levels, sources and spatial distribution. Chemosphere. 2016;**148**:248-254

[25] Liu Y, Xiao T, Ning Z, Li H, Tang J, Zhou G. High cadmium concentration in soil in the Three Gorges region: Geogenic source and potential bioavailability. Applied Geochemistry. 2013;**37**:149-156

[26] Nawab J, Khan S, Aamir M, Shamshad I, Qamar Z, Din I, Huang Q. Organic amendments impact the availability of heavy metal (loid) s in mine-impacted soil and their phytoremediation by *Penisitumamericanum* and *Sorghum bicolor*. Environmental Science and Pollution Research. 2016;**23**:2381-2390

[27] Nriagu JO. Global metal pollution: poisoning the biosphere?. Environment: Science and Policy for Sustainable Development. 1990;**32**:7-33

[28] Burke MP, Hogue TS, Kinoshita AM, Barco J, Wessel C, Stein ED. Pre-and post-fire pollutant loads in an urban fringe watershed in Southern California. Environmental Monitoring and Assessment. 2013;**185**:10131-10145

[29] Pacyna JM, Pacyna EG. An assessment of global and regional emissions of trace metals to the atmosphere from anthropogenic sources worldwide. Environmental Reviews. 2001;**9**:269-298

[30] Richardson GM, Garrett R, Mitchell I, Mah-Poulson M, Hackbarth T. Critical review on natural global and regional emissions of six trace metals to the atmosphere. Prepared for the International Lead Zinc Research Organisation, the International Copper Association, and the Nickel Producers Environmental Research Association. 2001

[31] Lee K, Do Hur S, Hou S, Hong S, Qin X, Ren J, Liu Y, Rosman KJR, Barbante C, Boutron CF. Atmospheric pollution for trace elements in the remote high-altitude atmosphere in central Asia as recorded in snow from Mt. Qomolangma (Everest) of the Himalayas. Science of the Total Environment. 2008;**404**:171-181

[32] Ilyin I, Gusev A, Rozovskaya O, Strijkina I. Transboundary Pollution by Heavy Metals and Persistent Organic Pollutants in 2014- Germany. 2016;**33**

[33] Six L, Smolders E. Future trends in soil cadmium concentration under current cadmium fluxes to European agricultural soils. Science of the Total Environment. 2014;**485**:319-328

[34] Akbar KF, Hale WH, Headley AD, Athar M. Heavy metal contamination of roadside soils of Northern England.Soil Water Research. 2006;**1**:158-163

[35] Belon E, Boisson M, Deportes IZ, Eglin TK, Feix I, Bispo AO, Galsomies L, Leblond S, Guellier CR. An inventory of trace elements inputs to French agricultural soils. Science of the Total Environment. 2012;**439**:87-95

[36] Shah MT, Begum S, Khan *S. pedo* and biogeochemical studies of mafic and ultramfic rocks in the Mingora and Kabal areas, Swat, Pakistan. Environmental Earth Sciences. 2010;**60**:1091-1102

[37] Smolders E, Mertens J. Cadmium. In: Alloway JB (Ed.) Heavy Metals in Soils - Trace Metals and Metalloids in Soils and Their Bioavailability, 3rd ed. Springer, Dordrecht. 2013;283-299

[38] He ZL, Xu HP, Zhu YM, Yang XE, Chen GC. Adsorption-desorption characteristics of cadmium in variable charge soils. Journal of Environmental Science and Health. 2005;**40**:805-822

[39] Tabelin CB, Igarashi T, Villacorte-Tabelin M, Park I, Opiso EM, Ito M, Hiroyoshi N. Arsenic, selenium, boron, lead, cadmium, copper, and zinc in naturally contaminated rocks: A review of their sources, modes of enrichment, mechanisms of release, and mitigation strategies. Science of the Total Environment. 2018;**645**:1522-1553

[40] Gnandi K, Tobschall H. Heavy metals distribution of soils around mining sites of cadmium-rich marine sedimentary phosphorites of Kpogame and Hahotoe (southern Togo). Environmental Geology. 2002;**41**:593-600

[41] Tóth G, Hermann T, Da Silva MR, Montanarella L. Heavy metals in agricultural soils of the European Union with implications for food safety. Environment International. 2016;**88**:299-309

[42] Eggleton J, Thomas KV. A review of factors affecting the release and bioavailability of contaminants during sediment disturbance events. Environment International. 2004;**30**:973-980

[43] Bi X, Feng X, Yang Y, Qiu G, Li G. Quantitative assessment of cadmium emission from zinc smelting and its influences on the surface soils and mosses in Hezhang County, Southwestern China. Atmospheric Environment. 2006;**40**:4228-4233

[44] Sprynskyy M, Kowalkowski T, Tutu H, Cozmuta LM, Cukrowska EM, Buszewski B. The adsorption properties

of agricultural and forest soils towards heavy metal ions (Ni, Cu, Zn, and Cd). Soil and Sediment Contamination. 2011;**20**:12-29

[45] Bui AT, Nguyen HT, Nguyen MN, Tran THT, Vu TV, Nguyen CH, Reynolds HL. Accumulation and potential health risks of cadmium, lead and arsenic in vegetables grown near mining sites in Northern Vietnam. Environmental Monitoring and Assessment. 2016;**188**:525

[46] Cadmium Compounds, Technology Transfer Network Air Toxics Web Site, Environmental Protection Agency, Washington, DC, USA, 2007

[47] Cannino G, Ferruggia E, Luparello C, Rinaldi AM. Cadmium and mitochondria. Mitochondrion. 2009;**9**:377-84

[48] Cappuyns V, Van Herreweghe S, Swennen R, Ottenburgs R, Deckers J. Arsenic pollution at the industrial site of Reppel-Bocholt (north Belgium). Science of the Total Environment. 2002;**295**:217-240

[49] Chen X, Zhu G, Jin T, Zhou Z, Gu S, Qiu J, Xiao H. Cadmium stimulates the osteoclastic differentiation of RAW264. 7 cells in presence of osteoblasts. Biological trace element research. 2012;**146**:349-53

[50] Beyer WN, Stafford C. Survey and evaluation of contaminants in earthworms and in soils derived from dredged material at confined disposal facilities in the Great Lakes Region. Environmental Monitoring and Assessment. 1993;**24**:151-165

[51] Borah P, Singh P, Rangan L, Karak T, Mitra S. Mobility, bioavailability and ecological risk assessment of cadmium and chromium in soils contaminated by paper mill wastes. Groundwater for Sustainable Development. 2018;**6**:189-199

[52] Chen X, Zhu GY, Jin TY. Progress of the study on toxic effects of cadmium on kidney and bone. Journal of Occupational and Environmental Medicine. 2008;**25**:412-5

[53] Cheng S, Liu G, Zhou C, Sun R. Chemical speciation and risk assessment of cadmium in soils around a typical coal mining area of China. Ecotoxicology and Environmental Safety. 2018;**160**: 67-74

[54] Clemens S, Aarts MGM, Thomine S, Verbruggen N. Plant science: the key to preventing slow cadmium poisoning. Trends in Plant Science.2013;**18**,92-99

[55] Clemens S. Evolution and function of phytochelatinsynthase.Journal of Plant Physiology. 2006;**163**:319-332

[56] Czarnecki LA, Moberly AH, Turkel DJ, Rubinstein T, Pottackal J, RosenthalMC, McCandlishEF, BuckleyB, McGann JP. Functional rehabilitation of cadmium-induced neurotoxicity despite persistent peripheral pathophysiology in the olfactory system. Toxicological Sciences. 2012;126:534-44

[57] Dala-Paula BM, Custódio FB, Knupp EA, Palmieri HE, Silva JBB, Glória MBA. Cadmium, copper and lead levels in different cultivars of lettuce and soil from urban agriculture.Environmental Pollution. 2018;**242**:383-389

[58] Deepali KK, Gangwar K. Metals concentration in textile and tannery effluents, associated soils and ground water. New York Science Journal. 2010;**3**:82-89

[59] Dwivedi AK, Vankar PS. Source identification study of heavy metal contamination in the industrial hub of Unnao, India. Environmental Monitoring and Assessment. 2014;**186**:3531-3539

[60] Eiswirth M, Hötzl H. The impact of leaking sewers on urban groundwater.

Groundwater in the urban environment. 1997;**1**:399-404

[61] El-Mahrouk ES, Eisa EA, Hegazi MA, Abdel-Gayed ME, Dewir YH, El-Mahrouk ME, Naidoo Y. Phytoremediation of cadmium-, copper-, and lead-contaminated soil by Salix mucronata (Synonym Salix safsaf). HortScience. 2019;**54**:1249-57

[62] EU. European Union Risk Assessment Report – Cadmium Oxide and Cadmium Metal. Part I – Environment Office for Official Publications of the European Communities, Luxembourg. 2007;608

[63] European Commission. Commission Regulation (EC) No 1881/2006 of 19 December 2006 setting maximum levels for certain contaminants in foodstuffs. Official Journal of European Union. 2006;**24**:364

[64] Everett CJ, Frithsen IL. Association of urinary cadmium and myocardial infarction. Environmental research. 2008;**106**:284-6

[65] Fan KC, Hsi HC, Chen CW, Lee HL, Hseu ZY. Cadmium accumulation and tolerance of mahogany (Swieteniamacrophylla) seedlings for phytoextraction applications. Journal of environmental management. 2011;**92**:2818-22

[66] Farid M, Ali S, Shakoor MB, Bharwana SA, Rizvi H, Ehsan S, Tauqeer HM, Iftikhar U, Hannan F. EDTA assisted phytoremediation of cadmium, lead and zinc. Internationa Journal of Agronomy and Plant Production. 2013;**4**:2833-46

[67] Gebrekidan A, Weldegebriel Y, Hadera A., Van der Bruggen B. Toxicological assessment of heavy metals accumulated in vegetables and fruits grown in Ginfelriver near Sheba Tannery, Tigray, Northern Ethiopia. Ecotoxicology and Environmental Safety. 2013; **95**, 171-178.

[68] Goher ME, AM AE, Abdel-Satar AM, Ali MH, Hussian AE, Napiórkowska-KrzebietkeA. Biosorption of some toxic metals from aqueous solution using non-living algal cells of *Chlorella vulgaris*. Journal of Elementology. 2016;**21**

[69] Golui D, Datta SP, Dwivedi BS, Meena MC, Trivedi VK, Jaggi S, Bandyopadhyay KK. Assessing Geoavailability of Zinc, copper, nickel, lead and cadmium in polluted soils using short sequential extraction scheme. Soil and Sediment Contamination: An International Journal. 2020;1-18

[70] Gupta N., Khan DK, Santra SC. Heavymetal accumulation in vegetables grown in a long-term wastewater-irrigated agricultural land of tropical India. Environmental Monitoring and Assessment. 2012;**184**:6673-6682

[71] Li F, Qiu Z, Zhang J, Liu W, Liu C, Zeng G. Investigation, pollution mapping and simulative leakage health risk assessment for heavy metals and metalloids in groundwater from a typical brownfield, middle China. International journal of environmental research and public health. 2017;**14**:768

[72] Hédiji H, Djebali W, Belkadhi A, Cabasson C, Moing A, Rolin D, Chaïbi W. Impact of long-term cadmium exposure on mineral content of Solanumlycopersicum plants: consequences on fruit production. South African Journal of Botany. 2015;**97**,176-181

[73] Idrees N, Tabassum B, Abd Allah EF, Hashem A, Sarah R, Hashim M. Groundwater contamination with cadmium concentrations in some West UP Regions, India. Saudi Journal of Biological Sciences. 2018;**25**:1365-1368

[74] Irfan M, Hayat S, Ahmad A, Alyemeni MN. Soil cadmium enrichment: allocation and plant physiological manifestations. Saudi Journal of Biological Sciences.2013;**20**,1-10

[75] Islam MA, Al-Mamun A, Hossain F, Quraishi SB, Naher K, Khan R, Das S, Tamim U, Hossain SM, Nahid F. Contamination and ecological risk assessment of trace elements in sediments of the rivers of Sundarban mangrove forest, Bangladesh. Marine Pollution Bulletin. 2017;**124**:356-366

[76] Järup L, Rogenfelt A, Elinder CG, Nogawa K, Kjellström T. Biological half-time of cadmium in the blood of workers after cessation of exposure. Scandinavian journal of work, environment and health. 1983:327-31

[77] Jordán MM, Montero MA, Pina S, García-Sánchez E. Mineralogy and distribution of Cd, Ni, Cr, and Pb in biosolids-amended soils from Castellon Province (NE, Spain). Soil Science. 2009;**174**:14-20

[78] Tamaddon F, Hogland W. Review of cadmium in plastic waste in Sweden. Waste Management and Research. 1993;**11**:287-295

[79] Khan A, Khan S, Alam M, Khan MA, Aamir M, Qamar Z, Rehman ZU, Perveen S. Toxic metal interactions affect the bioaccumulation and dietary intake of macro-and micro-nutrients. Chemosphere. 2016;**146**:121-8

[80] Brown TJ, Idoine NE, Raycraft ER, Shaw RA, Deady EA, Hobbs SF, Bide T. World mineral production 2011-15. British Geological Survey. 2017

[81] Kabir E, Ray S, Kim KH, Yoon HO, Jeon EC, Kim YS, Cho YS, Yun ST, Brown RJ. Current status of trace metal pollution in soils affected by industrial activities. The Scientific World Journal. 2012

[82] Azzi V, Kazpard V, Lartiges B, Kobeissi A, Kanso A, El Samrani AG. Trace metals in phosphate fertilizers used in Eastern Mediterranean countries. CLEAN–Soil, Air, Water. 2017 ;45

[83] White PJ, and Brown PH. Plant nutrition for sustainable development and global health. Annals of botany, 2010;**105**, 1073-1080

[84] Pan J, Plant JA, Voulvoulis N, Oates CJ, Ihlenfeld C. Cadmium levels in Europe: implications for human health. Environmental Geochemistry and Health.2010;**32**, 1-12.

[85] Kirkham M.B. Cadmium in plants on polluted soils: Effects of soil factors, hyperaccumulation, and amendments. Geoderma. 2006;**137**,19-32.

[86] Phillips CJ, and Tudoreanu L. A model of cadmium accumulation in the liver and kidney of sheep derived from soil and dietary characteristics. Journal of the Science of Food and Agriculture. 2010; **91**, 370-376

[87] Tudoreanu L, and Phillips CJC. Modeling cadmium uptake and accumulation in plants.Advances in Agronomy. 2004; **84**, 121-157

[88] Adams SV, Passarelli MN, Newcomb PA. Cadmium exposure and cancer mortality in the Third National Health and Nutrition Examination Survey cohort. Occupational and environmental medicine. 2012;**69**:153-6

[89] Sauvé S, Manna S, Turmel M C, Roy AG, Courchesne F. Solidsolution partitioning of Cd, Cu, Ni, Pb, and Zn in the organic horizons of a forest soil. Environmental Science and Technology. 2003;**37**:5191-5196

[90] Uraguchi S, and Fujiwara T. Cadmium transport and tolerance in

rice: perspectives for reducing grain cadmium accumulation. Rice, 2012; **5**, 5

[91] Zhao FJ, Hamon RE, Lombi E, McLaughlin MJ, McGrath SP. Characteristics of cadmium uptake in two contrasting ecotypes of the hyperaccumulatorThlaspicaerulescens. Journal of Experimental Botany. 2002;**53**, 535-543.

[92] Lugon-Moulin N, Zhang M, Gadani F, Rossi L, Koller D, Krauss M., and Wagner G. Critical review of the science and options for reducing cadmium in tobacco (Nicotianatabacum L.) and other plants. Advances in agronomy. 2004;**83**,112-181

[93] Vesey DA. Transport pathways for cadmium in the intestine and kidney proximal tubule: focus on the interaction with essential metals. Toxicology letters. 2010;**198**:13-9

[94] M Valko, H.Morris, andM. T. D. Cronin, "Metals, toxicity and oxidative stress, "Current Medicinal Chemistry. 2005;**12**:1161-1208

[95] Abernethy DR, DeStefano AJ, Cecil TL, Zaidi K, Williams RL, Panel UM. Metal impurities in food and drugs. Pharmaceutical research. 2010;**27**:750-5.

[96] Åkesson A, Lundh T, Vahter M, Bjellerup P, Lidfeldt J, Nerbrand C, Samsioe G, Strömberg U, Skerfving S. Tubular and glomerular kidney effects in Swedish women with low environmental cadmium exposure. Environmental health perspectives. 2005;**113**:1627-31

[97] Julin B, Wolk A, Johansson JE, Andersson SO, Andrén O, Åkesson A. Dietary cadmium exposure and prostate cancer incidence: a population-based prospective cohort study. British journal of cancer. 2012;**107**:895-900

[98] Nordberg GF, Nogawa K, Nordberg M, and Friberg L. "Cadmium," in Chapter 23 in Handbook of the Toxicology of Metals. F. Nordberg, B. F. Fowler, M.Nordberg, and L. Friberg, Eds., pp. 445-486, Elsevier, Amsterdam, The Netherlands, 3rd edition, 2007

[99] Abdulla M, Chmielnicka J. New aspects on the distribution and metabolism of essential trace elements after dietary exposure to toxic metals. Biological Trace Element Research. 1989;**23**:25-53

[100] R A Bernhoft. "Mercury toxicity and treatment: a review of the literature," Journal of Environment and land Public Health. 2012

[101] Whittaker MH, Wang G, Chen XQ, Lipsky M, Smith D, Gwiazda R, Fowler BA. Exposure to Pb, Cd, and As mixtures potentiates the production of oxidative stress precursors: 30-day, 90-day, and 180-day drinking water studies in rats. Toxicology and Applied Pharmacology. 2011;254:154-66

[102] Wang L, Gallagher EP. Role of Nrf2 antioxidant defense in mitigating cadmium-induced oxidative stress in the olfactory system of zebrafish. Toxicology and applied pharmacology. 2013;**266**:177-86

[103] Kjellstrom T. "Mechanism and epidemiology of bone effects of cadmium," IARC Scientific Publications.1992; **118**:301-310

[104] Kido S, Fujihara M, Nomura K, Sasaki S, Shiozaki Y, Segawa H, Tatsumi S, Miyamoto KI. Fibroblast growth factor 23 mediates the phosphaturic actions of cadmium. Nihon eiseigakuzasshi. Japanese journal of hygiene. 2012;**67**:464-71

[105] Ogawa T, Kobayashi E, Okubo Y, Suwazono Y, Kido T, Nogawa K. "Relationship among prevalence of patients with Itai-itai

disease, prevalence of abnormal urinary findings, and cadmium concentrations in rice of individual hamlets in the Jinzu River basin, Toyama prefecture of Japan," InternationalJournal of Environmental Health Research. 2004; **14**:243-252

[106] Bernhard D, Rossmann A, Henderson B, Kind M, Seubert A, Wick G. Increased serum cadmium and strontium levels in young smokers: effects on arterial endothelial cell gene transcription. Arteriosclerosis, thrombosis, and vascular biology. 2006;**26**:833-8

[107] Pacini S, Fiore MG, Magherini S, Morucci G, Branca JJ, Gulisano M, Ruggiero M. Could cadmium be responsible for some of the neurological signs and symptoms of Myalgic Encephalomyelitis/Chronic Fatigue Syndrome. Medical hypotheses. 2012;**79**:403-7

[108] Shargorodsky J, Curhan SG, Henderson E, Eavey R, Curhan GC. Heavy metals exposure and hearing loss in US adolescents. Archives of Otolaryngology–Head & Neck Surgery. 2011;**137**:1183-9

[109] Argonne National Laboratories, Cadmium, Human Health FactSheet, Argonne National Laboratories, Lemont, Ill, USA, 2001.

[110] Paulson AJ. The transport and fate of Fe, Mn, Cu, Zn, Cd, Pb and SO4 in a groundwater plume and in downstream surface waters in the Coeur d'Alene Mining District, Idaho, USA.Applied Geochemistry. 1997;**12**:447-464

[111] USEPA (US Environmental Protection Agency).Region 9, Preliminary Remediation Goals. 2012

[112] Vidali M: Bioremediation: An overview. (2001) Pure Applied Chemistry; **73**:1163-172.

[113] Roy M, Giri AK, Dutta S, Mukherjee P. Integrated phytobial remediation for sustainable management of arsenic in soil and water. Environment international. 2015;**75**:180-98

[114] Schlegel HG, Cosson JP, and Baker AJM. 1991. Nickel hyperaccumulating plants provide a niche for nickel resistant bacteria. BotanicaActa. 1991;104:18-25

[115] Yu G, Liu J, Long Y, Chen Z, Sunahara GI, Jiang P, You S, Lin H, Xiao H. Phytoextraction of cadmium-contaminated soils: Comparison of plant species and low molecular weight organic acids. International Journal of Phytoremediation..2020; **22**:383-391

[116] Silva JR, Fernandes AR, Junior MS, Santos CR, Lobato AK. Tolerance mechanisms in *Cassia alata* exposed to cadmium toxicity–potential use for phytoremediation. Photosynthetica. 2018;**56**:495-504

[117] Ali SY, Banerjee SN, Chaudhury S. Phytoextraction of cadmium and lead by three vegetable-crop plants. Plant Science Today. 2016 Aug 15;**3**:298-303

[118] Liu L, Li Y, Tang J, Hu L, Chen X. Plant coexistence can enhance phytoextraction of cadmium by tobacco (Nicotianatabacum L.) in contaminated soil. Journal of environmental Sciences. 2011;**23**:453-60

[119] Hassan MS, Dagari MS, Muazu AA, Sanusi KA. Effect of Citric Acid on Cadmium Ion Uptake and Morphological Parameters of Hydroponically Grown Jute Mallow (Corchorusolitorius). International Journal of Chemical, Material and Environmental Research. 2016;**3**:14-9

[120] Srivastava S, Agrawal SB, Mondal MK. A review on progress of heavy metal removal using adsorbents of microbial and plant origin.

Environmental Science and Pollution Research. 2015;**22**:15386-415

[121] Mosa KA, Saadoun I, Kumar K, Helmy M, Dhankher OP. Potential biotechnological strategies for the cleanup of heavy metals and metalloids. Frontiers in Plant Science. 2016;7:303

[122] Abioye OP, Oyewole OA, Oyeleke SB, Adeyemi MO, Orukotan AA. Biosorption of lead, chromium and cadmium in tannery effluent using indigenous microorganisms. Brazilian Journal of Biological Sciences. 2018;**5**:25-32

[123] Lakkireddy K, Kües U. Bulk isolation of basidiospores from wild mushrooms by electrostatic attraction with low risk of microbial contaminations. AMB Express. 2017;7:28

Chapter 5

Use of Swine Manure in Agriculture in Southern Brazil: Fertility or Potential Contamination?

Lucas Benedet, Guilherme Wilbert Ferreira, Gustavo Brunetto, Arcângelo Loss, Paulo Emílio Lovato, Cledimar Rogério Lourenzi, Sérgio Henrique Godinho Silva, Nilton Curi and Jucinei José Comin

Abstract

A major challenge in agricultural production systems is the maximization of resources used to promote the development of crops with a minimum of environmental impact. In this sense, the use of fertilizers of animal origin has great potential to promote the improvement of soil properties. In southern Brazil, swine manure (SM) is widely used in agricultural areas, allowing nutrient cycling within pig units and reducing costs for chemical fertilizers. Much of this manure is applied in liquid form (PS), but other strategies are often used, such as PS compost and swine bedding (DL). The use of these SMs improves the chemical, biological, and physical attributes of the soil, contributing to increased fertility and productivity of crops. However, prolonged use or applications with high doses of SM can result in the accumulation of metals and phosphorus in soils, representing a risk of contamination of soils and surface water resources, mainly due to losses by runoff, and subsurface, by leaching. Therefore, the adoption of criteria and the rational use of PMs need to be adopted to avoid dangerous effects on the environment, such as plant toxicity and water contamination. The potentialities and risks of SM applications are discussed in this chapter.

Keywords: pig slurry, pig deep litter, fertilization, copper, zinc, phosphorus

1. Introduction

The southern region of Brazil, represented by the states of Santa Catarina (SC), Paraná (PR) and Rio Grande do Sul (RS), stands out in swine production, having almost 50% of the total Brazilian herd, with a total of 20,594,238 heads [1]. However, due to the intensive production system in the swine units, the volume of waste produced daily is high. In these conditions, the use of swine manure as fertilizer in crop and pasture areas is an established strategy, promoting a targeting use of this material and cycling of nutrients [2].

Due to the large volume of swine manure (SM) produced in these swine units in southern Brazil, much work has been carried out to evaluate its effects on soils, as well as promoting strategies for its use, improving its efficiency [3, 4]. Also, alternatives have been proposed in the transformation of waste that is generated in these swine units, mainly in liquid form. These transformations ranged from anaerobic digestion, composting, and swine production in a deep litter production system. These strategies allow pig slurry (PS) transformation into a more stable material, with a higher concentration of nutrients and in a solid form, such as the PS compound and the deep litter (DL). SM transformation also has the advantage of allowing the waste transport and commercialization, becoming another source of income in the swine units [5–8].

Studies carried out in these Brazilian states have shown that SM are rich sources in macro and micronutrients, and their beneficial use for plant growth and development, serving as an interesting alternative in the partial or total replacement of mineral fertilizers, promoting the maintenance of productivity and decrease in external inputs in agricultural production areas [2, 5, 9, 10]. In addition, SM applications in agricultural areas promotes benefits to different soil properties, increasing soil fertility, improving soil structure and microbiological activity [4, 6, 9, 11, 12].

However, SM also has factors limiting its use, such as its variable concentration of nutrients and unbalanced to the needs of the plants, which can compromise the quality and functionality of the environment when excessive and/or continuous fertilization is carried out in the same areas [13–17]. The main concerns raised by researchers, and according to the publications of international academic society, are related to the phosphorus (P) and heavy metals, such as copper (Cu) and zinc (Zn), accumulation, availability, and losses in areas submitted a long history of applications [3, 18–23]. The results of the studies developed to emphasize the importance of adopting technical criteria for the use of SM in agricultural areas, proposing limits of doses used, applications based on soil attributes and composition of manure, the recommendation for crops, determination of limiting elements for the applied dose (mainly P), supplementation of the application of SM with chemical fertilizers and adequate treatments for the storage and stabilization of waste, for example [10, 17, 24, 25].

This chapter aims to survey studies carried out in the states of Santa Catarina, Paraná and Rio Grande do Sul about the use of swine manure in soils, covering its effects on soil fertility, crop productivity, and also on its potentials effects on the contamination of soils and water resources, mainly related to P, Cu, and Zn. Approximately 92 scientific publications conducted in these states were used to carry out this study.

2. Use of swine manure in agricultural areas: soil fertility and crop yield

The use of swine manure (SM) for grain and pasture production is an alternative to mineral fertilizers and allows the disposal of this residue within agricultural units, promoting nutrient cycling and reducing production costs [9, 10, 26].

Swine manure is used as fertilizer because it is rich in nutrients in readily available forms (mineral) to crops, in addition to increasing nutrients levels in organic forms that tend to have less availability, but which they can become available from their mineralization [4, 27, 28]. While pig slurry (PS) can present amounts close to mineral (49.6%) and organic (50.4%), for example, deep litter (DL) presents much higher proportions of organic N (90.6%) compared to mineral N (9.4%) [27]. Changes in these proportions are also observed in soils fertilized with SM, where applications, mainly with higher doses, increase the P labile forms in the soil. However, a large part of the P present in the soil is in more recalcitrant forms, with emphasis on the mineral forms (Pi), corresponding to up to 80% of the total value in soil superficial layers, mainly associated with Fe oxides. In deeper layers, this percentage is even higher. Additionally, the contribution of P organic forms (Po) can reach up to 20%, decreasing with the increase of soil depth and soil organic matter (SOM) content [29]. This presence of organic and mineral components in organic fertilizers favors, in addition to improving soil fertility, microbiological activity, and subsequent crop nutrition, with organic components mineralization over time (residual effect) [4, 26–28, 30–32].

SM applications promote nutrients added to the soil, increasing its contents and, consequently, its availability to plants [6, 33–38]. Naturally, SM applications allow a more complete nutrients fertilization, not limited to N, K, and P, present in high concentrations in the manure, but also adding other macro and micronutrients, such as Ca, Mg, Cu, Zn, and Mn [7, 9, 12]. In a study conducted by Pessotto et al. [35] in the municipality of Três Passos, RS, in a Entisol, PS applications (150 m^3 ha^{-1}) for four and 16 years in pasture areas (Cynodon spp.) promoted an increase in Cu, Zn, P, K, Ca, Mg and Na contents compared to mineral fertilization in an area of annual grain cultivation under no-tillage, mainly with the longest application time. A similar result was observed by Lourenzi et al. [6] evaluating the effect of PS compost applications in a no-tillage system with maize (*Zea mays* L.), black beans (*Phaseolus vulgaris* L.), in the off-season, and black oats (*Avena strigosa* L.) succession in a Oxisol, located in Chapecó, SC. The authors determined that the dose of 4 Mg ha^{-1} promoted nutrients contents similar to mineral fertilization (NPK), but with an increase in Cu, Zn and Ca contents, besides the SOM, in the soil surface layer.

However, unlike mineral fertilizers, the amounts of nutrients added by SM are difficult to control, as their chemical composition is variable [4, 9, 26]. This was evident in a study conducted by Da Ros et al. [36] in PS samples used for fertilizing sunflower (Oct/2009), canola (May/2010), beans (Jan/2011), and maize (Sep/2011) in Santa Maria, RS. During this period, four applications were carried out, with N, P_2O_5, and K_2O contents ranging from 0.83–2.21 g L^{-1}, 0.37–1.75 g L^{-1}, and 0.69–1.25 g L^{-1}, respectively. The different diets offered, the animal ages, the volume of water used to clean slatted floor stalls, and the shape and time of storage, for example, are components that promote variation in the levels of nutrients present in manure [34]. Additionally, the breeding systems and the SM stabilization processes, such as the deep litter system, composting, and anaerobic digestion, promote changes in the structure and composition of manure and in its soil dynamics [5, 7, 8]. Admittedly, the incorporation of substrates with high lignification, such as rice husks, shavings, and sawdust, in the preparation of deep litter and compost increases its C/N ratio. Also, the stabilization process of these swine fertilizers promotes the degradation of more labile organic compounds, reduces the carbon (C) content, increases the levels of nitrate (from the transformation of ammoniacal N) and other nutrients, producing a more stable material [8]. Consequently, its mineralization in the soil can be delayed, as well as the release of nutrients, also possibly causing greater immobilization of nutrients [4]. This was demonstrated by Giacomini et al. [27] evaluating the dynamics of N in the incubation of an Ultisol with DL, PS, and PS + oat straw application, in Santa Maria, RS. DL and PS additions increased soil mineral N contents during incubation, with DL promoting an increase of 17 mg kg^{-1}, while PS of 66 mg kg^{-1}, compared to the control treatment. Due to the nature of these wastes, the amount of mineralized N was differentiated, where 34.9% of the total N of the PS was mineralized, whereas for DL this percentage was only 14.6%. Additionally, the authors observed that the addition of PS to the soil covered with oat straw reduced the amount of mineral N by 30%, compared to the soil with only PS, due to the N microbial immobilization in response to the high C/N ratio of the straw (46.5/1), also to the lower mineralization of organic N from PS that remained on the straw, without making contact with the soil.

In addition to nutrients, SM applications also promote changes in the SOM contents, through the addition of organic compounds present in the manure, but mainly by increasing the crops biomass production [7, 9, 26, 37]. The maintenance of these residues in the soil, along with soil conservation systems, such as no-till system and integrated crop-livestock-forest (ICLF), favors the accumulation of MOS [35, 38]. This positive effect can be seen in **Figure 1**, which shows the number of scientific articles developed with SM in the Southern region of Brazil, starting in 2008, which showed changes in the SOM contents. This compilation reveals that in most studies (n = 21) the addition of SM increases the SOM contents compared to soils without fertilization, corresponding to approximately 60% of the studies. In the other studies (40%) there was no change in the levels. This distribution changes when assessing the effect of manure in comparison to applications with mineral fertilizers, changing to close to 40 those in which the MOS content increased and 60% in those in which there was no effect. The positive effect of SM on SOM was observed by Scheid et al. [12] with an increase in the organic C content of a Hapludox in pasture area (Cynodon spp.) after 15 years of PS applications with the dose of 200 m^3 ha^{-1} $year^{-1}$, in Nova Candelária, RS. The authors attributed these results to the direct effect of the C added by the manures (even in low concentration) and indirectly to the C added to the soil by crop residues and roots, which showed an increase in their biomass due to PS fertilization. The same mechanism was used by Lourenzi et al. [6] to justify the increase in SOM contents after six

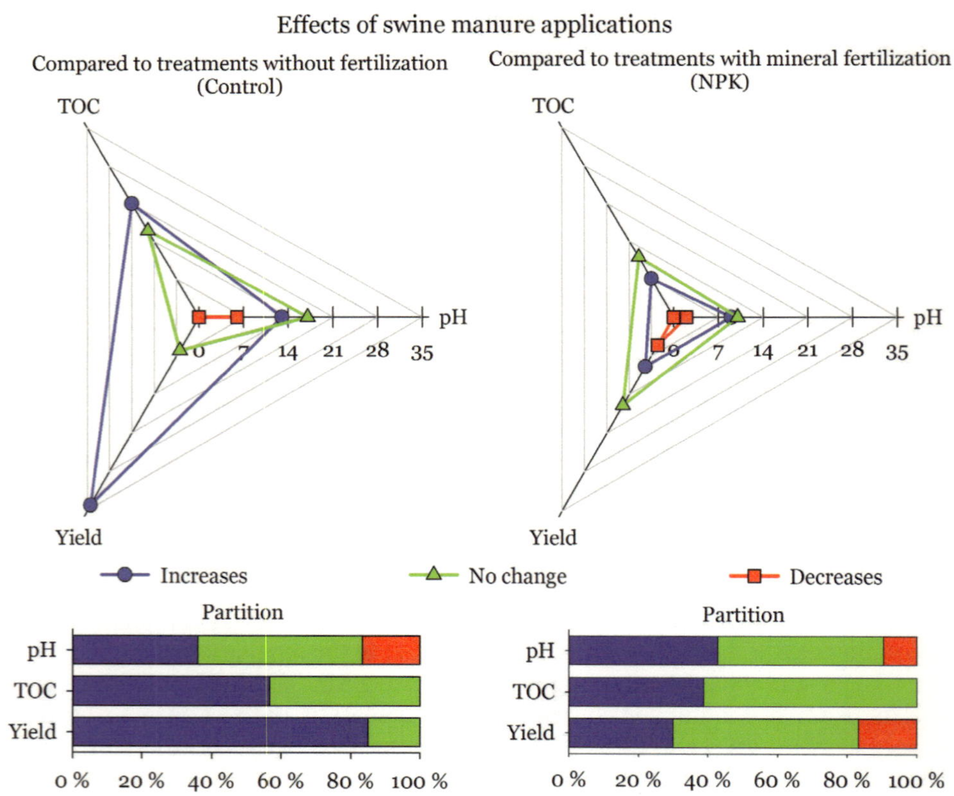

Figure 1.
Number of scientific articles using swine manure that show an increase, reduction, and without change in total organic carbon (TOC) content, pH values, and crop yield compared to treatments without fertilization (control) and with mineral fertilization (NPK).

years of PS composting with shavings applications in a no-till system in Chapecó, SC. The authors observed an increase of 22, 26, and 23% in SOM contents, with the application of 4, 8, and 16 Mg ha^{-1}, respectively. However, it is important to notice that both studies found greater effects on soil superficial layers, promoting changes up to 10 and 4 cm, respectively. Additionally, the addition of solid material, such as PS compost or DL, tends to promote the greater accumulation of organic C than PS applications, due to the higher C content in these solid manures [6, 7]. The increase in SOM contents in soils fertilized by SM is a factor of great importance in increasing soil fertility, contributing to its CEC, microbiological activity, and soil structure, also is an attribute associated with soil quality [11, 12]. However, it is important to notice that some studies have also shown the null effect of SM application on SOM contents [7, 16, 34]. In samples of Entisol, Inceptisol, and Oxisol, obtained from rural units in Western Santa Catarina under a no-tillage system, Scherer et al. [38] observed that SOM contents were similar between areas with PS application and with mineral fertilization, submitted to both managements for 15 and 20–25 years.

Another important component for soil fertility and crop growth influenced by SM fertilization is pH. However, as for SOM, the increase or decrease in pH values depends on the type of manure applied, dose and management used, and also to environmental conditions [26, 36]. However, the positive effect of SM on soil pH tends to be greater compared to mineral fertilization (**Figure 1**). Da Ros et al. [36], for example, observed an increase of 0.3 units in the pH values in the 0–20 cm layer, on average, with the application of 100 m^3 ha^{-1} of PS, being higher than the values observed in treatments without fertilization and with NPK fertilization, in Santa Maria, RS. In the same region, a similar result was found by Lourenzi et al. [37] whose observed an increase in pH values up to 8 cm deep in soil fertilized with PS, mainly with the dose of 80 m^3 ha^{-1}, compared to the treatment without fertilization. According to these authors, the addition of carbonates by the manure and the complexation of Al and H^+ present in soil by organic compounds from SM and crop residues are the main mechanisms responsible for the positive effect of SM on pH, pointed out by these authors. Therefore, the use of soil improvers may not be necessary or reduced. However, the absence of changes in pH values may occur due to the use of liming, as observed by Brunetto et al. [11], when evaluating the effect of DL and PS applications in an Typic Hapludalf in Braço do Norte, SC. Also, NH_4^+ nitrification from SM applications can promote pH values reduction by increasing the H^+ concentration, as observed by Veiga et al. [33] in a Rhodic Hapludox in Campo Novos, SC.

This improvement in soil fertility promoted by SM applications favors the increase of crop productivity, due to the greater supply of nutrients, root growth, and absorption [9, 32]. In comparison to unfertilized soils, SM application has a great effect on crop yield, due to the supply of nutrients in the soil (**Figure 1**). Freitas Alves et al. [28] observed that PS application at a dose of 40 m^3 ha^{-1} promoted an average yield of maize grains from the 2011/12 (Oxisol) and 2012/2013 (Inceptisol) similar to the soil fertilized with NPK (130, 185 and 70 kg ha^{-1} of N, P_2O_5, and K_2O, respectively) and higher than the soil without fertilization, in a study carried out in Lajes, SC. Also, Brugnara et al. [32] also observed the positive effect of PS compost application on the production of passion fruit seedlings, in a study carried out in Chapecó, SC. The authors established PS compost concentrations up to 64% (v/v) (maximum tested dose) and 38.87 to 95.8% (v/v) in mixtures with coconut fiber and conventional substrate, respectively, to obtain the best responses of leaf area, height, dry mass of the aerial part and root and number of leaves. Additionally, Fey et al. [5] observed that doses of 60 and 150 m^3 ha^{-1}, PS applications promoted greater production of maize biomass, and also a higher concentration of N, P, and Mn in the plant tissues, than PS obtained from manure and biodigester, mainly with the highest dose, both in a Oxisol and in an Ultisol, in Marechal Cândido Rondon, PR. The higher concentration of nutrients in the PS without treatment compared to the other wastes promoted this result. These results show that the dose used also needs to consider the nature and composition of the SM to be used and that the amount of nutrients added to the soil can vary considerably according to the type and management adopted about this manure.

However, in studies that compare crops yield with SM fertilization and mineral fertilizers, the results are more controversial (**Figure 1**). Under these conditions, soil natural fertility, manure doses used and the crop type are factors that have great influences on the results. An example is a study by Pandolfo and Veiga [2] on a Hapludox soil, in Campos Novos, SC, with increasing doses of PS in different management systems related to the maintenance of crop residues on the soil surface. First, the authors observed that winter pasture biomass production and maize grain yield in the summer were higher after the second year of evaluation, due to the improvement of soil fertility with PS applications. Second, the effects of PS applications were greater on the management that removed the winter and summer crop residues than on those where these residues were kept (either in winter or in summer). This revealed that in soils with less cycling or greater withdrawal of nutrients, the response in crop yield to PS applications is higher, although higher doses are necessary to maintain soil fertility and crop yield. Additionally, although the use of increasing doses of SM promotes a gradual increase in crop yields, balancing with the results obtained with the mineral fertilizers application, many studies also demonstrate that very high doses of SM are necessary to promote higher yields than those observed in crops with mineral inputs [9, 10, 28]. In a study conducted in Chapecó, SC, for example, Miranda et al. [31] observed that PS applications promoted greater production of dry matter (DM) of giant missionary grass only with the dose of 500 kg of N ha^{-1}, which corresponded to 275 m^3 ha^{-1} $year^{-1}$, in comparison to the application of 200 kg of N with mineral fertilizer. The lower PS doses promoted lower or similar yields to mineral fertilizer treatment. In other cases, some studies have also revealed that after reaching a certain dose, productivity increases are very low or zero [9]. The lower PS doses promoted lower or similar yields to mineral treatment. The lower effect of SM applications on crop yield compared to mineral fertilization is also documented (**Figure 1**). This was observed by Locatelli et al. [10] in maize grain yield in the 2014/15 and 2015/2016 harvests, in Curitibanos, SC. Even with the addition of 140 kg of N ha^{-1}, the yield was lower with PS application (12,778 and 7886 kg ha^{-1}, respectively) compared to the treatment with mineral fertilization (19,348 and 11,411 kg ha^{-1}, respectively) which in total added 122.7 kg of N ha^{-1}.

However, in these conditions with high SM application the fertilizer efficiency index is reduced, promoting greater addition of nutrients in the soil, but with less use by the crops [26]. Consequently, there is a greater accumulation of nutrients in the soil and greater potential for soil pollution and surface and subsurface water resources through losses from runoff and leaching, respectively [5, 23, 39–43]. In terms of runoff losses, the volume of the runoff and the amount of nutrients added to the soil (high doses of manure) are mainly responsible for the increase in losses. Also, the time between the SM application and the first rainfall event has a strong effect on the amount of nutrient lost in the runoff. Therefore, it is recommended that applications are not carried out when there is a risk of rain. On the other hand, the increase in the production of biomass from crops, promoted by its applications can offer a beneficial effect by increasing soil cover, increasing water infiltration, and reducing the volume of runoff [19, 39]. In two parallel studies carried out by Ceretta et al. [3] and Basso et al. [18], the authors assessed the losses by runoff and leaching, respectively, from the application of increasing doses of PS (0, 20 40, and 80 m^3 ha^{-1}). Basso et al. [18] observed that PS applications promoted in the early stages of crop development result in greater N losses by leaching, due to the low root volume and low capacity to absorb nutrients. Also, the greatest losses occur from the greatest addition of nutrients to the soil promoted by high SM doses. These applications, especially in high doses and in the initial stages of plant development, deserved special attention, as they caused concentrations of N in percolated water above 10 mg L^{-1}, considered as a critical level in water for human consumption [44], which can represent a risk to subsurface water quality. However, PS applications promote low changes in P concentrations in the percolated water. This is justified due to the greater interaction between this element and soil mineral components, especially clays [30]. In parallel, Ceretta et al. [3] showed that the P and N concentrations are much higher in runoff waters. The authors calculated that the P and N losses by flow with the doses of 20 40 and 80 m^3 ha^{-1} were 49, 21, and 20 times (available P) and 4.0, 3.7, and 1.5 times (nitrate) greater than those observed by percolation. The highest concentrations of N and P in the runoff waters were obtained after the first dates after the applications and with the highest doses of manure. Furthermore, in many evaluation periods, nitrate concentrations above 10 mg L^{-1} were found mainly with the highest PS doses, which was also observed by Ceretta et al. [19]. This strengthens the greatest concern in areas with a long history of SM application and with more sloping reliefs, which have a greater accumulation of nutrients, especially P, and which are more susceptible to runoff losses. Management systems without carrying out the soil overturning may even present an accumulation of nutrients in the soil's most superficial layers, increasing the risk of losses [40, 41]. As alternatives, the maintenance of vegetation cover, level and/or band cultivation, and the construction of terraces are important strategies to avoid and control these losses and their potential negative effect on the quality of water resources.

In swine units with continuous SM applications, the importance of adopting technical criteria and strategies for the management and application of manure is increasingly emphasized, including the recommendation of nutrients for crops for the choice of dose, application time, manure treatment, soil management, soil moisture, and nutrient concentration in SM. These measures prevent or delay nutrients accumulation in soil and its losses, whether through volatilization (NH_3), percolation, or runoff. The absence of measures and monitoring of applications can compromise the quality of water resources present in the swine units, as observed by the studies by Cadoná et al. [23] and Loss et al. [42]. In the first study, these authors observed in the supply wells of four swine units located in Braço do Norte, SC, an increase in N and P concentrations of in the water. For N, there was the pollution of these water resources with NH_4^+ and NO_3^- in all collections, which were carried out between July 2015 and June 2016. The highest NH_4^+ and NO_3^- concentrations percolated occurred mainly in the month with the highest rainfall, with a greater volume of percolated water and greater transport capacity. These data were worrying, since the concentrations were above the parameters regulated by CONAMA Resolution n° 357/2005 [44] for Total Ammoniacal Nitrogen and NO_3^-, with maximum limits of 3.7 mg L^{-1} and 10 mg L^{-1} for freshwater, respectively. The authors also evidenced that total P concentrations in the water were above that recommended by this Resolution (0.030 mg L^{-1}) in all collections performed, presenting a variation of 0.07 mg L^{-1} to 1.42 mg L^{-1} in collections for the four swine units. Also, a relationship was verified between soil available P contents and total P concentrations in waters. Loss et al. [42] also observed that the water resources (supply well, weir, and spring) of a swine unit (Braço do Norte) showed an increase in NH_4^+ and NO_3^- concentration in all collections, and in certain periods the values were above recommended by CONAMA [44], mainly for NH_4^+. These data strengthen the attention that must be given to the volume of applied manure, favoring the subdivision of applications (mainly in the initial stages of crop development) to reduce the accumulation of nutrients in the soil, especially P.

Therefore, due to the varied positive effects of SM on soil properties and crops yield, the addition of these residues to the soil is no longer considered as disposal, but a recommended direction in the swine production units, since followed by technical criteria [24, 25, 45]. This concern is fundamental for the protection of natural resources and soil functionality maintenance, maintaining attention to the high concentration of nutrients present in SM and its variable composition, being often incompatible with the nutritional requirements of the crops to be fertilized [7, 13, 26].

3. Soil contamination risk from swine manure applications

Although the use of SM contributes considerably to the improvement of soil fertility, its continuous use has been rising concern in academic and environmental society, especially with the use of high doses [22, 46]. Greater attention is paid to P and heavy metals accumulation in soils, which can cause water eutrophication and plant toxicity, respectively, representing risks to the quality and functionality of soils and nearby water resources [7, 14, 15, 17, 24, 25]. For this reason, many studies continue to evaluate the total contents and availability of these elements in different types of soil, applied SM, and management. The main objective is to develop strategies applicable to the management of SM fertilizers that promote soil fertility, without compromising its quality or with the least possible impact. The results obtained by these studies on P and heavy metals accumulations in soils are presented below.

3.1 Phosphorus

In areas subjected to constant application of swine manure, especially at high doses, there may be an increase in the levels of P in the soil. This is justified due to the variable composition of these nutrients in the manure, being predominantly out of balance with the recommendations for different cultures. This makes it difficult to control the quantities added in each application. As noted by **Figure 2**, in studies with SM applications where doses are based on the amount added, that is, by volume (m^3 ha^{-1} for PS) or weight (t ha^{-1} for DL or compost), the higher doses tend to promote greater increases in the availability of P in soils, due to the greater addition of this element. Also, management systems that aim to meet N demands through the exclusive use of SM have resulted in more excessive P and K additions to the soil. In these situations, to reach the recommended N dose, the amount of manure used promotes an addition higher than what will be absorbed by the plants, resulting in accumulation in the soil. And as the number of applications increases, the availability and total P content in the soil also tends to increase (**Figures 2** and **3**) [7, 40, 41]. However, the rate of increase in these levels will depend on the amount of P added in each application, the type of culture and management adopted, and the soil's attributes, such as the type and content of clay, pH, and MOS (**Figure 4**) [15, 46].

In the soil, P has low mobility and is highly reactive to the functional groups of clay minerals and Fe and Al oxides. For this reason, soils with high clay content and mineralogical composition with a greater abundance of hematite, goethite, gibbsite, and some minerals of 1: 1 clay, such as kaolinite, have a greater capacity to adsorb P [30].

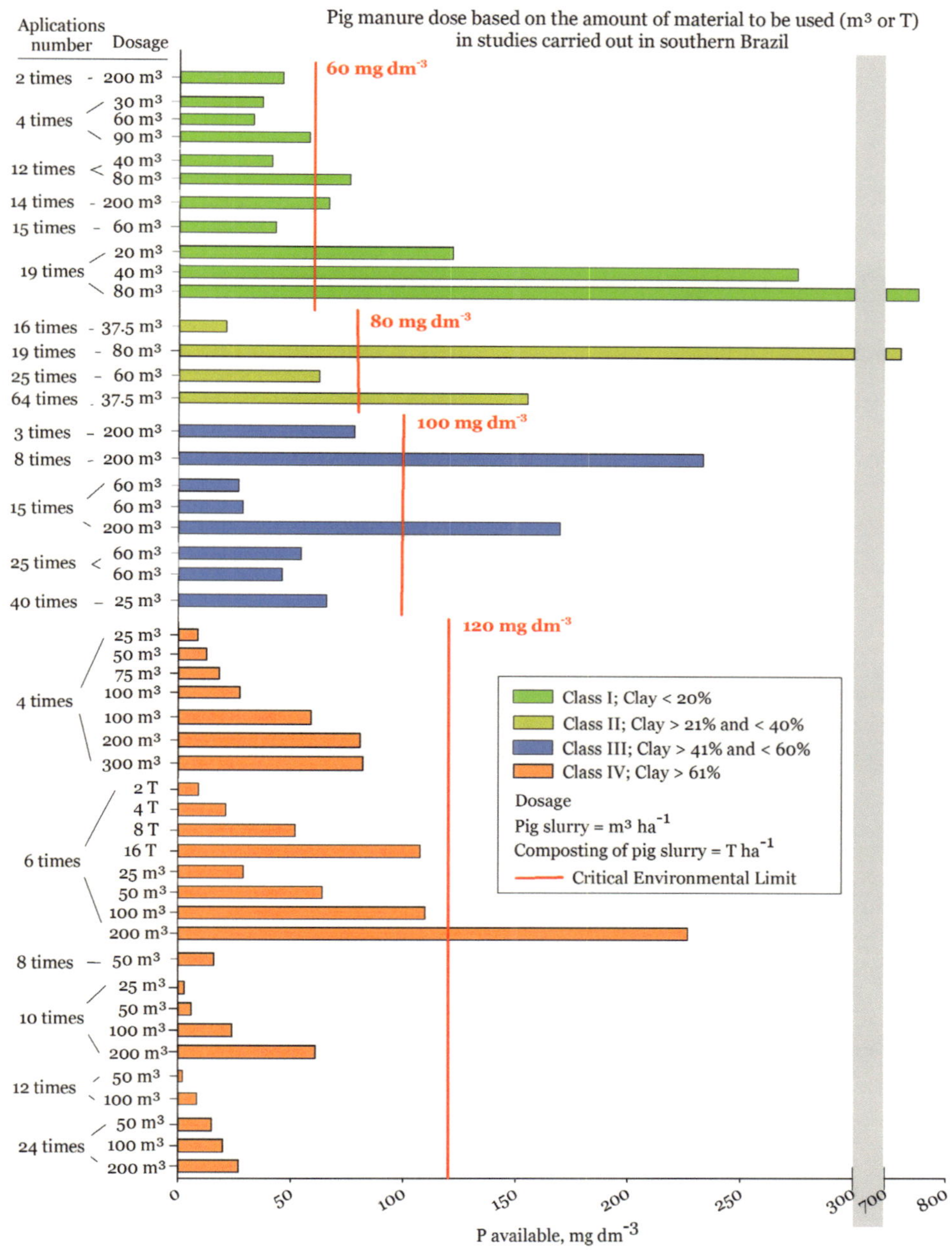

Figure 2.
Available P contents in soils with different clay contents obtained from studies with application of SM conducted in the states of SC, PR, and RS, using the volume (m^3 ha^{-1}) or weight (t ha^{-1}) of the applied waste (PS or PS compost).

Thus, the P that is added by the SM applications tends to be adsorbed, mainly by the more avid adsorption sites, with a greater binding energy of these mineral components [13, 24, 25]. Naturally, the weathering process, the chemical reactions of the soil, and the action of the roots can make available this adsorbed P, as well as releasing the one that is present in the composition of the minerals, but in less quantity and when the available levels are low [39, 47].

Also, the effect of microbiological activity and mineralization of MOS allows P release associated with organic soil components. In this sense, in studies performing P fractionation in the soil with SM applications, it is observed that most of the

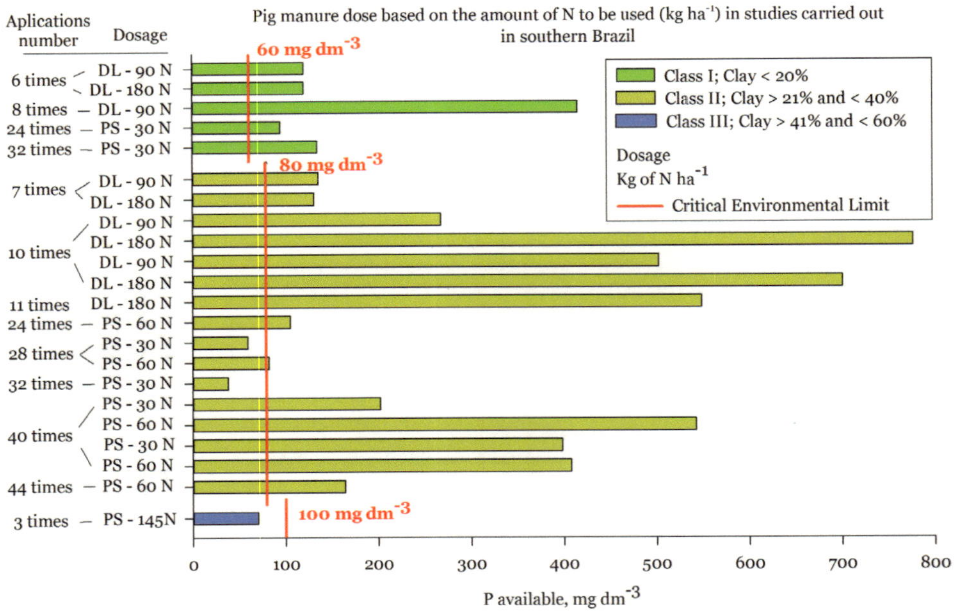

Figure 3.
Available P contents in soils with different clay contents obtained from studies with application of PS and DL conducted in the states of SC, PR, and RS, using as a dose the recommendation for N (kg ha^{-1}).

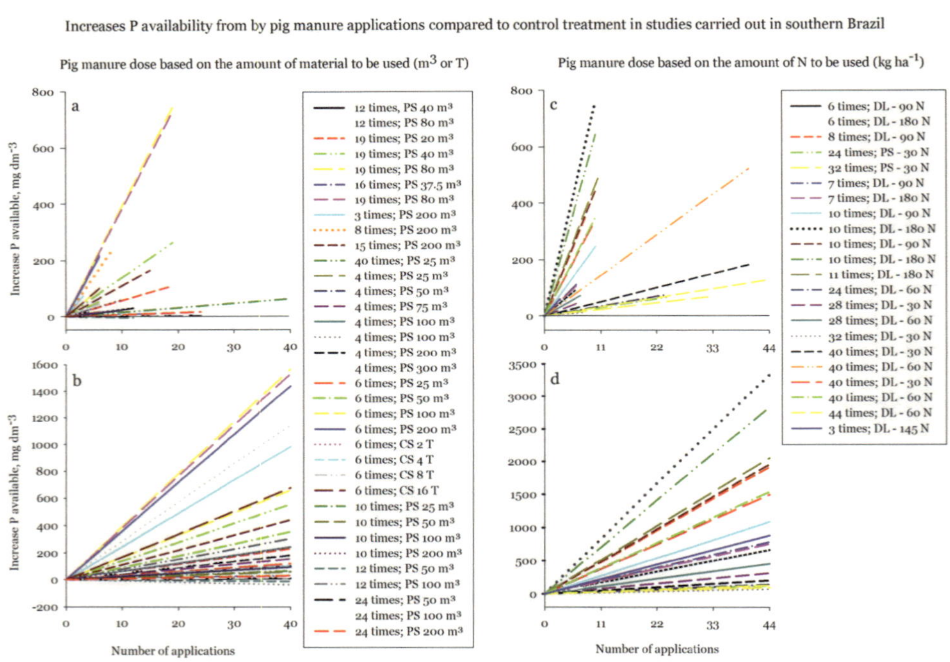

Figure 4.
Increase in soil available P contents based on the number of SM applications used in the studies (a and c) and projected up to 44 applications (b and d) from studies conducted in the states of SC, PR, and RS, using as dose the volume (m^3 ha^{-1}) or weight (t ha^{-1}) of the applied manure (a and b) and the recommendation by N (c and d).

added P is in inorganic forms, associated with minerals. However, it is also possible to observe an increase in P organic form, mainly with an increase in SOM promoted by SM application or an increase in cultivated crops DM. The little effect is observed on the P residual contents, a highly recalcitrant form and not available, but there is an increase in the more labile forms, highly related to the applied SM dose [13, 15, 30, 34, 40, 41, 46]. In a study with increasing SM doses in Santa Maria, RS, De Conti et al. [47] also observed that most of the chemical species of P in the soil solution were in the form of orthophosphates ($H_2PO_4^-$ and HPO_4^{2-}). These type of studies, together with the evaluation of the maximum adsorption capacity, demonstrate that the SM applications are saturating the functional groups of the soil colloids, reducing the binding energy, and contributing to the increase of P availability and mobility (migration) in the soil profile (**Figure 4**) [25, 40, 41]. This behavior is observed mainly in the no-tillage system, where applications are made superficially and there is no soil disturbance. This behavior is observed mainly in the no-tillage system, in which applications are carried out superficially and the soil tilling is restricted to the planting lines or seedling cradles. This promotes a marked accumulation of P in the most superficial layers, and a gradual decrease with increasing depth. In the case of systems with overturning, it is expected that P availability will be more evenly distributed over the overturned layer, which may promote a reduction in P contents [13]. Additionally, Gatiboni et al. [34] determined that in samples of ground soil the maximum P adsorption capacity of soils was 1.4 to 14 times greater than samples with preserved structure. The authors pointed out that in samples with the preserved structure, P can migrate through the soil profile through preferential flows, which does not happen with samples with ground soil. Therefore, in samples with the destruction of the soil structure, the area and the contact time between the water containing P that percolates in the soil and the adsorption bands with soil colloids, especially clays, can increase. However, soil losses in systems with tumbling tend to be greater, which can contribute to the losses of P (particulate P) by runoff [14, 15, 39, 46].

The major problem of P accumulation in soils is related to its potential for water resources eutrophication [48]. Although P has low mobility in the soil, the saturation of the adsorption sites by P accumulation can favor its migration in the profile through water infiltration. This migration can occur through preferential flows, through a continuous network of pores and bio pores, and fissures present in the soil structure [18–20, 47]. Additionally, Tiecher et al. [43] also observed that the amount of P added, but mainly the volume of percolated water, highly associated with the intensity of the rain, have the greatest effect on the amount of percolated P. These authors when taking a sample of percolated water at a depth of 0.6 m, observed in many instances the concentration of soluble and available P above 0.15 mg L^{-1}, the maximum limit allowed by Brazilian legislation [44], mainly using the dose of 80 m^3 (the highest dose assessed). However, as previously stated (item, 2), the greatest P loss occurs through its transport through surface runoff water [3, 39, 46]. Soil surface P accumulation, the amount of this element added by the applications, the interval between application, and the occurrence of rain are factors with a high correlation with the amount of P present in the runoff water. Also, the clay content has a great influence on how much the soil can retain P, reducing its losses by the flow. This was evident in the study conducted by Gatiboni et al. [25], who observed that in soils with higher clay content (80%) the increase in the concentration of P in the drained solution only occurs when P availability (determined by Mehlich-1) is very high (147 mg dm^{-3}). On the other hand, soils with lower clay content (20%) already show an increase in P in the drained solution with lower values of available P (74 mg dm^{-3}).

Therefore, the increase in P availability in soils can increase its losses and its direction to nearby watercourses, mainly rivers, springs, dams, and artificial wells, compromising its quality. This is even more worrying due to a large number of watercourses present in the studies of Southern Brazil and the very diversified relief present in this region, with many sloping areas, which favors soils and nutrients loss, especially in the absence of more conservationist soil management [15]. In a study conducted by Boitt et al. [46] in a Typic Hapludox from Campos Novos, Santa Catarina, counting the amount of P added by PS (25, 50, 100 and 200 $m^3\ ha^{-1}\ year^{-1}$) and by the crops residues, the amount accumulated in the soil and extracted by the crops, the authors evaluated that from 6 to 38% of the added P was lost, mainly by runoff. The amount of P that was considered lost was 45 to 1550 kg P ha^{-1}. This study reveals, therefore, that a considerable amount of the added P can be taken from the area of application and, potentially, reach the water resources. However, in a study evaluating the water quality of 13 watercourse points at Taquari Antas Watershed, Rio Grande do Sul State, which featured 861 swine farmers, Schneider et al. [29] determined that the Water Quality Index (WQI) and the trophic state class of the rivers were classified as "good" and "mesotrophic" (P concentration, $137 < P \leq 296$ mg m^{-3}), although they found an increase in the concentration of P at some points. Also, the authors warned about the proximity of swine units to water resources and the flow of rivers, where the closer the facilities are to the rivers, the greater their contaminating potential, and the greater the river's flow should be to promote a dilution effect of polluting agents. The potential of swine units to contaminate water resources was also addressed by Couto et al. [14, 15]. In the first study, the authors used data from estimated soil loss, the distance between the P source and a water body, and soil P concentrations (using P chemical fractionation data) to assess the vulnerability of P contamination in the soils of swine units submitted for different times (years) of PS application in Braço do Norte, SC. In summary, areas with higher levels of P in more labile forms, higher accumulation of P in the soil, lower clay content, and greater erosion caused by water showed greater vulnerability in losing P, which can cause it to be directed to water streams. In parallel, in the same region, Couto et al. [15] separated areas with different years of PS application and different types of land use and used P fractionation data and the universal equation of soil loss (USLE) to create a map representing the vulnerability to P losses for the entire Upper Coruja/Bonito River watershed. The different classes of vulnerability developed by this approach showed an excellent correlation with the P losses assessed in runoff water. For example, areas with the cultivation of black oats and rotating maize showed greater losses of P due to soil tillage and poor surface protection. With these results, the authors recommended the adoption of soil conservationist practices, such as not disturbing with PS incorporation and soil cover.

Therefore, more and more work involving the use of SM as fertilizer reveals that its prolonged and excessive use can pose great risks to environmental quality [48]. From this, strategies for storage, stabilization and application began to be devised in order to facilitate or increase control over what is applied to the soil or to prevent environmental problems in swine areas. In prominence, they became a reference for the assessment of critical P levels in soils, which became known as Environmental Critical Limits (ECL), which are related to the clay content of the soil and the concentration of P in the drained solution and allowed to regulate the amount of SM to be applied in the cultivation areas [24]. These studies were so relevant that they became the basis for the new environmental regulation of the State of Santa Catarina, based on the formulation of Normative Instruction No. 11 [45], which establishes criteria for the implementation of activities related to pig farming, including treatment of swine manure (liquid and solid) and its disposal in the soil. Until that time, Santa Catarina legislation only recommended PS applications in doses of up to 50 $m^3\ ha^{-1}\ year^{-1}$ [49]. Briefly, the new legislation determines that the maximum allowed content (Environmental Critical Limit) of P (ECL-P) extracted by Mehlich-1 in the 0–10 cm layer will be defined by the equation (Eq. (1)) prepared by Gatiboni et al. [24]:

$$ECL - P = 40 + clay(\%) \tag{1}$$

where clay (%) is the soil clay content present in the 0–10 cm layer.

From this equation, it is possible to identify the maximum allowed P content according to soil clay content, being a simple assessment, aided by soil analysis. With the values obtained by Eq. (1), it is then possible to determine the SM dose to be used. In cases where P content is up to 20% above the ECL-P, it is recommended to reduce the recommended maintenance dose by 50% for the crop to be fertilized. And in cases where P content exceeds by more than 20%, the ECL-P is recommended to temporarily suspend any type of fertilization that contains P. Mitigating measures are also recommended in both situations, such as the use of crops with high capacity of extraction and exportation of P and the revolving of the soil aiming at the dilution of P in deeper layers (which must be associated with soil erosion control practices) [45].

Based on this Normative Instruction, the authors of this chapter performed surveyed the scientific articles published in the states of SC, PR, and RS using SM to assess the impact of these applications on the P contents present in the different soils and management. The result of this survey is shown in **Figures 2–4**. As can be seen in **Figure 2**, in the studies where the SM applications used their quantity, that is, their volume (m^3 ha^{-1} for PS) or weight (t ha^{-1} for PS compost) to formulate the doses, there are some cases where the levels of available P are already above the ECL-P, especially with the highest doses. At the same time, the increase in the number of applications also tends to contribute to a higher P availability. However, it is also noticeable that in studies with soils with higher clay contents (> 61%) the increase in P availability is more subtle, due to the greater capacity of these soils to adsorb P [24, 25]. Also, in only one case were the available P content was above the ECL-P. On the other hand, when considering the studies that used N as the reference nutrient to determine the dose (**Figure 3**), it is clear that the PS and DL applications promoted an increase in the available P contents above the ECL-P in its vast majority (although the number of studies on soils with clay content>41% is very small). This reveals that the added amount of P in this management system is much higher, increasing its contents in the soil more quickly (**Figure 4**), which tends to cause greater problems in the accumulation and losses of P in the swine units.

The survey of these studies also reveals that if rates of increase in P availability are maintained until 44 applications have been made, approximately 53% of the cases that used the SM volume or weight as the dose will have available P contents above the CEL- P (**Figure 4**). In cases where N is used as a reference element, this number rises to approximately 95%.

Therefore, although SMs are rich and advantageous sources of supplying P to the crops of interest, their use must be carried out with technical criteria, keeping up with P availability in the soil. In swine units, where SM applications were and still are carried out periodically, the choice of dose should take into account the soil fertility, the chemical composition of the manure, and the nutritional requirements of the crops. With these criteria, the application must be carried out until it meets what was recommended for a given nutrient, which is considered the criterion of the critical nutrient. And the other nutrients must be supplemented with mineral fertilizer [45, 50]. Also, the use of N as a reference element in fertilizers should be avoided, as it promotes excessive P addition, especially in richer materials of this element, such as DL. The use of highly demanding plants in P (phytoaccumulators) and the use of conservation practices are also important measures to be taken. These precautions are important to ensure adequate targeting of the SM, avoiding P accumulation in the soil and potential losses, which can affect the quality of water resources present in the swine units or nearby.

3.2 Heavy metals

Continuous SM application and in excessive doses, mainly carried out in swine units also presents the risk of promoting metals accumulation and contamination in soils, which also enhances its transfer to the adjacent water bodies from these areas [51, 52]. Studies carried out in the southern states of Brazil demonstrate this increase in Cu and Zn total and available contents in areas subjected to SM applications, with emphasis on treatments with a greater volume of application or with doses based on the N recommendation (**Figures 5–7**). In parallel, some studies have also shown an increase in Mn, Fe, Ni, Cd, Pb, and Cr contents. However, the greatest focus of studies on metals in the states of SC, PR, and RS was given to Cu and Zn, which tended to have high levels in some studies [57].

SMs present variable but significant concentrations of Cu and Zn in their composition due to the addition of excessive amounts of these elements in the swines' diets about the physiological requirements of these animals [58]. In general, 6.08 to 11.68 mg of Cu kg^{-1}, and from 39.67 to 76.15.0 mg of Zn kg^{-1} of feed is added to the pig's diets between the pre-initial and termination phases [59]. In a study published by the National Research Council (NRC) [60], it was emphasized that Cu requirements can be from 3.0 to 6.0 mg kg^{-1} and Zn from 50.0 to 100.0 mg kg^{-1} during the animals breeding and 5.0 and 50.0 mg kg^{-1} of Cu and Zn, respectively, in the lactation and gestation phases. However, as highlighted by the NRC [60] and Ribeiro et al. [61], it is common to use doses of 100 to 250 mg of Cu kg^{-1} and 1000 to 3000 mg of Zn kg^{-1} of feed to promote pharmacological effects in pigs.

The use of these metals in feed formulations is due to the functions they present in swine metabolism. Copper sulfate ($CuSO_4$) is used as a growth promoter

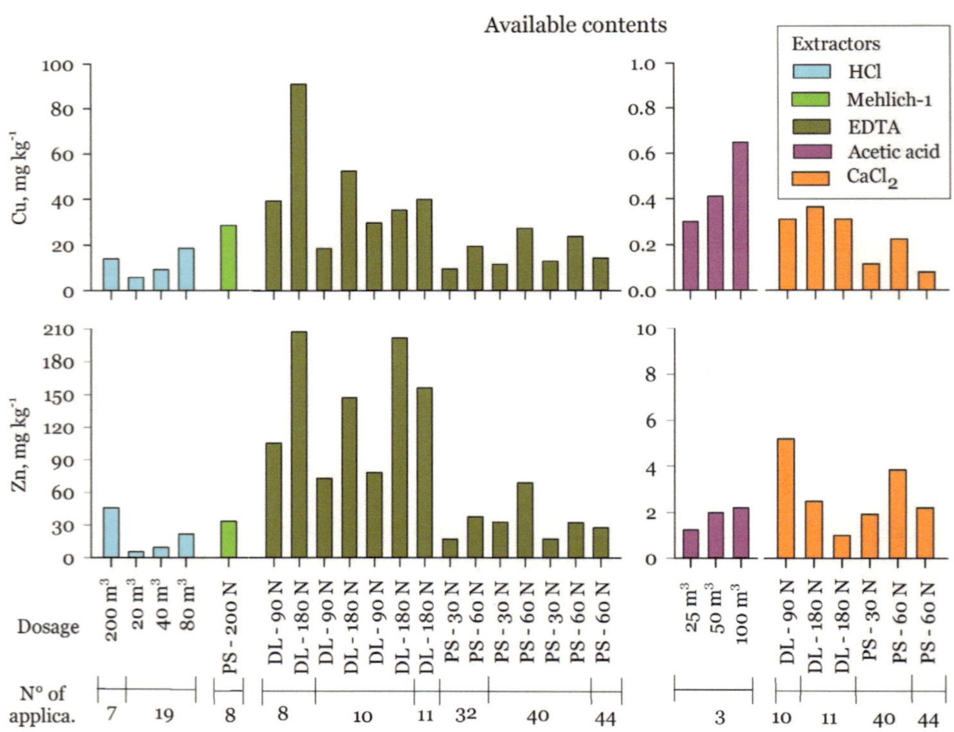

Figure 5.
Available Cu and Zn contents determined by different extractors (HCl, EDTA, Mehlich-1, CaCl2, and acetic acid) in study soils in the states of Santa Catarina, Paraná, and Rio Grande do Sul.

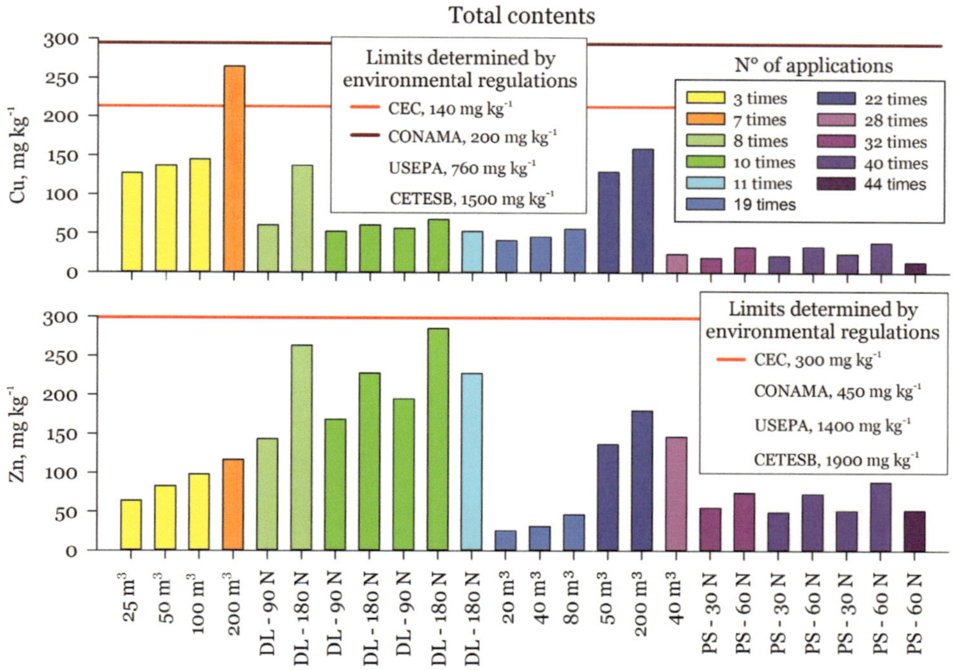

Figure 6.
Total Cu and Zn contents in soils from studies carried out in the states of Santa Catarina, Paraná and Rio Grande do Sul and the values recommended by environmental agencies [53–56].

in animals, while zinc oxide (ZnO) acts as a preventive and curative product for diarrhea [60, 61]. As a result, much of the Cu and Zn added to the feed is eliminated by the animals' feces. At the same time, the adoption of certain manure management systems, such as swine production in a deep litter system and the composting of manure, can increase the proportion of these elements in the final product. The high temperatures during the process of stabilizing the DL or the PS compost and the microbial activity favor the mineralization of OM and the reduction of the C content at the end of this process, increasing the concentration of metals [8].

Cu and Zn when added to the soil via SM are distributed in pre-existing forms in the soil, such as mineral and organic, being adsorbed by the most avid adsorption sites. This ability of soils to adsorb Cu and Zn will depend on the nature of the soil source material, the content, and composition of the clay fraction, the content and composition of the MOS, and the geochemical conditions of the soils, such as pH and ionic strength [17, 57, 62]. However, the continuous addition of Cu and Zn to the soil by SM applications promotes bonds with less adsorption energy and precipitates with greater solubility [57]. In work developed by Girotto et al. [63] in Santa Maria, RS, the authors found that after 17 applications with different doses of DL (0, 20, 40, and 80 $m^3 ha^{-1}$) Cu and Zn are accumulated in the soil mainly in bioavailable forms, in which the highest levels of Cu were found in organic and mineral form, and Zn, in mineral form. Also, successive manure additions tend to increase the presence of Cu and Zn in more labile forms, such as soluble and exchangeable, due to the saturation of the adsorption sites, reducing the bond stability. Similar results were obtained by other authors in soils in the municipalities of Campos Novos and Braço do Norte, performing the chemical fractionation of Cu and Zn in the soil [21, 57].

Therefore, as Cu and Zn are added to the soil, they are distributed among soil mineral and organic components. However, as the applications are maintained, the bonding groups are saturated, increasing the availability of these metals in the

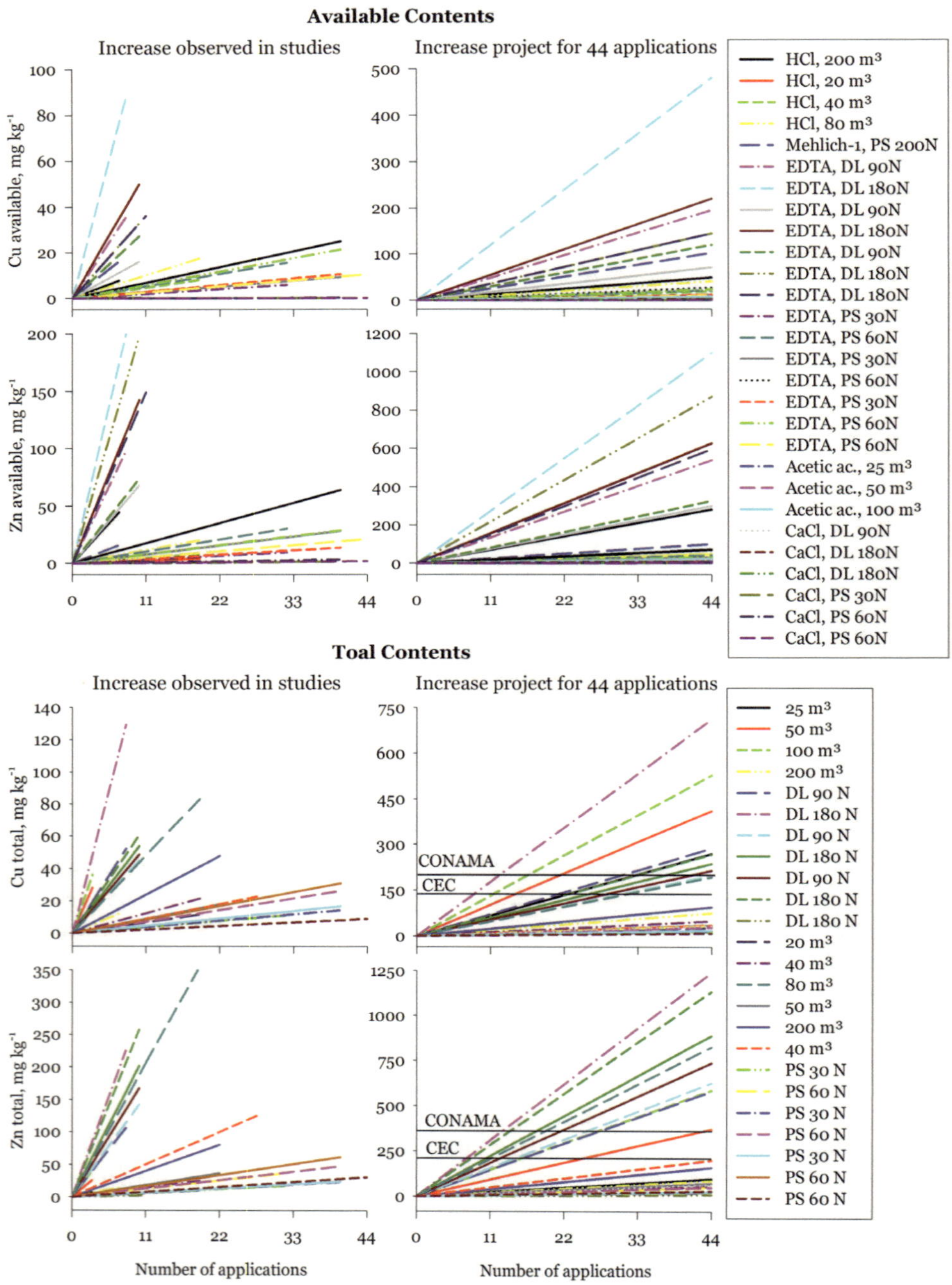

Figure 7.
Increase in available and total Cu and Zn contents in the soil based on the number of SM applications used in the studies and projected up to 44 applications from studies conducted in the states of SC, PR, and RS.

soil. This availability is influenced by pH, redox potential, ionic strength, competing ions, and soil constituents (minerals and organics), in addition to the metal's ionic potential, that is, the ratio of its charge and ionic radius (z^2/r) [62]. Thus, the magnitude in increasing the availability of metals in areas with SM application is not only related to how much is added by the manure, but also by soil attributes. In a studie carried out on swine units in Santa Catarina, Couto et al. [21] observed an increase in Cu and Zn total and available contents, but that this increase was dependent on the type and attributes of the soil, application time and SM composition.

When surveying studies carried out in the southern states of Brazil, we also observed a trend in increasing Cu and Zn availability (determined by HCl, EDTA, Mehlich-1, $CaCl_2$, and Ac. Acetic extractors) and in total contents in the trials, mainly with the highest applied volumes (m^3 of PS ha^{-1}) and with the use of N as a reference element (**Figures 5** and **6**). However, greater attention should be paid to observing that in some cases the Cu total contents are already above the reference values of the Council of the European Communities (CEC) [57] and CONAMA [59]. This represents a risk to the quality of this environment, requiring monitoring and possible mitigating actions in the future. In the case of Zn, no study had levels above the recommended. However, when projecting an increase in the total contents in these areas, up to a number of 44 applications, it is observed that the number of situations where the levels were above that recommended by CEC [57] and CONAMA [59] will be higher (**Figure 7**). Additionally, it is observed that the increase in availability with the 44 applications will provide very high levels, which can represent potential toxic effects in these areas. Therefore, it is important to exercise caution in doses used excessively so as not to compromise the quality of the environment and increase Cu and Zn available contents.

With the increase in Cu and Zn saturation and availability in the soil promoted by SM, consequently, there is an increase in the concentration of these metals in the soil solution, which potentiates their toxic effects on plants and organisms and their movement in the soil profile [7, 16, 17, 20, 63, 64]. The increase in the concentration of these metals in the soil solution after continuous applications of PS was demonstrated by De Conti et al. [65] in soil samples obtained in Santa Maria, RS. In addition to the increased concentration, mainly with the highest dose (80 m^3 ha^{-1}), these authors also observed that the main species of Cu and Zn in the soil solution were free species (Cu^{2+} and Zn^{2+}) and species complexed with dissolved organic compounds (CuDOC and ZnDOC). The increase in free species can pose a risk to crops, as these are forms absorbed by the roots, while the CuDOC and ZnDOC species have greater mobility. However, the authors themselves revealed that the crops could complex these free species with exudates release, which increased the concentrations of CuDOC and ZnDOC in the soil solution. This mobility of Cu and Zn, referring to the soil solution, is dependent on its concentration in the solution, the nature of the association with other soluble ionic species, and the ability of the soil to provide them for the solution [64].

The increase in soil Cu and Zn availability in areas subjected to SM application has raised concerns about the possibility of toxic effects on crop cultivation. This concern is justified because, even though these metals play an important role in many physiological processes in crops, it is expected that in the swine units there may be a steady growth in the contents of these metals in soils, which can reach critical levels, affecting the productive capacity of these areas [22]. In Santa Maria, RS, Girotto et al. [20], when evaluating maize grown in soil samples (deformed and non-deformed) submitted to 19 applications of increasing doses of DL (20, 40, and 80 m^3 ha^{-1}), observed an increase in soil Cu and Zn available contents and physiological changes in the maize, with increased lipid peroxidation, several senescent leaves, and ascorbate peroxidase activity and decreased plant weight and superoxide dismutase activity, especially at higher doses and with the deformed soil sample. In a parallel study, evaluating the maize cultivation in a soil (obtained from Braço do Norte, SC) submitted to nine years of PS and DL applications with doses based on the recommendation of N (90 and 180 kg N ha^{-1}), Benedet et al. [7] observed an increase in Zn concentration in the aerial part and small changes in stomatal density, chlorophyll content and APX and CAT activity in maize grown in the soil with PS 180 kg N ha^{-1}, but this did not affect the crop development. With soil samples obtained in the same area (but with 10 years of application), Benedet et al. [16] observed that black oat plants showed an accumulation of Cu and Zn in the aerial part and Cu in the roots, especially when grown in the treatment with DL 180 kg N ha^{-1}. Additionally, at the beginning of crop development, leaves with chlorosis in DL 180 kg N ha^{-1} were observed. However, no negative effect was observed in the production of DM on plant growth. Although these studies have used high doses of SM, higher than that used in swine units, they indicate a potential negative effect of soil Cu and Zn accumulation of Cu and Zn, their translocation to crop tissues, the need to maintain control on the doses used of SM and monitoring of these areas [21]. Additionally, as presented by Benedet et al. [17], the maintenance of conservationist practices, especially those that favor the maintenance or increase of SOM contents, mainly in the most stable forms, is important to control the availability of metals, mainly Cu, due to their high affinity with the organic groups, increasing the adsorption of these metals and reducing their soluble forms.

In parallel, the accentuated additions of Cu and Zn in the soil by the successive SM applications present a high potential for environmental contamination [21, 61], as they can be submitted losses through runoff and leaching [51, 52, 64]. In an experiment conducted in Santa Maria, Girotto et al. [64] evaluated the Cu losses in the drained and percolated soil solutions submitted to PS applications and found that the greatest Cu losses occurred due to surface runoff. Additionally, the Cu losses in the drained solution, with a dose of 80 m^3 ha^{-1}, were 2.3 times greater than in soil without fertilization, the majority being transferred in particulate form. Smanhotto et al. [53], on the other hand, did not observe changes in the Cu and Zn concentrations in the percolated water collected at a depth of 60 cm in a soil fertilized with PS (100; 200, and 300 m^3 ha^{-1}) in Cascavel, PR. The authors justify the changes in the Cu and Zn concentrations in the percolated solution as dependent on the soil clay content present and the presence of pores and bio pores that favor percolation and alter the interaction between the percolated water and the surface of the clay minerals and Fe oxides and Al. Additionally, Dal Bosco et al. [66] also observed an increase in Cu and Zn concentration in water lost by runoff after application of increasing doses of PS (50, 100, 150, and 200 m^3 ha^{-1}) in Toledo, PR. The authors observed that with the increase of the dose and the first collection of drained water (48 days after sowing) there was a greater loss of Cu. Also, the Zn concentrations in the drained material were close to the maximum concentration of 5.0 mg L^{-1} provided by CONAMA Resolution 357/2005 [44] for the discharge of effluents. These losses due to runoff, although they do not seem significant, can cause contamination of water resources. In parallel studies, Capoane et al. [52] found low Cu and Zn concentrations in the water resources of the hydrographic basin of Arroio Caldeirão, Palmitinho, RS, formed by 124 swine units with the intensive swine production system, dairy cattle, and tobacco production, mainly. However, Capoane [51], when evaluating the Cu and Zn contents in the sediments present in the bottom of the water resources of this basin, observed that Zn concentrations were between the range Threshold Effect Level (TEL) and Probable Effect Level (PEL) (123.1–315.0 mg kg^{-1}), a range that represents the occasional occurrence of adverse effects for organisms. And the Cu concentrations were on average 2.3 times higher than the Canadian Council of Ministers of Environment (CCME) [67] toxicity values (> 197.0 mg kg^{-1}), indicating that adverse effects may already be occurring in the aquatic organisms. The data collected by these studies reinforce the need to use conservationist practices to prevent erosion caused by rains, as described in item 3.1.

Thus, although the use of SM promotes productive gains to agricultural crops and improves soil fertility, its excessive application can promote the accumulation of Cu and Zn above the soil's support capacity, contributing to its mobility in the soil profile and toxicity to plants, in addition to favoring surface and subsurface waters contamination, compromising the functionality of the environment [21, 57, 63, 64].

4. Conclusion

The use of SM as fertilizer in swine units in the states of Santa Catarina, Paraná and Rio Grande do Sul is an advantageous practice for producers, as it allows the use and recycling of a source rich in nutrients, reducing the need for external inputs. The application of SM in soils favors the increase of available contents of macro and micronutrients, but it also tends to promote improvements in various soil attributes. The addition of C by the manure, even in a lower concentration in the PS, and the increase in crops DM production favors the increase of SOM, which consequently contributes to the improvement of soil biological and physical attributes. In addition, many studies carried out in these states demonstrate that SM applications can also favor soil pH values. Reducing the use of limestone or agricultural plaster in these areas. The increase in soil fertility results in enhance crop productivity, and in many cases the gains obtained are equivalent to those promoted by mineral fertilizers. However, it was also common to observe through the studies raised that very high doses of SM are necessary to achieve yields proportional to mineral fertilizers. Therefore, it is important to carefully use the SM in agricultural areas, avoiding excessive applications, higher than what will be absorbed by the crops. In these swine units the formulation of the doses of application of SM must take into account soil fertility, crops recommendations to be cultivated and the chemical composition of the SM. From these criteria, the most limiting element can be determined, that is, the one that first contemplates what was recommended, and the other nutrients will be supplemented with mineral fertilizer. The transformation of PS into solid materials, such as the PS compound and DL, can be an interesting strategy to be used in these swine units, reducing the volume of SM and allowing the transport and commercialization of a product rich in nutrients. However, as these solid manures have a higher concentration of nutrients, their use must present greater care and criteria. Additionally, it was observed by the studies survey that the use of N as a reference element for the recommendations of doses is a practice that should be avoided, as it causes the addition of very high amounts of other elements, such as P, Cu and Zn. Potentializing the accumulation of these nutrients in the soil. In this sense, P is an element of greatest concern in these swine units, as studies already show high levels in the soil and potential losses due to runoff, especially in areas subject to water erosion. Due to the provisions of the new normative instruction in Santa Catarina, the increase in the P content in these areas will make it impossible to continue with SM applications in these areas. And knowing that the extension of the areas of the pig units is limited, this can cause a shortage of options for farmers to target the MPs. Finally, the adoption of conservationist practices is important to maintain the soil structure and avoid losses by erosion and mineralization of the SOM. These practices contribute to increasing the adsorption capacity of these soils, contributing to the retention of P, Cu, and Zn in mineral and organic components.

Acknowledgements

The authors thank the National Committee of Technological and Scientific Development (CNPq), College Education Improvement Coordination (CAPES), Foundation for Support of Research and Innovation in the State of Santa Catarina (FAPESC) and Social Technologies for Water Management (TSGA II)—Petrobrás Environmental.

Author details

Lucas Benedet[1,2*], Guilherme Wilbert Ferreira[2], Gustavo Brunetto[3], Arcângelo Loss[2], Paulo Emílio Lovato[2], Cledimar Rogério Lourenzi[2], Sérgio Henrique Godinho Silva[1], Nilton Curi[1] and Jucinei José Comin[2]

1 Department of Soil Science, Federal University of Lavras, University Campus, Lavras, MG, Brazil

2 Department of Rural Engineering, Center of Agricultural Sciences, Federal University of Santa Catarina, Rod. Admar Gonzaga 1346, Florianópolis, SC, Brazil

3 Department of Soil Science, Center of Rural Science, Federal University of Santa Maria, Av. Roraima 1000, University City, Santa Maria, RS, Brazil

*Address all correspondence to: lucas_benedet@hotmail.com

References

[1] Instituto Brasileiro de Geografia e Estatistica – IBGE. Pesquisa da Pecuária Municipal – PPM [internet]. Available from: https://sidra.ibge.gov.br/pesquisa/ppm/quadros/brasil/2018 [Accessed: 2020-08-29].

[2] Pandolfo CM, Veiga MD. Crop yield and nutrient balance influenced by shoot biomass management and pig slurry application. Revista Brasileira de Engenharia Agrícola e Ambiental. 2016; 20(4): 302-307. DOI: 10.1590/1807-1929/agriambi.v20n4p302-307

[3] Ceretta CA, Basso CJ, Vieira FC, Herbes MG, Moreira IC, Berwanger AL. Dejeto líquido de suínos: I-perdas de nitrogênio e fósforo na solução escoada na superfície do solo, sob plantio direto. Ciência Rural. 2005; 35(6): 1296-1304. DOI: 10.1590/S0103-84782005000600011

[4] Giacomini SJ, Aita C, Miola EC, Recous S. Mineralização do carbono da palha de aveia e dejetos de suínos aplicados na superfície ou incorporados ao solo. Revista Brasileira de Ciência do Solo. 2008; 32(SPE): 2661-2668. DOI: 10.1590/S0100-06832008000700008

[5] Fey R, Lana MD, Zoz T, Richart A, Luchese EB. Relações entre lixiviação de nitrato e produção de biomassa do milho com dejetos suínos provenientes de diferentes tratamentos. Revista Brasileira de Ciências Agrárias. 2010; 5(2): 212-218. DOI: 10.5039/agraria.v5i2a715

[6] Lourenzi CR, Scherer EE, Ceretta CA, Tiecher TL, Cancian A, Ferreira PA, Brunetto G. Atributos químicos de Latossolo após sucessivas aplicações de composto orgânico de dejeto líquido de suínos. Pesquisa Agropecuária Brasileira. 2016; 51(3): 233-242. DOI: 10.1590/S0100-204X2016000300005

[7] Benedet L, Comin JJ, Pescador R, de Oliveira PA, Belli Filho P, De Conti L, da Rosa Couto R, Lovato PE, Cesco S, Mimmo T, Brunetto G. Physiological changes in maize grown in soil with copper and zinc accumulation resulting from the addition of pig slurry and deep litter over 10 years. Water, Air, & Soil Pollution. 2016; 227(11): 401. DOI: 10.1007/s11270-016-3103-9

[8] Wong JW, Karthikeyan OP, Selvam A. Biological nutrient transformation during composting of pig manure and paper waste. Environmental technology. 2017; 38(6): 754-761. DOI: 10.1080/09593330.2016.1211747

[9] Cassol PC, Costa AC, Ciprandi O, Pandolfo CM, Ernani PR. Disponibilidade de macronutrientes e rendimento de milho em Latossolo fertilizado com dejeto suíno. Revista Brasileira de Ciência do Solo. 2012; 36(6): 1911-1923. DOI: 10.1590/S0100-06832012000600025

[10] Locatelli JL, Bratti F, Ribeiro RH, Besen MR, Turcatel D, Piva JT. Uso de dejeto líquido de suínos permite reduzir a adubação mineral na cultura do milho?. Revista de Ciências Agrárias. 2019; 42(3): 31-40. DOI: 10.19084/rca.17538

[11] Brunetto G, Comin JJ, Schmitt DE, Guardini R, Mezzari CP, Oliveira BS, Moraes MP, Gatiboni LC, Lovato PE, Ceretta CA. Changes in soil acidity and organic carbon in a sandy typic hapludalf after medium-term pig-slurry and deep-litter application. Revista Brasileira de Ciência do Solo. 2012; 36(5): 1620-1628. DOI: 10.1590/S0100-06832012000500026

[12] Scheid DL, Silva RF, Silva VR, Ros CO, Pinto MA, Gabriel M, Cherubin MR. Changes in soil chemical and physical properties in pasture fertilised with liquid swine manure. Scientia Agricola. 2020; 77(5). DOI: 10.1590/1678-992x-2019-0017

[13] Berwanger AL, Ceretta CA, Santos DR. Alterações no teor de fósforo no solo com aplicação de dejetos líquidos de suínos. Revista Brasileira de Ciência do Solo. 2008; 32(6): 2525-2532. DOI: 10.1590/S0100-06832008000600029

[14] Couto RR, Santos MD, Comin JJ, Pittol Martini LC, Gatiboni LC, Martins SR, Filho PB, Brunetto G. Environmental vulnerability and phosphorus fractions of areas with pig slurry applied to the soil. Journal of environmental quality. 2015; 44(1): 162-173. DOI: 10.2134/jeq2014.08.0359

[15] Couto RR, Martini LC, Gatiboni LC, Belli Filho P, Martins SR, Lazzari CJ, Júnior VM, Comin JJ, Withers PJ, Schenato RB, Brunetto G. Vulnerability to contamination by phosphorus in a zero-order basin with a high density of pigs and a history of slurry addition: extrapolation of an index. Environmental earth sciences. 2018; 77(4): 144. DOI: 10.1007/s12665-018-7301-1

[16] Benedet L, De Conti L, Lazzari CJ, Júnior VM, Dick DP, Lourenzi CR, Lovato PE, Comin JJ, Tiecher TL, Ricachenevsky FK, Brunetto G. Copper and Zinc in Rhizosphere Soil and Toxicity Potential in White Oats (*Avena sativa*) Grown in Soil with Long-Term Pig Manure Application. Water, Air, & Soil Pollution. 2019; 230(8): 209. DOI: 10.1007/s11270-019-4249-z

[17] Benedet L, Dick DP, Brunetto G, dos Santos Júnior E, Ferreira GW, Lourenzi CR, Comin JJ. Copper and Zn distribution in humic substances of soil after 10 years of pig manure application in south of Santa Catarina, Brazil. Environmental Geochemistry and Health. 2020; 25: 1-21. DOI: 10.1007/s10653-020-00572-9

[18] Basso CJ, Ceretta CA, Durigon R, Poletto N, Girotto E. Dejeto líquido de suínos: II-perdas de nitrogênio e fósforo por percolação no solo sob plantio direto. Ciência Rural. 2005; 35(6): 1305-1312. DOI: 10.1590/S0103-84782005000600012

[19] Ceretta CA, Girotto E, Lourenzi CR, Trentin G, Vieira RC, Brunetto G. Nutrient transfer by runoff under no tillage in a soil treated with successive applications of pig slurry. Agriculture, ecosystems & environment. 2010a; 139(4): 689-699. DOI: 10.1016/j.agee.2010.10.016

[20] Girotto E, Ceretta CA, Rossato LV, Farias JG, Tiecher TL, De Conti L, Schmatz R, Brunetto G, Schetinger MR, Nicoloso FT. Triggered antioxidant defense mechanism in maize grown in soil with accumulation of Cu and Zn due to intensive application of pig slurry. Ecotoxicology and environmental safety. 2013; 93:145-155. DOI: 10.1016/j.ecoenv.2013.03.021

[21] Couto RR, Lazzari R, Conti D, Belli F. Accumulation and distribution of copper and zinc in soils following the application of pig slurry for three to thirty years in a microwatershed of southern Brazil. Archives of Agronomy and Soil Science. 2016; 62(5): 593-616. DOI: 10.1080/03650340.2015.1074183

[22] Feng Z, Zhu H, Deng Q, He Y, Li J, Yin J, Gao F, Huang R, Li T. Environmental pollution induced by heavy metal(loid)s from pig farming. Environmental Earth Sciences. 2018; 77(3): 103. DOI: 10.1007/s12665-018-7300-2

[23] Cadoná EA, Lourenzi CR, DE Souza EL, Rampinelli EC, Dos Santos ML, Sete PB, Soares CR. Contaminação por nitrogênio e fósforo de águas destinadas ao consumo humano em região com intensa atividade suinícola. Geociências (São Paulo). 2019; 37(4): 883-891.

[24] Gatiboni LC, Smyth TJ, Schmitt DE, Cassol PC, Oliveira CMB, editors. Proposta de Limites Ambientais de

Fósforo para Solos de Santa Catarina. Lages: UDESC/CAV; 2014. 38 p.

[25] Gatiboni LC, Smyth TJ, Schmitt DE, Cassol PC, Oliveira CM. Soil phosphorus thresholds in evaluating risk of environmental transfer to surface waters in Santa Catarina, Brazil. Revista Brasileira de Ciência do Solo. 2015; 39(4): 1225-1234. DOI: 10.1590/01000683rbcs20140461

[26] Lourenzi CR, Ceretta CA, Brunetto G, Girotto E, Tiecher TL, Vieira RC, Cancian A, Ferreira PA. Pig slurry and nutrient accumulation and dry matter and grain yield in various crops. Revista Brasileira de Ciência do Solo. 2014; 38(3): 949-958. DOI: 10.1590/S0100-06832014000300027

[27] Giacomini SJ, Aita C, Barbosa SP, Miola EC. Transformações do nitrogênio no solo após adição de dejeto líquido e cama sobreposta de suínos. Pesquisa Agropecuária Brasileira. 2013; 48(2): 211-219. DOI: 10.1590/S0100-204X2013000200012

[28] Freitas Alves CT, Cassol PC, Sacomori W, Gatiboni LC, Ernani PR, Aita C, Panisson J, Ferreira AK. Influência da adubação com dejeto suíno e adubo mineral adicionada de inibidor de nitrificação sobre a produtividade e a nutrição do milho. Revista de Ciências Agroveterinárias. 2017; 16(1): 2-10. DOI: 10.5965/223811711612017002

[29] Schneider VE, Marques RV, Bortolin TA, Cemin G, Dos Santos GM. Monitoring and assessment of surface water quality in Taquari-Antas Watershed, South Brazil—region with intensive pig farming. Environmental monitoring and assessment. 2016; 188(11): 617. DOI: 10.1007/s10661-016-5635-2

[30] Abdala DB, Gatiboni LC, Schmitt DE, Mumbach GL, Dall'Orsoletta DJ, Bonfada EB, Veiga M. Phosphorus speciation and iron mineralogy in an oxisol after 11 years of pig slurry application. Science of The Total Environment. 2020; 743: 140487. DOI: 10.1016/j.scitotenv.2020.140487

[31] Miranda M, Scheffer-Basso SM, Escosteguy PA, Lajús CR, Scherer EE, Denardin RB. Dry matter production and nitrogen use efficiency of giant missionary grass in response to pig slurry application. Revista Brasileira de Zootecnia. 2012; 41(3): 537-543. DOI: 10.1590/S1516-35982012000300009

[32] Brugnara EC, Nesi CN, Verona LA. Cama de aviário e composto de dejetos suínos em substratos para mudas de maracujazeiro-amarelo. Científica. 2014; 42(3): 242-251. DOI: 10.15361/1984-5529.2014v42n3p242-251

[33] Veiga MD, Pandolfo CM, Balbinot Junior AA, Spagnollo E. Chemical attributes of a Hapludox soil after nine years of pig slurry application. Pesquisa Agropecuária Brasileira. 2012; 47(12): 1766-1773. DOI: 10.1590/S0100-204X2012001200013

[34] Gatiboni LC, Schmitt DE, Cassol PC, Comin JJ, Heidemann JC, Brunetto G, Nicoloso RD. Samples disturbance overestimates phosphorus adsorption capacity in soils under long-term application of pig slurry. Archives of Agronomy and Soil Science. 2019; 65(9): 1262-1272. DOI: 10.1080/03650340.2018.1562274

[35] Pessotto PP, da Silva RF, Da Ros CO, da Silva VR, da Silva DA, Grolli AL. Atributos químicos de um Neossolo Regolítico sob uso de dejeto de suínos. Revista de Ciências Agroveterinárias. 2018; 17(3): 408-416. DOI: 10.5965/223811711732018408

[36] Da Ros CO, da Silva VR, Silvestrin TB, da Silva RF, Pessotto PP. Disponibilidade de nutrientes e acidez do solo após aplicações sucessivas de água residuária de suinocultura. Revista Brasileira de Tecnologia Agropecuária. 2017; 1(1): 35-44.

[37] Lourenzi CR, Ceretta CA, Silva LS, Trentin G, Girotto E, Lorensini F, Tiecher TL, Brunetto G. Soil chemical properties related to acidity under successive pig slurry application. Revista Brasileira de Ciência do Solo. 2011; 35(5): 1827-1836. https://doi.org/10.1590/S0100-06832011000500037

[38] Scherer EE, Nesi CN, Massotti Z. Atributos químicos do solo influenciados por sucessivas aplicações de dejetos suínos em áreas agrícolas de Santa Catarina. Revista Brasileira de Ciência do Solo. 2010; 34(4): 1375-1384.

[39] Lourenzi CR, Ceretta CA, Cerini JB, Ferreira PA, Lorensini F, Girotto E, Tiecher TL, Schapanski DE, Brunetto G. Available content, surface runoff and leaching of phosphorus forms in a typic hapludalf treated with organic and mineral nutrient sources. Revista Brasileira de Ciência do Solo. 2014; 38(2): 544-556. DOI: 10.1590/S0100-06832014000200019

[40] Guardini R, Comin JJ, Santos DR, Gatiboni LC, Tiecher T, Schmitt D, Bender MA, Belli Filho P, Oliveira PA, Brunetto G. Phosphorus accumulation and pollution potential in a hapludult fertilized with pig manure. Revista Brasileira de Ciência do Solo. 2012; 36(4): 1333-1342. DOI: 10.1590/S0100-06832012000400027

[41] Guardini R, Comin JJ, Santos DR, Gatiboni LC, Tiecher T, Schmitt D, Bender MA, Belli Filho P, Oliveira PA, Brunetto G. Phosphorus accumulation and pollution potential in a hapludult fertilized with pig manure. Revista Brasileira de Ciência do Solo. 2012; 36(4): 1333-1342. DOI: 10.1590/S0100-06832012000400027

[42] Loss A, Soares CR, Lourenzi CR, de Souza EL, Cadoná EA, Sete PB. Teores de Nitrogênio no Solo e na Água em Propriedade Suinícola de Braço do Norte/SC. In: Tullio LA, editor. Formação, Classificação e Cartografia dos Solos [recurso eletrônico]. Ponta Grossa, PR: Atena Editora; 2019. p. 99-106. DOI: 10.22533/at.ed.3481923099

[43] Tiecher TL, Lourenzi CR, Girotto E, Tiecher T, De Conti L, Marques AC, Silva LO, Marchezan C, Brunetto G, Ceretta CA. Phosphorus forms leached in a sandy Typic Hapludalf soil under no-tillage with successive pig slurry applications. Agricultural Water Management. 2020; 242: 106406. DOI: 10.1016/j.agwat.2020.106406

[44] Conselho Nacional do Meio Ambiente - CONAMA. Resolução CONAMA N°357 [Internet]. 2005. Available from: http://www.crq4.org.br/downloads/resolucao357.pdf [Accessed: 2020-08-27].

[45] Fundação de Meio Ambiente de Santa Catarina - FATMA. Instrução Normativa N° 11 - Suinocultura, de 14/11/2014 [internet]. 2014. Available from: https://static.fecam.net.br/uploads/273/arquivos/654746_in_11_Suinocultura.pdf [Accessed: 2020-08-25].

[46] Withers PJ, Vadas PA, Uusitalo R, Forber KJ, Hart M, Foy RH, Delgado A, Dougherty W, Lilja H, Burkitt LL, Rubæk GH. A global perspective on integrated strategies to manage soil phosphorus status for eutrophication control without limiting land productivity. Journal of environmental quality. 2019; 48(5): 1234-1246. DOI:10.2134/jeq2019.03.0131

[47] Boitt G, Schmitt DE, Gatiboni LC, Wakelin SA, Black A, Sacomori W, Cassol PC, Condron LM. Fate of phosphorus applied to soil in pig slurry under cropping in southern Brazil. Geoderma. 2018; 321: 164-172. DOI: 10.1016/j.geoderma.2018.02.010

[48] De Conti L, Ceretta CA, Ferreira PA, Lorensini F, Lourenzi CR, Vidal RF, Tassinari A, Brunetto G. Effects of pig slurry application and

crops on phosphorus content in soil and the chemical species in solution. Revista Brasileira de Ciência do Solo. 2015; 39(3): 774-787. DOI: 10.1590/01000683rbcs20140452

[49] Fundação do Meio Ambiente de Santa Catarina - FATMA. Instrução normativa para licenciamento ambiental. FATMA: Instrução normativa n° 11, de 13/12/2004. [Internet]. 2004. Available from:<http://www.fatma.sc.gov.br> [Accessed: 2020-08-26].

[50] Silva LSda, Gatiboni LC, Anghinoni I, editors. Manual de Calagem e Adubação para os Estados do Rio Grande do Sul e Santa Catarina. 11th ed. Porto Alegre: Sociedade Brasileira de Ciência do Solo, Núcleo Regional Sul, Comissão de Química e Fertilidade do Solo - RS/SC; 2016. 376 p.

[51] Capoane V. Concentrações de cobre e zinco no sedimento de fundo de uma bacia hidrográfica com suinocultura intensiva. Revista de Ciências Ambientais. 2016; 10(1): 43-58.

[52] Capoane V, Tiecher T, Schaefer GL, dos Santos DR. Impactos da suinocultura na qualidade da água de uma bacia hidrográfica agrícola do Rio Grande do Sul. Brazilian Geographical Journal: Geosciences and Humanities research medium. 2014; 5(2): 494-509.

[53] Smanhotto A, Sousa AD, Sampaio SC, Nóbrega LH, Prior M. Cobre e zinco no material percolado e no solo com a aplicação de água residuária de suinocultura em solo cultivado com soja. Engenharia Agrícola. 2010; 30(2): 347-357. DOI: 10.1590/S0100-69162010000200017

[54] Rostagno HS, Albino LFT, Hannas MI, Donzele JL, Perazzo FG, Saraiva A, Teixeira ML, Rodrigues PB, Oliveira RF, Barreto SLT, Brito CO. Tabelas brasileiras para aves e suínos - Composição de alimentos e exigências nutricionais. 4th ed. Viçosa: Universidade Federal de Viçosa; 2017. 488 p.

[55] National Research Council. Nutrient Requirements of Swine. 10th ed. Washington, DC: The National Academies Press. 1998. 221p. DOI: 10.17226/6016

[56] Ribeiro AM, Pinheiro CC, Gianfelici MF. Nutrientes que afetam a imunidade dos leitões. Acta scientiae veterinariae. Porto Alegre, RS. 2008.

[57] Council of the European Communities - CEC. Council Directive 86/278/EEC of 12 June 1986 on the protection of the environment, and in particular of the soil, when sewage sludge is used in agriculture. [Internet]. 1986. Official Journal of European Communities. Available from: https://data.europa.eu/eli/dir/1986/278/oj [Accessed: 2020-08-16]

[58] United State Environmental Protection Agency - USEPA. 40 CFR-Standards for the use and disposal of sewage sludge. Part 503. [Internet]. 1993. Washington, DC: United State Environmental Protection Agency. Available from: https://www.govinfo.gov/content/pkg/CFR-2018-title40-vol32/xml/CFR-2018-title40-vol32-part503.xml [Accessed: 2020-08-16]

[59] National Council for the Environment. Resolução N° 460/2013. [internet]. 2013. Available from: http://www2.mma.gov.br/port/conama/legiabre.cfm?codlegi=702. [Accessed: 2020-08-16].

[60] Companhia Ambiental do Estado de São Paulo - CETESB. Values oriented to soil and groundwater in the State of São Paulo. [Internet]. 2016. Available from: https://cetesb.sp.gov.br/aguas-subterraneas/wp-content/uploads/sites/13/2013/11/dd256_2016_e_valoresorientadoresdioxinasefuranos2016.pdf. [Accessed: 2020-08-16].

[61] Tiecher TL, Ceretta CA, Comin JJ, Girotto E, Miotto A, Moraes MPD, Benedet L, Ferreira PAA, Lourenzi CR, Couto RdR, Brunetto G. Forms and accumulation of copper and zinc in a sandy typic hapludalf soil after longterm application of pig slurry and deep litter. Revista Brasileira de Ciência do Solo. 2013; 37: 812-824. DOI: 10.1590/S0100-06832013000300028

[62] Shaheen SM, Tsadilas CD, Rinklebe J. A review of the distribution coefficients of trace elements in soils: Influence of sorption system, element characteristics, and soil colloidal properties. Advances in Colloid and Interface Science. 2013; 201: 43-56. DOI: 10.1016/j.cis.2013.10.005

[63] Girotto E, Ceretta CA, Brunetto G, Santos DR, Silva LS, Lourenzi CR, Lorensini F, Vieira RC, Schmatz R. Acúmulo e formas de cobre e zinco no solo após aplicações sucessivas de dejeto líquido de suínos. Revista Brasileira de Ciência do Solo. 2010; 34(3): 955-965. DOI: 10.1590/S0100-06832010000300037

[64] Girotto E, Ceretta CA, Santos DR, Brunetto G, Andrade JG, Zalamena J. Formas de perdas de cobre e fósforo em água de escoamento superficial e percolação em solo sob aplicações sucessivas de dejeto líquido de suínos. Ciência Rural. 2010; 40(9): 1948-1954. DOI: 10.1590/S0103-84782010005000141

[65] De Conti L, Ceretta CA, Ferreira PA, Lourenzi CR, Girotto E, Lorensini F, Tiecher TL, Marchezan C, Anchieta MG, Brunetto G. Soil solution concentrations and chemical species of copper and zinc in a soil with a history of pig slurry application and plant cultivation. Agriculture, Ecosystems & Environment. 2016; 216: 374-386. DOI: 10.1016/j.agee.2015.09.040

[66] Dal Bosco TC, Sampaio SC, Opazo MA, Gomes SD, Nóbrega LH. Aplicação de água residuária de suinocultura em solo cultivado com soja: cobre e zinco no material escoado e no solo. Engenharia Agrícola. 2008; 28(4): 699-709. DOI: 10.1590/S0100-69162008000400010

[67] Canadian Council of Ministers of the Environment - CCME. Protocol for the derivation of Canadian Sediment quality guidelines for the protection of aquatic life. 1995. Otawa: CCME, 35 p.

Chapter 6

Excessive and Disproportionate Use of Chemicals Cause Soil Contamination and Nutritional Stress

Nikita Bisht and Puneet Singh Chauhan

Abstract

Incredible achievements have been made in agricultural production worldwide, but many daunting challenges remain unresolved to ensure food security and environmental sustainability. Chemical fertilisers are used in excessive and disproportionate quantities to raise crop yields in order to combat certain circumstances. However, apart from being processed in crop plants, chemical fertilisers above the threshold level pollute the atmosphere. As the availability of nutrients becomes a constraint of plant growth and production, sustained crop productivity relies on constant renewal. To increase agriculture production and maintain soil fertility, the application of chemical fertilisers is indispensable. However, insufficient or unnecessary application of fertiliser does not guarantee consistently growing yields, which can result in low efficiency of nutrient usage. Today, the key goals are the study of the effective use of chemicals, the reduction of production costs and the efficient use of fertilisation.

Keywords: soil, chemicals, agriculture, toxicity, fertility

1. Introduction

Soil is a very important and sensitive resource of a nation. In order to meet increasing public needs and to promote crop products, the use of high inputs of chemicals in the soil in the form of fertilisers, pesticides, fungicides, insecticides, nematicides and weedicides, along with intensive irrigation practises, helped to achieve the target to a certain stage. However, the decrease in crop yield took place despite the application of fertiliser. The toxic chemicals influence the life of beneficial soil microorganisms, which are indeed responsible for maintaining soil fertility. Moreover, groundwater, air, and human and animal health have also been adversely affected by these chemicals directly and indirectly. Therefore, preserving the health of the soil is very essential. The avoidance of chemical fertilisers and the use of natural fertilisers such as biofertilizers, vermicompost, green manure and biopesticides, as well as the nourishing of the soil and the environment, can be a sustainable approach to crop productivity.

In order to boost crop quality and satisfy the global demand for food, chemical formulations being introduced as fertilisers and pesticides in appropriate amount are important for food management resources in agriculture. On the other hand, if used in excessive and disproportionate amount, there are harmful aspects of inorganic fertilisers and pesticides that can not be ignored. They persist for a long time in the soil and atmosphere and influence various biotic and abiotic factors. They negatively influence soil, microflora, other organisms, human health and the environment. The excessive quantities of agrochemicals, industrial chemicals, trace metals and urban waste enter the soil through atmospheric deposition, disposal of waste, industrial effluents and direct application, and pollute it [1–3]. Soil contamination is responsible for decreasing the soil biodiversity and fertility and hence, decrease soil health by obstructing the breakdown of soil organic matter and altering nutrient cycling. The contamination of soil, therefore reduces crop yield and affects food safety, especially when bioconcentrated pollutants enter organisms within food chains [4].

Through their roots, plants can also take up soil pollutants or absorb them through their leaves. The prolonged intake of infected foods, including human beings, can cause disease and lead to animal deaths [5]. In particular, urbanisation causes soil contamination in peri-urban areas, which have to deal with urban air pollution deposition and municipal solid waste disposal [1, 6]. Acid rain decreases soil aggregate stability, affects soil microorganisms and enzyme activities, increases soil erosion and mobility of nutrients, and in turn contribute to loss of nutrients [1, 7–9]. Soil pollution is also linked to the water quality used for irrigation purposes and to flooding events. Industrial and urban sewage is rapidly being adopted for irrigation to meet the rising demand for drinking water. This is particularly common in countries facing rapid urbanisation rates, such as China, where contaminated water and sewage have irrigated 3.62 million ha of agricultural land [1]. Due to atmospheric accumulation, industrial waste and the use of waste water for irrigation, soil contamination from trace metals is prevalent in peri-urban areas [1, 10, 11]. Trace metal supply is directly influenced by plant root exudates and by the activities of soil microorganisms. Owing to the high sensitivity of soil microorganisms to excessive trace metal concentrations, they are responsible for reducing soil biodiversity and fertility [12, 13]. Moreover, due to their close affinity with organic matter, trace metals accumulate in surface organic deposits, and passively taken up by plants by water flow [14]. Studies have documented the accumulation of trace metals in agricultural foods with high concentrations in stems and leaves rather than in fruits and seeds [15].

Intensive cultivation and mono-cropping have contributed to a large increase in the usage and reliance on agrochemicals. Fertiliser and pesticide soil contamination is due to inadequate management of nutrients in combination with pest and weed mismanagement, respectively [4]. In addition, because their behaviour in the environment and especially in the food chain is not comprehensively understood, the fate of pesticide metabolites is of concern [16]. The growth of the population raises the risk of pollution of the soil. Food protection is thus threatened by the presence of toxins and by the associated risks of bioaccumulation. Soil contamination is responsible for reducing crop yields and for turning productive agricultural land into unproductive areas by decreasing soil fertility and biodiversity. As a result, the food availability and stability dimensions of food security are affected by this. Food accessibility is challenged by the extent and spatial distribution of soil pollution, which, in particular in urban and peri-urban areas, restricts food access. Soil pollution is, therefore, a hazard to all dimensions of food safety.

2. Healthy soils: a prerequisite for sustainable food security

Soil health is defined as a living soil's ability to function within natural or regulated ecosystem boundaries, to preserve productivity of plants and animals, to conserve and enhance the quality of water and air, and to promote plant and animal health [17, 18]. Therefore, soil health is a multi-dimensional and holistically vital soil characteristic, and forms the basis for healthy food production, thereby contributing to local and global food security. By 2050, a 60 percent rise in global food production and related ecosystem services need to be accomplished. But, through soil erosion, nutrient loss, salinity, sealing and pollution, one-third of global soils are currently facing moderate to extreme degradation.To achieve sustainable soil management, evidence-based decisions and soil information are essential [19]. Soils impoverished by nutrients lead to systemic food and nutritional problems. Micronutrient deficiencies are significant cause of morbidity and mortality, and affect over two billion people [20–23]. Protein-energy malnutrition is due to food scarcity and ingestion of trace elements (i.e. iron, lithium, magnesium, zinc, copper, iodine) from crops with low tissue concentrations, which are directly attributable to nutrient-poor soils [24].

3. Impact of chemicals used in agriculture on environment

Since the chemical fertiliser increases the growth and vigour of the plant, it thus meets the world's food security, but the plants grown in this way do not develop good plant characteristics such as good root system, shoot system, nutritional characteristics and will not have time to grow and mature properly [25]. The deleterious effect of chemical fertilisers itself begins with the processing of chemicals whose products and by-products are certain harmful chemicals or gases that cause air pollution, such as NH_4, CO_2, CH_4, etc. And it will cause water pollution when the waste from industries is disposed of untreated in nearby water bodies. It also involves the most damaging impact of the accumulation of chemical waste in the bodies of water, i.e. water eutrophication. And its constant use, when applied to the soil, degrades the health and quality of the soil, thereby causing soil contamination. It is therefore high time to realise that our climate and biodiversity are depleted by this crop production input. Therefore, its continuous use without taking any remedial action to reduce or judicious use will one day deplete all natural resources and threaten the entire life of the earth. The adverse effects of these synthetic chemicals on human health and the environment can be reduced or eliminated by adopting new agricultural technological practises, including the use of organic inputs such as manure, biofertilizers, biopesticides, slow-release fertilisers and nanofertilizers, etc., and moving away from chemical intensive cultivation.

4. Influence on soil compaction and degradation

Soil compaction is an important component of the syndrome of land degradation and is a major problem for advanced agriculture, affecting soil resources adversely [26]. As the soil is compacted, its composition changes by crushing aggregate units, decreasing the size of pore spaces between the soil particles, decreasing compaction due to the use of heavy equipment, reducing the use of organic fertiliser, repeated use of chemical fertilisers, and ploughing for several years at the same depth [27]. One of the principal causes of compaction is the usage of fertilisers more than the recommended amount for long periods and intensive cropping. Soil compaction causes problems such as excessive soil strength, root growth restriction, poor aeration, poor drainage, runoff, erosion and deterioration of the soil, etc. [28]. Such modifications lead to permeability, hydraulic conductivity and groundwater recharge reductions [29]. Excessive soil compaction impedes root growth and this decreases the capacity of plants to absorb nutrients and total porosity, leading to an increase in the density of soil bulk and resistance to penetration. It is reported that compaction decreases both root growth and yield by more than 80 percent [30]. Nitrification decreases by 50 percent as the density of soil bulk increases and plants consume less N, P and Zn from soil [31]. A great concern is the reduction of biological activities in soil due to compaction [32]. The most significant element in soil structure stability is organic matter. Soil that has high organic matter content and thrives with soil species is more compaction-resistant and can recover much better from mild damage to compaction [33, 34]. Over-use of fertilisers has led the development of continuous monoculture cropping, accumulation of fertiliser mineral salts in soil that forms compaction layers in soil, and cause long-term soil degradation.

5. Disproportionate usage of chemicals and soil nutrients

The soil is a home for soil organisms which are a mechanism for nutrient recovery, and offers many other environmental services. Chemical fertiliser overuse can contribute to soil acidification and soil crust, thereby reducing the content of organic matter, humus content, beneficial species, stunting plant growth, altering the pH of the soil, growing pests, and even leading to the release of greenhouse gases. The acidity of the soil reduces crop phosphate intake, raises the concentration of harmful ions in the soil and inhibits crop growth [35]. The soil's loss of humus decreases its capacity to store nutrients. The atmosphere is polluted by greenhouse emissions resulting from the excess use of nitrogen fertiliser. Over the time nitrogen fertilisers added in large quantities to fields kills the balance between the three macronutrients, N, P and K, resulting in decreased crop yields. Sandy soils are much more vulnerable to soil acidification than clay soils. Clay soils have the potential to buffer excess chemical fertilisation effects. Repeated chemical fertiliser applications may lead to a toxic build-up in the soil of heavy metals such as arsenic, cadmium, and uranium. Not only do these toxic heavy metals pollute the ground, but they also accumulate in food grains, fruits and vegetables. Fertilisers such as triple superphosphate, have trace elements such as cadmium and arsenic that accumulate in plants and enter humans via food chains that can cause health problems [36]. Application of fertilisers without the recommendation of soil testing can lead to implications such as soil degradation, nutrient imbalance, soil structure destruction, bulk density increase [37].

When crop plants are harvested, soil nutrient levels are reduced over time, and these nutrients are replenished either by natural decomposition or by adding fertilisers. Therefore, the basic component of modern agriculture thesedays is fertiliser. However, while chemical fertilisers are the main cause of adequate crop production for the world's population, their overuse presents serious challenges for present and future generations, such as contaminated air, water and soil, degraded land, soils and increased greenhouse gas emissions. Not only are these synthetic fertilisers being harmful to our climate, but also to humans, livestock, and microbial forms of life. It is high time that everyone realises the detrimental effects of using excess chemical fertilisers and take steps to minimise the usage of chemical fertilisers and pesticides by substituting other organic modifications such as organic manures that not only provide plants with essential nutrients, but also preserve soil quality for subsequent crops. There are so many other technologies that are being developed, such as slow or controlled released fertilisers, prilled or granulated fertilisers, inhibitors of nitrification, nano-fertilisers, etc., all of which are the promising alternatives that can be used to solve these serious challenges and save both our environment and the ecosystem [35].

6. Microbial community structure

Soil microorganisms play an important role in the conservation of soil fertility and ecosystem work [38, 39]. The plant roots secrete carbon-containing organic material in the rhizosphere which is the source of carbon, nitrogen and energy needed for the growth and reproduction of soil microorganisms. A large number of microbes gather around plant roots, which results in a distinction between the state of soil nutrients and the composition of the soil microbial population [40]. The region with the greatest contact between plant roots, soil and microorganisms is the rhizosphere. Microbes of the rhizosphere play an important role in the cycling of soil material and the transfer of energy. Fertiliser application is an important management measure in agricultural production that not only promote crop growth and yield but negatively influence the soil microorganisms as well [41]. The widespread use of chemical fertilisers currently leads to a decline in soil fertility and a number of environmental problems, while bioorganic fertiliser not only improves soil fertility through the contribution of beneficial microorganisms and organic materials, but also eliminates many of the environmental problems caused by chemical fertilisers. Studies have shown that various fertilisation treatments have a significant effect on the structure of soil microbial biomass and the community. Different applications of fertilisers change the physical and chemical properties of the soil, which in turn affects the structure of the soil bacterial community. Previous studies have found that pH, nitrate, and available phosphate and potassium are significant soil factors that influence the structure of the microbial community [9, 42]. By direct effects on the quality of soil nutrients, fertilisation affects soil microbial diversity. In conjunction with other mineral fertilisers, the long-term application of nitrogen fertiliser influences the nitrogen cycle and associated bacterial populations. Repeated overuse of chemical fertiliser may have a detrimental impact on the quality of soil and the composition of the soil microbial population. Long-term use of chemical fertilisers can dramatically decrease soil pH, which is closely related to reduced bacterial diversity and major changes in the composition of the bacterial population [43].

7. Potential of biofertilizers to replace chemical fertilisers

As the land for agriculture is restricted and even diminished over time, the worldwide increase in the human population poses a major threat to the food security [44]. It is therefore important that agricultural productivity should be dramatically improved over the next few decades in order to meet the high demand for food from the emerging population. Furthermore, too much reliance for crop production on chemical fertilisers ultimately affects both environmental ecology and human health with great severity. A biofertilizer is a material that contains living microorganisms that colonise the rhizosphere or the interior of plants when applied to seeds, plants or soil and encourage plant growth by increasing the host plant's supply of nutrients [45]. The use of microbes as biofertilizers in the agricultural sector is considered an alternative to chemical fertilisers because of their wide potential to increase crop production and food safety [38]. Extensive work on biofertilizers has revealed their ability to supply the crop with the requisite nutrients in sufficient quantities to increase crop yield. Biofertilizers are widely used to accelerate certain microbial processes that increase the availability of nutrients that can be easily assimilated by plants. By fixing the atmospheric nitrogen and solubilising insoluble phosphates, biofertilizers increase soil fertility and produce plant growth-promoting substances in the soil [45].

The naturally accessible biological system of nutrient mobilisation, which greatly increases soil fertility and ultimately crop yield, has been encouraged by biofertilizers. Biofertilizers are expected to be a healthy alternative to chemical inputs and to a great extent mitigate ecological disruption. Biofertilizers are cost-effective in nature, eco-friendly, and their extended usage greatly increases soil fertility. It has been stated that the use of biofertilizers increase the protein content, essential amino acids, vitamins, and nitrogen fixation, thereby increases crop yield by about 10–40 percent [46]. The advantages of using biofertilizers include low-cost nutrient sources, excellent microchemical and micronutrient suppliers, organic matter suppliers, growth hormone secretion, and the counteraction of chemical fertiliser adverse effects. Microbes are important soil components and play a crucial role in the different biotic activities of the soil ecosystem that make the soil dynamic for the mobilisation of nutrients and sustainable for the production of crops [47].

8. Improving soil fertility

Physical fertility refers to the soil's physical properties, its composition, texture, water holding properties, the way water flows to the roots of plants, and how the soil is penetrated by those roots. Biological fertility refers to the species and their capacity to play important roles that live in the soil. A soil's composition, its acidity or alkalinity, and its ambient temperature are only a sample of the several variables that decide the degree to which plants have access to nutrients [9]. The relative value of these variables depends on the nutrients, the soil and the plant. Most notably, soil structure determines how well the soil holds nutrients and water. Organic matter-containing clays and soils retain nutrients and water much better than sandy soils. The microbial community of the soil would also be highly influenced by the soil structure. If the soil does not allow these species to survive, plants that rely on bacteria or fungal species for nutrient uptake will not grow. Until recently, the application of fertiliser was the most commonly used treatment for nutrient deficiency. As plant nutrient requirements vary over the plant life cycle, timing is also important. The effect of Liebig 's Law may obscure the identification of genuinely deficient nutrients, as the correct scarce nutrient may not be directly recognised by deficiency symptoms. Additional fertiliser would be of little to no assistance if the requisite structural and biological conditions are not present. The improved vitality of plants would rely on improving the structural and biological fertility of the soil. Inappropriate application of fertiliser is a waste of time and money, but it can also have dire environmental implications as well [19].

9. Conclusion

All living beings in some way are vulnerable to the widespread long-term use of chemicals in agriculture in any form such as fertilisers, pesticides, etc. Agricultural soil has been disrupted by the extensive and disproportionate use of chemicals and putting it back into order will take time and transition. While we cannot fully prevent the adverse effects of chemical fertilisers at an instant of time, we can definitely reduce the impact by minimising their use and promoting the use of biofertilizers. Biofertilizers will not reduce the use chemical fertilisers but they will improve the soil quality in various ways such as maintaining soil nutrient cycles, soil microbial communities, etc.

Acknowledgements

The authors are grateful to Director CSIR- National Botanical Research Institute for providing facilities and support during the study. This work is supported by the CSIR-Network project (MLP0049; OLP 0109) funded by Council of Scientific and Industrial Research, New Delhi, India.

Author details

Nikita Bisht and Puneet Singh Chauhan*
Microbial Technologies Division, CSIR-National Botanical Research Institute, Lucknow, India

*Address all correspondence to: puneetnbri@gmail.com; puneet@nbri.res.in

References

[1] Chen, J. 2007. Rapid urbanization in China: a real challenge to soil protection and food security. Catena 69:1-15.

[2] Ashraf, M.A., Maah, M.J., Yusoff, I. 2014. Soil contamination, risk assessment and remediation. In: Hernandez-Soriano MC (ed) Environmental risk assessment of soil contamination. INTECH, Rijeka, 3-56.

[3] Masindi, V., Muedi, K. 2018. Environmental contamination by heavy metals, heavy metals. Hosam El-Din M. Saleh & Refaat F. Aglan, IntechOpen.

[4] Edwards, C.A. 2002. Assessing the effects of environmental pollutants on soil organisms, communities, processes.

[5] Liu, H., Probst, A., Liao, B. 2005. Metal contamination of soils and crops affected by the Chenzhou lead/zinc mine spill (Hunan, China). Sci Total Environ 339:153-166.

[6] Mclaughlin, M.J., Tiller, K.G., Naidu, R., Stevens, D.P. 1996. Review: the behaviour and environmental impact of contaminants in fertilizers. Aust J Soil Res 34:1-54.

[7] Calace, N., Fiorentini, F., Petronio, B.M., Pietroletti, M. 2001. Effects of acid rain on soil humic compounds. Talanta. 54:837-846

[8] Zhang, J.E., Ouyang, Y., Ling. D.J. 2007. Impacts of simulated acid rain on cation leaching from the Latosol in south China. Chemosphere. 67:2131-2137

[9] Neina, D. 2019. The Role of Soil pH in Plant Nutrition and Soil Remediation. *Appl. Environ. Soil Sci.* 5794869.

[10] Wang, Q.R., Cui, Y.S., Liu, X.M., Dong, Y.T., Christie, P. 2003. Soil contamination and plant uptake of heavy metals at polluted sites in China. J Environ Sci Health 38:823-838.

[11] Zhao, Y.F., Shi, X.Z., Huang, B., Yu, D.S., Wang, H.J., Sun, W.X., Öboern, I., Blombäck. K. 2007. Spatial distribution of heavy metals in agricultural soils of an industry-based peri-urban area in Wuxi, China. Pedosphere.17:44-51.

[12] Giller, K.E., Witter, E., Mcgrath, S.P. 1998. Toxicity of heavy metals to microorganisms and microbial processes in agricultural soils: a review. Soil Biol Biochem. 30:1389-1414.

[13] Lenart-Boroń, A., and Boroń, P. 2014. The effect of industrial heavy metal pollution on microbial abundance and diversity in soils–a review. Actinomycetes, 1012, 107-108.

[14] Bradl, H.B. 2004. Adsorption of heavy metal ions on soils and soils constituents. J Colloid Interface Sci. 277(1), 1-18

[15] Mclaughlin, M.J., Tiller, K.G., Naidu, R., Stevens, D.P. 1996. Review: the behaviour and environmental impact of contaminants in fertilizers. Aust J Soil Res 34:1-54.

[16] Arias-Estévez, M., López-Periago, E., Martínez-Carballo, E., Simal-Gándara, J., Mejuto, J.C., García-Río, L. 2008. The mobility and degradation of pesticides in soils and the pollution of groundwater resources. Agric Ecosyst Environ 123:247-260.

[17] Doran, J.W. 2002. Soil health and global sustainability: translating science into practice. Agric Ecosyst Environ 88:119-127.

[18] Tahat, M., Alananbeh, K., Othman, Y., Leskovar, D. 2020. Soil Health and Sustainable Agriculture. Sustainability. 12, 4859.

[19] Rojas, R.V., Achouri, M., Maroulis, J. Caon, L. 2016. Healthy soils: a prerequisite for sustainable food security. Environ Earth Sci 75, 180.

[20] Black, R. 2003. Micronutrient deficiency—an underlying cause of morbidity and mortality. No. 81 (2) World Health Organization.

[21] Shetty, P. 2009. Incorporating nutritional considerations when addressing food insecurity. Food Security 1:431-440.

[22] Charles, H., Godfray, J., Beddington, J.R., Crute, I.R., Haddad, L. 2010. Food security: the challenge of feeding 9 billion people. Science 327:812-817.

[23] FAO, WHO (2014) Rome declaration on nutrition. second international conference on nutrition. Rome, Italy.

[24] Clair, S.B., Lynch, J.P. 2010. The opening of pandora's box: climate change impacts on soil fertility and crop nutrition in developing countries. Plant Soil 335:101-115.

[25] Li, D. P., and Wu, Z. J. 2008. Impact of chemical fertilizers application on soil ecological environment. Chinese Journal of Applied Ecology, 19, 1158-1165.

[26] Weisskopf, P., Reiser, R., Rek, J., Oberholzer, H.R. 2010. Effect on different compaction impacts and varying subsequent management practices on soil structure, air regime and microbiological parameters. Soil Till. Research111,65-74.

[27] Mari, G. R., Ji. Changying, Jun Zhou. 2008. Effects of soil compaction on soil physical properties and nitrogen, phosphorus, potassium uptake in wheat plants. J. Transactions of the CSAE, 24(1): 74-79

[28] Batey, T. 2009. Soil compaction and soil management -a review. Soil Use and Management. 25(4): 335-345

[29] Blanco, C. H., C. H. Gantzer, S. H. Anderson, E. E. Alberts, and F. Ghidey. 2002. Saturated hydraulic conductivity and its impact on simulated runoff for claypan soils. Soil Sci. Soc. Am. J. 66, 1596-1602.

[30] Rannik, K. 2009. Soil compaction effects on soil bulk density and penetration resistance and growth of spring barley (*Hordeum vulgare* L.). J. Acta Agriculturae Scandinavica. Plant Soil Science. 59 (3): 265-272

[31] Barzegar, A. R., Nadian, H., Heidari, F., Herbert, S. J., Hashemi, A. M. 2006. Interaction of soil compaction, phosphorus and zinc on clover growth and accumulation of phosphorus. Soil & Tillage Res. 87: 155-162

[32] Beylich, A., Oberholzer, H. R., Schrader, S., Hoper, H., Wilke. B. M. 2010. Evaluation of soil compaction effects on soil biota and soil biological processes in soils. Soil & Tillage Res. 109(2): 133-143.

[33] Dexter, A. R., Richard, G., Arrouays, D., Czyz, E. A., Jolivet, C., Duval, O. 2008. Complexed organic matter controls soil physical properties. Geoderma. 144, 620-627.

[34] Celik, I., Gunal, H., Budak, M., Akpinar, C. 2010. Effects of long-term organic and mineral fertilizers on bulk density and penetration resistance in semiarid Mediterranean soil conditions. Geoderma. 160(2): 236-243.

[35] Chandini, K. R., Kumar, R., Prakash, O. 2019. The impact of chemical fertilizers on our environment and ecosystem. Research trends in environmental Sciences, 2nd edition, pp 69-86

[36] Sonmez Kaplan, M., Sonmez, S. 2007. An investigation of seasonal changes in nitrate contents of soils and irrigation waters in greenhouses located

in Antalya-Demre region. Asian Journal of Chemistry. 19(7):5639

[37] Savci, S. 2012. Investigation of effect of chemical fertilizers on environment. *Apcbee Procedia*.*1*, 287-292

[38] Bisht, N., Tiwari, S., Singh, P.C., Niranjan, A., Chauhan, P.S., 2019. A multifaceted rhizobacterium Paenibacillus lentimorbus alleviates nutrient deficiency-induced stress in *Cicer arietinum* L. Microbiol. Res. 223-225, 110-119.

[39] Bisht, N. and Chauhan, P. S. 2020. Comparing the growth-promoting potential of Paenibacillus lentimorbus and *Bacillus amyloliquefaciens* in *Oryza sativa* L. var. Sarju-52 under suboptimal nutrient conditions. Plant Physiology and Biochemistry, 146, 187-197.

[40] Jacoby, R., Peukert, M., Succurro, A., Koprivova, A.,Kopriva, S. 2017. The role of soil microorganisms in plant mineral nutrition-current knowledge and future directions. Front. Plant Sci. 8:1617.

[41] Bargaz, A., Lyamlouli, K., Chtouki, M., Zeroual, Y., Dhiba, D. 2018. Soil microbial resources for improving fertilizers efficiency in an integrated plant nutrient management system. Front. Micribiol. 9:1606.

[42] Liang, R., Hou, R., Li, J., Lyu, Y., Hang, S., Gong, H., Ouyang, Z. 2020. Effects of Different Fertilizers on Rhizosphere Bacterial Communities of Winter Wheat in the North China Plain. Agronomy.10, 93.

[43] Wu, L., Jiang, Y., Zhao, F. He, X., Liu, H., Yu, K. 2020. Increased organic fertilizer application and reduced chemical fertilizer application affect the soil properties and bacterial communities of grape rhizosphere soil. Sci Rep 10, 9568.

[44] Bisht N., Chauhan P.S. 2020. Microorganisms in Maintaining Food and Energy Security in a World of Shifting Climatic Conditions. In: Singh P., Singh R., Srivastava V. (eds) Contemporary Environmental Issues and Challenges in Era of Climate Change. Springer, Singapore

[45] Ramasamy,M., Geetha, T., Yuvaraj, M. 2020. Role of Biofertilizers in Plant Growth and Soil Health, Nitrogen Fixation, Everlon Cid Rigobelo and Ademar Pereira Serra, IntechOpen, DOI: 10.5772/intechopen.87429.

[46] Bhardwaj, D., Ansari, M. W., Sahoo, R. K., Tuteja, N. 2014. Biofertilizers function as key player in sustainable agriculture by improving soil fertility, plant tolerance and crop productivity. Microb. Cell Fact. 13:66. doi: 10.1186/1475-2859-13-66

[47] Ahemad, M., Kibret, M. 2014. Mechanisms and applications of plant growth promoting rhizobacteria: current perspective. JKSUS 26, 1-20.

Chapter 7

Increasing Yields and Soil Chemical Properties through the Application of Rock Fines in Tropical Soils in the Western Part of Cameroon, Africa

Samuel Tetsopgang

Abstract

Local tropical soils were amended with pulverized rock fines such as trachyte, basalt, pyroclastic materials, limestone and gneiss with ± manure in different proportions in Cameroon. And soil textures and chemicals were assessed after harvesting. Cabbage and potatoes as test crops treated with fines of pyroclastic materials and basalts, portrayed highest and lowest productivities, respectively. The early loamy sand texture of controls changes to clay; and clay textures remained unchanged after treatments. This indicates the decrease of sand proportion and gain in clay particles after treatments. The pH of local soils was strongly to slightly acidic ($4.8 \leq pH \leq 6.5$) and rose up to slightly acidic and slightly alkaline affinity ($6.6 \leq pH \leq 7.2$). A significant pH increase from 5.9 to 6.9 was observed on a treated sample with pulverized pyroclastic materials. Organic carbon and Organic matter show parallel oscillated tendencies from controls to treated soils. There is a general increase of Mg and Ca after treatments while Na and K remain constant. Rock fines from trachyte, limestone and basalt as treatments significantly increase phosphorus in soils with contents of 96.0, 51.5 and 50.9 ppm, respectively.

Keywords: tropical soils, rock fines, yield, soil texture, chemicals, pH and phosphorus

1. Introduction

1.1 Soil erosion and infertility in sub-Sahara Africa

Soil erosion is a process acting over millions of years. It is known as "geologic" or natural when caused by factors such as climate, soil type and topography [1]. Human also induced soil erosion through activities such as overgrazing, deforestation and agriculture which are the major factors of the soil erosion accounting for 92% of all activities destroying the soil structure [2]. Then, soil erosion is a leading cause of soil infertility with a detrimental impact on the agricultural productivity. There are a number of soil physical, chemical and biological parameters on which depends the soil productivity. Then, available amount of nutrients in a soil on which depends the soil fertility is directed by the soil pH, organic matter content and other

IntechOpen

physico-chemical soil parameters. And, the loss of chemical fertility causes soil erosion favored by the leaching of some nutrients with a major consequence being the decrease of the crop productivity.

In fact, the consequence of the decreasing of the agricultural productivity can be felt at the level of common farmers in sub-Sahara Africa, where the major part of the population still depends on this sector to earn an income. Then, the serious consequence is the economic damage with alarming damage at the level of a nation. For example, nitrogen and phosphorus are three times lost by erosion that their application as fertilizers in soils in Zimbabwe alone. Then, the agricultural productivity is negatively impacted by the soil erosion which has depleted soils with essential nutrients needed for crop growth in Africa [3]. In 1983, [3] demonstrated the negative values of -22, -2.5 and -15 kgha^{-1} yr.$^{-1}$ were the N, P and K balances in Africa, respectively. In fact, these detrimental balances are originated through the depletion due to water runoff and wide spread eroding sediments by the harvested products and wide spread erosive processes. Then, the common method to solve the problem of soil depletion used by farmers is adding NPK chemical fertilizers in soils in sub-Sahara Africa.

This work focuses on the application of pulverized rock fines from different rock types to replenish soil chemicals' contents as an alternative to the common use of chemical fertilizers for soil amendment in Africa.

2. Site locations, geology, and experiments

Different sites namely Befang (06°20′18″N, 10°02′47″E), Foumbot (05°32′25″N, 10°35′30″E), Batibo (05°45′10″N, 09°45′35″E), Santa (05°47′58″N, 10°09′46″E) Kalong (04°47′30″N, 11°03′53″E) and Bonandale (04°09″36″N, 09°34′34″E) are the locations where the experiments were carried out and soil samples collected for different analyses. They are located in the northwest, west, center and littoral regions of Cameroon (**Table 1**) comprising between the latitudes 04°30′ 00″ N and 06°20′ 00″ N and the longitudes 09°30′00″E and 11°04′00″E. This region is characterized with average temperatures of 22°C in Santa and 28°C in Bonandale. The climate is hot and humid tropical type comprising one to two rainy seasons alternating with dry seasons on the savannah to partly equatorial forest vegetation. The highest rainfall of about 4000 mm/year is encountered in the area of Befang. Soil horizons are thick and can reach more than 20 m in some the equatorial forest. They are brown to reddish lateritic soils and may be also black in color depending on the content of organic matter. They are developed on the metamorphic or granitic substratum as observed in the area of Kalong. However, undifferentiated sedimentary rocks make the substratum in Bonandale. In other regions such as Befang, Foumbot, Santa and Batibo soils were mostly developed on volcanic rocks.

Each site is represented by an experimental plot made up of a control and treated soils replicated three times when growing a test crop. The test crops were chosen based on its growth capacity on a specific site. They were mostly made up of maize (*Zea mays*). However cabbage (*Brassica oleracea*), carrots (*Daucas carota*) and Irish potatoes (*Solanum tuberosum*) were also used as test crops in some sites.

Then, the 1 × 1 mm mesh sieve was used several times on crushed and pulverized rock samples devoid of any sign of weathering to collect fines used as fertilizers. Fines used as fertilizers for these experiments are common rocks encountered in abundance in each respective site. They are made up of volcanic pyroclastic materials, basalt, trachyte, gneiss and limestone. Fines of dried *Tithonia diversifolia* were also used as green manure. Poultry manure or cow dump were also added to some treatments. After harvesting and yields determined for each crop, soil samples were collected from all the controls and treated soils in the studied sites. On each

Samples	Soil types	Composition	Crop	Yield (kg/ha)	Yi	Growth period (months)	Localities (basement)	Coordinates
T01	Control	Local soil	Maize	833	1	3	Foumbot (basalt)	05°32′25″N 10°35′30″E
T41	Treated soil	T01 + 600 g basalt fines +600 g poultry manure		2500	3	3		
T02	Control	local soil	Maize	3200	1	3	Bonandale (sediments)	04°09″36″N 09°34′34″E
T12	Treated soil	T02 + 2 kg basalt fines		5400	1.70	3		
T12B				/	/	6		
T62		T02 + 1 kg limestone fines		4000	1.25	3		
T62B				/	/	6		
T22		T02 + 2 kg limestone fines		8300	2.59	3		
T22B				/	/	6		
T03	Control	Local soil	Maize	4000	1	3	Kalong (gneiss)	04°47′30″N 11°03′53″E
T13	Treated soil	T03 + 3 kg basalt fines		8000	2			
T23		T03 + 3 kg gneiss fines		11,000	2.75			
T04	Control	Local soil	Cabbage	2444	1	3	Befang (volcanic pyroclastic materials)	06°20′18″N 10°02′47″E
T14	Treated soil	T04 + 200 g lapilli fines		3578	1.50			
T24		T04 + 200 g fines from pyroclastic bombs		15,000	6.13			
T34		T04 + 200 g fines from highly vesicular pyroclastic materials		6444	2.63			
T44		T04 + 200 g fines from less vesicular pyroclastic materials		11,666	4.77			

Samples	Soil types	Composition	Crop	Yield (kg/ha)	Yi	Growth period (months)	Localities (basement)	Coordinates	
T05	Control	Local soil	Carrots	500	1	3	Santa (basalt)	05°47′58″N	10°09′46″E
T15	Treated soil	T05 + 1 kg basalt fines		150	0.3				
T25		T05 + 1 kg basalt fines +10 ml LMO		506	1.01				
T35		T05 + 1 kg basalt fines +10 ml LMO + 0.5 kg Tithonia		925	1.85				
T45		T05 + 1 kg basalt fines +0.5 kg Tithonia		525	1.05				
T06	Control	Local soil	Potatoes	14,815	1	3	Batibo (basalt)	05°45′10″N	09°45′35″E
T26	Treated soil	T06 + 2 kg basalt fines +2 kg coal fines		20,741	1.4				
T46		T06 + 2 kg trachyte fines + 2 kg coal fines		13,333	0.89				

LMO: light organic matrix; Yi = performance index (=yield per treatment/yield per control).

Table 1.
Localities of soil samples with their compositions, test crops and yields in the west, Centre and littoral regions of Cameroon.

plot unit, soil samples were collected at depth ≤25 cm, mixed, dried and stored in clean plastic bags and taken for additional description and analysis to the Laboratory of Soil Sciences, Faculty of Agronomy, University of Dschang, Cameroon. For the analysis, at room temperature and for a week, soil samples were air dried and passed through a 2 mm polyethylene sieve to eliminate any contamination with plant debris or/and pebbles. Then, later on, an agate mortar was used to lightly crush these soil samples into fine powders before passing them through a 0.149 mm nylon sieve and stored them under ambient conditions into glass containers pending analysis. The standard laboratory procedure for soil analyses (AFNOR) were undertaken for the physiological analysis of these soil samples. Soil reaction was determined in soil water suspension 1:2.5 using a glass electrode. The analysis method by wet digestion according to reference [4] was utilized to determine the content of organic matter in these soils. The modified analysis method of reference [5] was used to determine the total nitrogen content. The percolation with 1 M ammonium acetate was used to determine exchangeable cations and exchange capacity (CEC). The determination of Ca, K and Na was possible using a flame photometer and Mg with an atomic absorption spectrophotometer. In addition, pH_{water} as is in [6] was measured with a pH meter at 1:2.5 soil/water.

3. Results

The results are made up of the yields of each test crops (**Table 1**) in addition to the textures and chemical compositions obtained from the analyses of controls and treated soils as presented in **Table 2**. The percentage composition in sand, silt and clay was used for the textural determination of each sample. Chemical compositions in this study are made up of parameters such as pH, OM and OC (%), N (g/kg), Ca, Mg, Na and K (meq/100 g) and P (ppm). The variations in chemical contents between controls and treated soils were determined to appraise chemicals' contents in soils after treatments.

3.1 Crop yields on controls and treated soils

Maize was used as the test crop in the localities of Foumbot, Bonandale and Kalong. However cabbage, carrots and potatoes were also used in Befang, Santa and Batibo, respectively.

Maize as the test crops yielded (after 03 months of growth) 833, 3200 and 4000 kg/ha for the controls T01, T02 and T03 in the localities of Foumbot, Bonandale and Kalong, which are made up of basalt, sediments and gneiss as country rocks, respectively. Out of the different treated soils where maize was used as the test crop, the best yield was obtained from T23 (=T03 + 3 kg gneiss fines) in the locality of Kalong. This is followed by higher yields of 8300 and 8000 kg/ha obtained on T22 (=T02 + 2 kg limestone fines) and T13 (=T03 + 3 kg basalt fines) in Bonandale and Kalong, respectively. Yields as low as 5400 and 3200 kg/ha were obtained on treatments T12 (=T02 + 2 kg basalt fines) and T62 (=T02 + 1 kg limestone fines) in the same locality (Bonandale). The lowest yield comes from T41 (=T01 + 600 g basalt fines +600 g poultry manure) cultivated in Foumbot.

Cabbage yielded productivities as high as 15,000 and 11,666 kg/ha for treatments T24 (=T04 + 200 g fines from pyroclastic bombs) and T44 (=T04 + 200 g fines from less vesicular pyroclastic materials), respectively. However, the lowest yield of 2444 kg/ha was obtained from the control soil (T04). Intermediate yields of 3578 and 6444 kg/ha were obtained with T14 (=T04 + 200 g lapilli fines) and T34

	Control soils						Soils treated with different rock fines ± manure																		
	T01	T02	T03	T04	T05	T06	T41	T12	T22	T62	T22B	T12B	T62B	T13	T23	T14	T24	T34	T44	T15	T25	T35	T45	T26	T46
Texture																									
Sand	86.00	82.50	18.00		5.00	40.00	83.00	75.00	70.00	77.50	62.00	76.00	75.00	27.00	38.00					5.00	8.00	6.00	4.00	40.00	39.00
Silt	10.00	5.00	24.00		40.00	28.00	13.00	10.00	20.00	15.00	31.00	10.00	10.00	28.00	34.00					40.00	34.00	40.00	36.00	25.00	27.00
Clay	4.00	12.50	58.00		55.00	32.00	4.00	15.00	10.00	7.50	7.00	14.00	15.00	45.00	43.00					55.00	58.00	54.00	60.00	35.00	34.00
	LS	LS	C		SC	CL	LS	SL	SL	C	SL	SL	C	C	C					SC	SC	SC	C	CL	CL
Soil reaction																									
pH water	7.10	6.40	5.60	5.92	4.60	5.80	7.20	6.00	6.60	6.50	6.71	6.98	7.17	5.70	6.40	6.18	6.90	6.45	6.72	4.80	4.90	4.90	4.80	5.20	6.10
Organic matter																									
CO(%)	5.52	1.30	0.95	2.30	3.81	7.22	6.39	0.21	0.34	2.05	1.14	1.13	3.10	1.59	2.06	4.10	6.80	5.40	5.60	3.16	3.61	4.15	3.77	7.03	6.74
MO(%) 9.25	2.24	1.64	4.00	6.57	12.45	11.29	0.35	0.59	3.54	1.97	1.95	5.35	2.74	4.12		6.20	9.11	6.91	7.10	5.45	6.23	7.15	6.57	12.12	11.62
N (g/kg)	4.62	0.56	0.06		3.06	0.16	5.46	0.98	0.70	0.85	0.36	0.35	0.37	0.14	0.26					2.91	3.06	2.92	2.77	0.05	0.16
C/N	11.60	23.00	15.00		12.00	11.62	2.00	5.00	24.00	32.00	32.00	88.00	11.00	9.00						11.00	12.00	14.00	14.00		
Exchangeable cations (meq/ 100 g)																									
Ca	1.76	1.64	3.84	0.90	2.52	3.20	1.96	1.92	2.96	2.44	0.16	2.32	1.78	6.64	9.12	0.96	1.70	1.50	1.36	2.60	2.56	3.04	2.54	10.24	3.20
Mg	0.88	0.84	2.27	0.30	0.52	3.20	0.96	0.40	0.44	0.24	0.08	0.40	0.60	3.67	4.26	0.40	1.02	0.81	0.86	0.03	0.42	0.36	0.36	45.76	24.00
K	0.70	0.80	0.73		0.07	0.01	1.08	0.80	0.20	0.62	0.07	0.07	0.45	0.73	0.67					0.07	0.09	0.09	0.07	0.01	0.01
Na	0.04	0.01	0.06		0.02	0.00	0.06	0.04	0.02	0.06	0.76	0.76	0.76	0.04	0.08					0.03	0.02	0.02	0.02	0.01	0.00
Sum of exchangeable bases (meq/ 100 g)	3.38	3.29	6.90	1.90	3.13	6.40	4.06	3.16	3.62	3.36	1.07	3.55	3.59	11.08	14.13	2.07	3.64	3.09	2.94	3.08	3.09	3.51	2.99	56.00	27.20
Capacity of cationic	7.40	15.00	15.27	22.00	8.80	20.40	8.80	15.20	13.60	14.20	18.25	19.25	19.50	33.18	48.89	23.00	28.00	24.40	23.60	10.70	8.80	8.80	8.80	24.16	21.20

DOI: http://dx.doi.org/10.5772/intechopen.93969

	Control soils						**Soils treated with different rock fines ± manure**																		
	T01	**T02**	**T03**	**T04**	**T05**	**T06**	**T41**	**T12**	**T22**	**T62**	**T22B**	**T12B**	**T62B**	**T13**	**T23**	**T14**	**T24**	**T34**	**T44**	**T15**	**T25**	**T35**	**T45**	**T26**	**T46**
exchange (meq/ 100 g)																									
Phosphorus assimilable (ppm) Bray II	6.82	7.78	26.48	19.60	20.52	19.13	7.24	26.65	51.50	50.91	71.37	75.80	2.07	27.95	32.70	21.49	30.11	26.38	28.25	23.56	13.82	17.90	18.45	24.30	95.96

T01, T02, T03, T04, T05 and T06 are control soils collected in the localities of Foumbot, Bonandale, Kalong, Befang, Santa and Batibo in Cameroon, respectively. Treated soils were collected in the same localities, respectively.
T41 = T01 + 600 g basalt fines +600 g poultry manure; T12 = T02 + 2 kg basalt fines; T22 = T02 + 2 kg limestone fines; T62 = T02 + 1 kg limestone; T22B = T22 after 06 months of growth period; T12B = T12 after 06 months of growth period; T62B = T62 after 06 months of growth period; T13 = T03 + 3 kg basalt fines; T23 = T03 + 3 kg gneiss fines. Abbreviations: LS = loamy sand; C = Clay; SC = silty clay; CL = Clay loam; SL = silty loam. pHw = pH water. The letter "B" indicates soil samples collected after 06 months of plant growth. T14 = T04 + 200 g lapilli; T24 = T04 + 200 g fines from volcanic pyroclastic materials; T34 = T04 + 200 g highly vesicular pyroclastic materials; T44 = T04 + 200 g fines from less vesicular pyroclastic materials; T15 = T05 + 1 kg basalt fines; T25 = T05 + 1 kg basalt fines +10 ml LMO (light organic material); T35 = T05 + 1 kg basalt fines +0.5 kg green manure (Tithonia); T45 = T05 + 1 kg basalt fines +0.5 kg green manure; T26 = T06 + 2 kg basalt fines +0.75 kg green manure; T46 = T06 + 2 kg trachyte fines +0.75 kg green manure. Abbreviations: LS = loamy sand; C = Clay; SC = silty clay; CL = clay loam; SL = silty loam; pHw = pH water.

Table 2.
Physico-chemical properties of controls and soils treated with different rock fines ± manure collected in different localities in Cameroon.

(=T04 + 200 g fines from highly vesicular pyroclastic materials), respectively. More details on this work can be found in [7].

The highest yield of carrots with the value of 925 kg/ha was obtained from treatment T35 (=T05 + 1 kg basalt fines +10 ml LMO + 0.5 kg Tithonia). Lower yields of 525, 506 and 150 kg/ha were obtained from treatments T45 (=T05 + 1 kg basalt fines +0.5 kg Tithonia), T25 (=T05 + 1 kg basalt fines +10 ml LMO) and T15 (=T05 + 1 kg basalt fines), respectively. The control (T05) yielded intermediate productivity with 500 kg/ha. The highest yield of potatoes was from T26 (=T06 + 2 kg basalt fines +2 kg coal fines) with 20741 kg/ha, followed by the control (T06) with 14816 kg/ha. The lowest yield is found on treatment T46 (=T06 + 2 kg trachyte fines +2 kg coal fines) with 13,333 kg/ha.

The performance index (Yi = Yield per treatment/Yield per control) indicates the number of folds increase of each treatment in relation to its control. The highest performances (Yi = 6.13 and 4.77) are found in the locality of Befang with treatments T24 and T44, respectively. This locality also portrayed another high Yi (=2.63) for treatment T34. Other performances as high as 3, 2.75, 2.59 and 2 are found in treatments T41, T23 and T22 in the localities of Foumbot, Kalong and Bonendale. Other performances are $1 \leq Yi \leq 2$ and found sparse in all localities subject to this study. However, the localities of Santa and Batibo showed that treatments T15 and T46 yielded less than their controls with Yi (=0.30 and 0.89), respectively.

3.2 Textures and chemicals of controls and treated soils

The control soils show textures of loamy sand (T01 and T02) to clay (T03) passing through clay loam (T06) and silty clay (T05). The highest pH values (=7.10 and 6.40) were observed on loamy sandy samples (T01 and T02, respectively). The lowest pH (=4.60) is portrayed by the silty clayish sample (T05). Intermediate pH values (=5.92, 5.80 and 5.60) were observed on samples T04, T06 and T03 with clayish affinity. For the organic matters (CO, MO and N), the highest values of CO and MO (=6.39 and 11.29%, respectively) were observed on the control T06 collected on a clayish soil in Batibo (**Table 2**). Samples T03 and T02 show lowest values (=0.95 and 1.30%) of CO and MO. Intermediate values of CO and MO (=2.30; 3.81 and 4.00; 6.57%) came from samples T04 and T05, respectively. N also exhibits higher values (=4.62 and 3.06) on T01 and T05, respectively. Lower values (=0.06, 0.56 and 0.16) were found on T03, T02 and T06. For the exchangeable cations (Ca, Mg, K and Na), Ca exhibits highest values (=3.84 meq/100 g) while lowest values belong to Na and K (=0.01 meq/100 g). K and Mg exhibit values between 3.2 and 0.01 meq/100 g. The strongest capacity of cationic exchange belongs to T04 and T06 with values of 22.00 and 20.40. Available phosphorus (P) values are between 26.5 and 6.8 ppm for these controls.

The textures of most treated soils fall in the field of sandy loam (T12, T22, T12B, T22B) and clay (T62, T62B, T13, T23 and T45). However, some treated samples presented properties of clay loam (T26 and T46) and loamy sand (T41). This sample exhibits the highest pH (=7.2) while the lowest pH (=4.8) belong to T15 and T45. For CO, the highest values (=7.03, 6.80 and 6.74%) were encountered on samples T26, T24 and T46, respectively (**Table 2**). Values of CO as low as 0.21 and 0.34% were observed on samples T12 and T22, respectively. In fact, most samples show intermediate CO values with values within 1.13 and 6.39%. Treated soils with lowest values of CO also show lowest MO which are 0.35 and 0.59% for samples T12 and T22, respectively. Samples (T26, T46 and T41) with highest CO also portrayed highest values of MO (=12.12, 11.62, 11.29%), respectively. N portrays the highest concentration (=5.46 g/kg) on T41 followed by lower values (=3.06, 2.92 and 2.91) encountered on T25, T35 and T15, respectively. For the exchangeable cations (Ca,

Mg, K and Na), Mg exhibits the highest values (=45.76 and 24.00 meq/100 g) belonging to T26 and T46. These samples also present higher values of Ca (=10.24 and 3.20 meq/100 g), However samples T13 and T23 also present higher values (=6.64 and 9.12 meq/100 g, respectively) for Ca and Mg (=3.67 and 4.26 meq/100 g, respectively). Then the sum of exchangeable cations are higher on samples T26 (=56.00 meq/100 g), T46 (=27.20 meq/100 g), T23 (=14.13 meq/100 g) and T13 (=11.08 meq/100 g). The lowest values of exchangeable cations (=0.1–0.0 meq/100 g) are those of Na. K also exhibits low values (=1.08–0.0). The strongest and the weakest capacity of cationic exchangeable (=48.89 and 8.8 meq/100 g, respectively) were found on T23 and T25, T35, T45 and T41, respectively. For the available phosphorus (P), the highest value (=95.96 ppm) was found on sample T46. Samples T22B and T12B also present higher values of 71.37 and 75.80 ppm, respectively. Values as high as 51.50, 50.91 ppm were observed with samples T22 and T62. Other higher values of P are encountered on T23, T24, T44, T13, T12, T34, T26, T15 and T14 with 32.70, 30.11, 28.25, 27.95, 26.65, 26.38, 24.30, 23.56 and 21.49 ppm, respectively. Intermediate values are between 7.24 and 18.45 ppm and 2.07 ppm is the lowest values of P found on T62B (**Table 2**).

The variation of pH and the concentrations of various chemicals between values of different controls and those of corresponding treated soils are presented in **Table 3**. The highest positive pH variation (ΔpH = +0.98) was encountered on sample T24 (=T04 + 200 g fines from volcanic pyroclastic materials). This sample also exhibits the highest OC and OM variations (ΔOC = 4.50 and ΔOM = 5.11), respectively. A positive ΔpH as high as +0.80 was observed on a couple of samples, T44 (=T04 + 200 g fines from less vesicular pyroclastic materials) and T23 (=T03 + 3 kg gneiss fines). This couple of samples, T44 and T23 also show higher ΔOC (=+3.30 and =+1.11) and ΔOM (=+3.10 and =+2.48), respectively. Samples T26 (=T06 + 2 kg basalt fines +0.75 kg green manure) and T12 (=T02 + 2 kg basalt fines) both exhibit negative pH variations (=−0.60 and =−0.40), corresponding to negative ΔOC (=−0.19 and −1.09) and ΔOM (=−0.33 and −1.89), respectively. The highest ΔN (+0.84) belongs to sample T41 with manure in his composition.

Most exchangeable cations exhibit low variations with $-0.73 \leq \Delta K \leq 0.38$, $-0.02 \leq \Delta Na \leq 0.75$, $-1.48 \leq \Delta Ca \leq 2.80$ and $-0.76 \leq \Delta Mg \leq 1.99$. However some higher variations of ΔCa and ΔMg are encountered for a couple of samples T23 (=5.28 and 1.99) and T26 (=7.04 and 52.56), respectively. ΔMg is also higher for T46 (=20.80). The most remarkable variation of chemical concentrations was found on phosphorus contents. Although negative variations are observed on T45 (=−2.07), T25 (=−2.62), T62B (=−5.71) and T25 (=−6.70), other samples such as T12, T22, T62, T22B, T12B and T46 exhibit positive phosphorus variations with values as high as 18.87, 43.72, 43.13, 63.59, 68.02 and 76.83, respectively. Positive and low phosphorus contents between 0.42 and 8.65 are observed on samples T41, T13, T23, T14, T34, T15, T26 and T44 (**Table 3**).

4. Performance index (Yi) and physico-chemicals fluctuations of controls and treated soils

Treatments on different local soils in several localities indicated that the best yields as measured by the performance index (Yi) came from the locality of Befang where soils are underlain on volcanic pyroclastic materials. These geologic resources were also used for treatments. These performance indexes on volcanic pyroclastic materials (Yi = 6.13 and 4.77) are higher than those treated with volcanic lava (Yi $\leq$ 3). This suggests that the textures of volcanic rocks influence the yields on different treatments. For instance [8], have indicated that volcanic ash can be one

T41	T12	T22	T62	T22B	T12B	T62B	T13	T23	T14	T24	T34	T44	T15	T25	T35	T45	T26	T46
+0.10	−0.40	+0.20	+0.10	+0.31	+0.58	+0.77	+0.10	+0.80	+0.26	+0.98	+0.53	+0.80	+0.20	+0.30	+0.30	+0.20	−0.60	+0.30
+0.87	−1.09	−0.96	+2.05	−0.16	−0.17	+1.8	+0.64	+1.11	+1.80	+4.50	+3.10	+3.30	−0.65	−0.20	+0.34	−0.04	−0.19	−0.48
+2.04	−1.89	−1.65	+1.30	−0.27	−0.29	+3.11	+1.10	+2.48	+2.20	+5.11	+2.91	+3.10	−1.12	−0.34	+0.58	0.00	−0.33	−0.83
+0.84	+0.42	+0.14	+0.29	−0.20	−0.21	−0.19	+0.08	+0.20					−0.15	0.00	−0.14	−0.29	−0.11	0.00
+0.38	0.00	−0.60	−0.18	−0.73	−0.73	−0.35	0.00	−0.06					0.00	+0.02	+0.02	0.00	0.00	0.00
+0.02	+0.03	+0.01	+0.05	+0.75	+0.75	+0.75	−0.02	+0.02					+0.01	0.00	0.00	0.00	0.00	0.00
+0.20	+0.28	+1.32	+0.80	−1.48	+0.68	+0.14	+2.80	+5.28	+0.06	+0.80	+0.60	+0.46	+0.08	+2.55	+0.52	+0.02	+7.04	0.00
+0.08	−0.44	−0.40	−0.60	−0.76	−0.44	−0.24	+1.40	+1.99	+0.10	+0.72	+0.51	+0.56	−0.49	−0.10	−0.16	−0.16	+42.56	+20.80
+0.42	+18.87	+43.72	+43.13	+63.59	+68.02	−5.71	+1.47	+6.22	+1.89	+10.51	+6.78	+8.65	+3.04	−6.70	−2.62	−2.07	+5.17	+76.83

Δ: difference between 02 values. T41 = T01 + 600 g basalt fines +600 g poultry manure; T12 = T02 + 2 kg basalt fines; T22 = T02 + 2 kg limestone fines; T62 = T02 + 1 kg limestone; T22B = T22 after 06 months of growth period; T12B = T12 after 06 months of growth period; T62B = T62 after 06 months of growth period; T13 = T03 + 3 kg basalt fines; T23 = T03 + 3 kg gneiss fines; T14 = T04 + 200 g lapilli; T24 = T04 + 200 g fines from volcanic pyroclastic materials; T34 = T04 + 200 g highly vesicular pyroclastic materials; T44 = T04 + 200 g fines from less vesicular pyroclastic materials; T15 = T05 + 1 kg basalt fines; T25 = T05 + 1 kg basalt fines +10 ml LMO (Light Organic Material); T35 = T05 + 1 kg basalt fines +0.5 kg green manure (Tithonia); T45 = T05 + 1 kg basalt fines +0.5 kg green manure; T26 = T06 + 2 kg basalt fines +0.75 kg green manure; T46 = T06 + 2 kg trachyte fines + 0.75 kg green manure (Tithonia); Tithonia = Tithonia diversifolia.

Table 3.
Variation of the chemical parameters between controls and treated soils collected in the different localities of Cameroon.

of the best rock powder additives for soil amelioration since it contains a wide range of chemicals and weathers relatively fast to provide a natural fertilizer [9, 10]. Other treatments on metamorphosed terrain exhibit the best yields below those on volcanic rocks with the highest Yi < 3. It is the similar situation for treatments on sedimentary rocks related fines. According to [11], rocks found in crystalline terrains contain significant amounts of quartz which dilute the effectiveness of rocks as source of K, Ca and Mg from minerals such as feldspars and mica. The different experiments also indicate a general variation of textures and physico-chemical parameters of treated soils in relation to the controls: the initial loamy sand texture of the controls moves to sandy loam and clay loam textures; sandy clay texture moves to clay texture and initial clay textures remained unchanged. This suggests that a soil treated with rock fines ± manure losses sand proportion while increasing mostly in clay and somehow in silt particles. These clayish soils also exhibit the best yields. In respect to the pH, the slightly, moderately to strongly acidic properties of the local soils were shifted upwards in between the slightly acidic and the slightly alkaline soils (**Figure 1**). This suggests a general increase of pHs after treatments. For example, in the locality of Santa, a pH (=4.60) of a control T05 increases to pH (=4.80 and =4.90) corresponding to treated soils T15, T45 and T25, T35, respectively. In fact, there is a general positive increase of pHs between +0.10 and +0.98 (**Table 3**). However, a couple of samples exhibit negative pH variations (ΔpH = −0.40 and ΔpH = −0.60) on samples T12 (=T02 + 2 kg basalt fines) and T26 (=T06 + 2 kg basalt fines +2 kg coal fines) in the locality of Bonandale and Batibo, respectively. This implies the potential of these treatments to increase or decrease soil pHs. According to [12], the advantages for the application of rocks' fines in soils are in correcting the pHs with nutrient supply and its long residual effect.

The organic carbon (OC) and organic matter (OM) also showed variations after the application of different treatments. The trends of fluctuation of OM and OC are parallel throughout all control and treated soils (**Figure 2**). Some samples exhibit positive ΔOM and ΔOC while other show negative ΔOM and ΔOC (**Table 3**). This implies that these soils showed increasing or decreasing OC and OM after treated with rock fines ± manure. The highest values of ΔOM (=+5.11) and ΔOC (=+4.50) were found on sample T24 (=T04 + 200 g fines from pyroclastic bombs) which also showed the highest ΔpH (=+0.98) and yield index (Yi = 6.13). The second highest Yi (=4.77) from the sample T44 (=T04 + 200 g fines from less vesicular pyroclastic

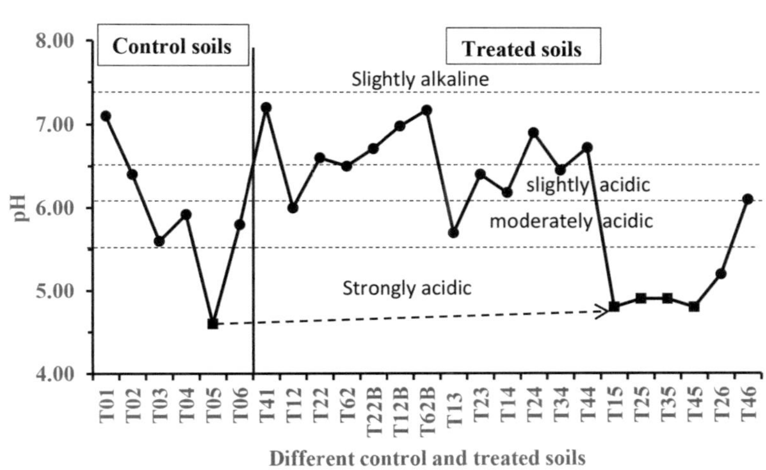

Figure 1.
pH of control and treated soils collected in some selected sites in Cameroon. Different acidic fields modified after [13–15]. The broken arrow indicates an increase of pHs from the controls to the treated soils.

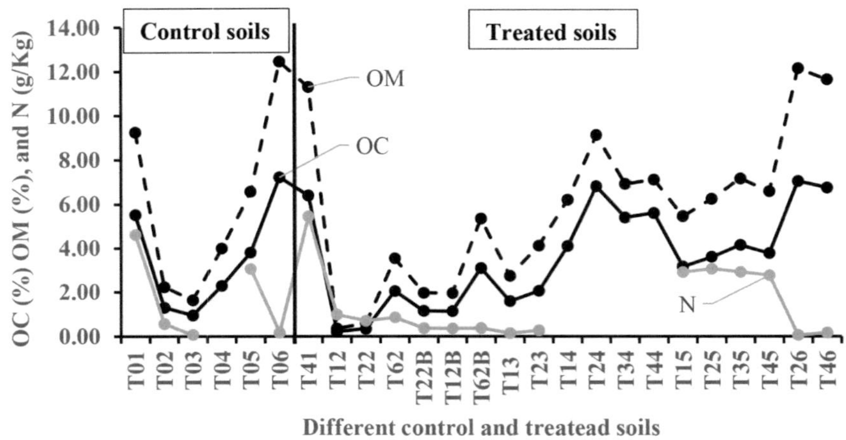

Figure 2.
OC (%), OM (%) and N (g/kg) of control and treated soils collected in some selected sites in Cameroon.

bombs) also exhibit higher ΔpH (=+0.80), ΔOM (=+3.10) and ΔOC (=+3.30). This suggests that fines from pyroclastic materials increase pH, OM and OC contents in different treatments. This is also observed on treatments with fines from basalt. However negative ΔpH is observed on samples T12 and T26 which were treated with fines from basalt. These samples also showed negative ΔOM and ΔOC. This suggests that a decrease of pH implies as decrease in OM and OC. Regarding other soil nutriments, N contents remained very weak ≤1.00 g/kg except some higher values between 2.77 and 3.06 g/kg encountered for sample T15, T25, T35 and T45. These samples were treated with basalt fines + green manure. Then, added N may come from the manure. There is a slight increase of K contents towards 0.8 and 1.1 meq/100 g while the values of Na remain monotone (**Figure 3**). However, some samples under the treatments of basalt and trachyte fines exhibit outstanding high contents of Mg and Ca. This indicates pulverized products of these rocks as potential sources of Ca and Mg in soils. Phosphorus contents after treatment (**Figure 4**) exhibits a broad increase in relation to the controls. Mostly, soils treated under fines from trachyte and limestone exhibit the highest values of phosphorus (=(96.0, 51.5) and 50.9 ppm), respectively. The fines from volcanic pyroclastic materials and gneiss indicated higher contents of phosphorus of 32.7 and 30.1 ppm, respectively.

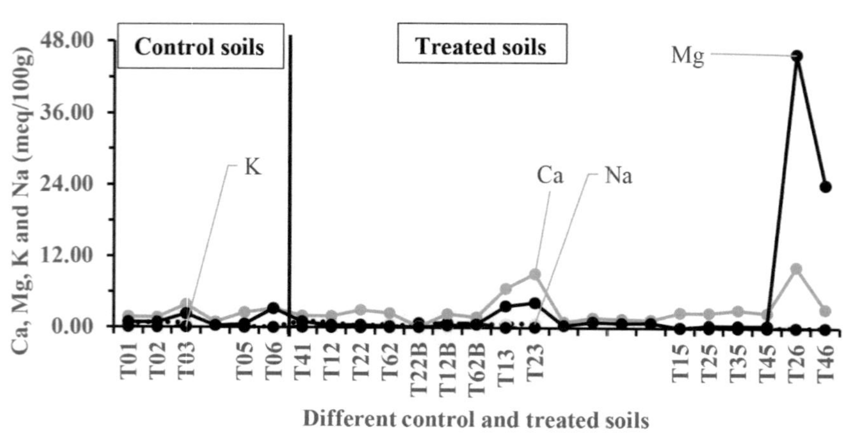

Figure 3.
Ca, Mg, K and Na (meq/100 g) of control and treated soils collected in some selected sites in Cameroon.

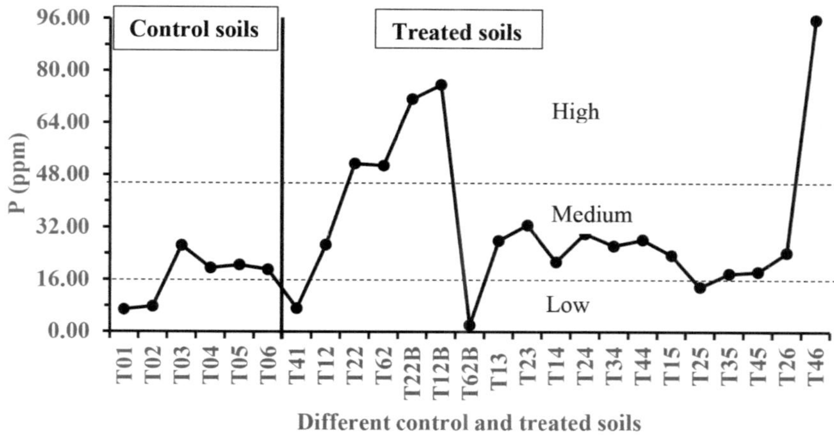

Figure 4.
Available phosphorus (ppm) of control and treated soils collected in some selected sites in Cameroon. Phosphorus limit modified after [13–15].

5. Conclusion

This work demonstrated that fines from different rock types such as basalt, trachyte, volcanic pyroclastic materials in addition to limestone and gneiss applied as fertilizers, indicate a slight increase of pH in all samples in several localities in Cameroon. However, the treatment under basalt fines showed a significant pH increase. Then, the required soil acidity may be accomplished using pulverized products from rocks of basaltic compositions, while this also simultaneously increasing soil contents in MO, CO, Mg and Ca. There is a good and parallel correlation of MO and CO contents in all treated soils. The higher values of phosphorus suggest that the application of the concerned rock fines as a good potential sources of this nutrients in these tropical soils.

Acknowledgements

This work was entirely carried with the assistance of some small scale local farmers in the areas of Befang, Santa, Batibo, Foumbot, Bonandale and Kalong in the northwest, west, littoral and center regions of Cameroon. They accepted to host these field trials on their land. The local agricultural technicians and some bypassing villagers assisted also during field works. Many thanks to all of them.

Author details

Samuel Tetsopgang
Department of Geology, Higher Teacher Training College, The University of Bamenda, Bamenda, Cameroon

*Address all correspondence to: tetsopgang@yahoo.com

References

[1] Bennett, H. H. & Chapline, W. R. (1928). Soil erosion a national menace. US Dep. Agric., Circ. 33.

[2] Mostafa, K. T. & Osama, A. E. K. (1992) (eds.). The world environment 1972–1992. Two decades of challenge. Chapman and Hall, London, England.

[3] Smaling, E. M. A., Nandwa, S. M. & Janssen, B. H. (1997) Soil Fertility Is at Stake, in Replenishing Soil Fertility in Sub-Saharan Africa. In: Buresh, R., Sanchez, P.A. and Calhoun, F., Eds., American Society of Agronomy and Soil Science Society of America, Madison Wisconsin, 47–61.

[4] Walkley, A. J. & Black, I. A. (1934). Estimation of soil organic carbon by the chromic acid titration method. Soil Sci. 37, 29–38

[5] Kjeldahl, J. (1883). "Neue Methode zur Bestimmung des Stickstoffs in organischen Körpern" (New method for the determination of nitrogen in organic substances), *Zeitschrift für analytische Chemie*, 22 (1) 366–383.

[6] Peech, M. (1965). Hydrogen-ion activity In C.A. Black et al., Eds., Methods of Soil Analysis. Part 2. American Society of Agronomy, Madison, WI; pp. 914–926.

[7] Tetsopgang, S. & Fonyuy, F. (2019) Enhancing growth quality and yield of cabbage (*Brassica oleracea*) while increasing soil pH, chemicals and organic carbon with the application of fines from volcanic pyroclastic materials on a tropical soil in Wum, Northwest Cameroon, Africa. Scientific African 6 (2019) e00199. https://doi.org/10.1016/j.sciaf.2019.e00199

[8] Blum WEH, Herbinger B, Mentler F, Ottner M, pollak E, Unger E and Wenzel WW (1989) Zur verwendung von gesternsmelen in der landwirtschaft. II. Wikung von gsteinsmehlen als bodenverbesserungsmittel. Z Pflanzenernahr Bodenk 152: 427–430.

[9] Kronberg B I (1977) The geochemistry of some Brazillain soils and geochemical consideration for agriculture on highly leached soils. PhD Thesis. The University of western Ontarion, London, Canada. 137 p

[10] Salinas JG, Saif SR, Kronberg BI, Fyfe WS and Walden DB (1986) Rock and glassy materials as nutrient conservative additives. In: memorias de IX congress Latin Americana de la Cienca del Suelo.

[11] T.J. Weerasuriya, S. Pushpakumara and P.I. Cooray (1993) Acidulated pegmatitic mica: a promising new multi-nutrient mineral fertilizer. Fertilizer Research 34: 67–77

[12] C.G. Ramos, A.G. de Mello and R. M. Kautzmann (2014) A preliminary study of acid volcanic rocks for stonemeal application, Environmental Nanotechnology, Monitoring and Management 1–2, 30–35.

[13] Cass, A. (1998). Measuring and managing chemical impediments to growth. Aust. Grapegrower Winemaker. 415:13–16

[14] Hazelton, P. and Murphy, B. (2007). Interpreting soil test results. What do all the numbers mean? Published by CSIRO Publishing. Collingwood Victoria – Australia. http://www.publish.CSIRO.

[15] Euroconsult (1989) Agricultural compendium for rural development in the tropics and the subtropics, Elsevier Amsterdam, pp. 778.

Section 2

Remediation Technologies

Chapter 8

Evaluation of Ornamental Plants for Phytoremediation of Contaminated Soil

Adeyela Ibironke Okunlola, Dotun Nathaniel Arije and Katherine Olayinka Olajugbagbe

Abstract

A completely randomized design with three replicates was conducted at the Screen house of the Department of Crop Soil and Pest Management, Federal University of Technology Akure, Ondo State, to examine the phytoremediation potential of *Codiaeum variegatum* and *Basella alba* on contaminated soils from four locations. Soils were collected from the Mechanic workshop, Dumpsite, Forest Topsoils, and Effluent site, and filled into the buckets. Initial soil analysis was conducted on all the soils to determine heavy metal concentration (Cu, Cd, Ni, Pb, and Zn). At 12 weeks after planting, soil and plant (root and shoot) samples were analyzed to determine the heavy metals accumulated. WHO permissible limit value for heavy metal concentration in soil and plant were used as a standard to evaluate plant phytoremediation potential. Results from the study confirm the phytoremediation potential of *C. variegatum* and its high tolerance for the accumulation of heavy metals. *B. alba* plant also shows its potential in removing heavy metals from the soil, but it was not as tolerant as *C. variegatum* as *B. alba* planted in soils from mechanic workshop and effluent site had stunted growth.

Keywords: ornamental plants, metals, toxics, phytoremediation

1. Introduction

Environmental pollution has been on the rise over the past decades because of the increasing human activities on energy reservoirs, unsafe agricultural practices, and rapid industrialization. The result of these technogenic and anthropogenic activities are the major sources of heavy metals in the environment [1]. In Nigeria, soil contaminations is caused by industrial and agricultural practices such as chemical fertilizer and pesticide application, wastewater irrigation, mining activities, and metal smelting. All these human activities have contributed to problems hindering the nation's agriculture from attaining food security.

However, agriculture in Nigeria has been facing two challenges for a long time, these challenges are; promoting environmental sustainability and enhancing food production. To ameliorate these challenges, there is a need to adopt management techniques that promote environmental sustainability. Phytoremediation has been identified as a cost-effective and easy way to sustain our environment by removing toxic elements from contaminated soils. Phytoremediation is a technique that relies on the use of plant interactions (physical, biochemical, biological, chemical, and microbiological) in polluted sites to mitigate the toxic effects of pollutants [2]. In Nigeria, heavy metals, pesticides, greenhouse gases, and hydrocarbons are pollutants that are of environmental and public health concerns.

The toxicity of heavy metals in the biota is because of their bioaccumulative nature and persistence in the environment thereby contaminating the food chains. The soil-to-plant transfer of heavy metals made it easy for metal transfer into the food chains. Metals are absorbed by plant roots and transferred to herbivorous animals along the food chain [3]. When plants like vegetables or cereals are planted in contaminated soils, the consumption of such food becomes a serious health issue to man [4].

However, because of the threat posed by the heavy metals on the growth and development of arable crops, scientists have warned against the use of crops as a phytoremediator because of their risk on human health after consumption. This is the reason for the shift to ornamental plants. The use of ornamentals continues to attract attention in recent years. In Nigeria, most ornamentals plants are not edible, therefore, the risk of contaminants entering the food chain is reduced.

The use of ornamental plants as a test plant in a phytoremediation experiment is because of their high biomass which means they can accumulate more heavy-metal concentration through their roots, into their tissues. Many studies have been conducted to evaluate the potentials of ornamental plants as in phytoremediation [5–8]. However, most of the selected ornamental plants used in all the studies were not indigenous and not commonly cultivated in Nigeria. Therefore, this study aims to evaluate the phytoremediation potential of two ornamental plants in common, although not indigenous in Nigeria. In addition, *Codiaeum variegatum* and *Basella alba* were used in this study because they grow well in heavy metal contaminated soils, but their mechanism to resist the heavy metals has not been reported. *B. alba* belongs to the family of Basellaceae and commonly refers to as Indian spinach, Malabar spinach, Ceylon spinach, and vine spinach. The plant is an underutilized vegetable in Nigeria compare to *Amaranthus* spp. and *Telfairia occidentalis.* In addition to being edible, *B alba* is also grown as an ornamental foliage vine. *Codiaeum variegatum* is an ornamental plant species that belongs to the genus Codiaeum, and the family Euphorbiaceae.

2. Objective

To examine the phytoremediation potential of *C. variegatum* and *B. alba* in heavy metal contaminated soils collected from four sites. The study also analyzed part of the plant with higher heavy metal concentration (shoots or roots) and the heavy metal concentration left in the soil after the experiment.

3. Materials and methods

A Completely Randomized Design with three replicates was conducted at the Screen house of the Department of Crop Soil and Pest Management, Federal University of Technology Akure, Ondo State located in the rain forest vegetation zone of Nigeria (7°16′N, 5°12′E). Soils were collected from four sites (Mechanic workshop, Dumpsite, Forest topsoil, and Effluent site) and filled into the buckets and transported to the screen house. The soils from the four locations served as the treatments. A total of 12 plastic buckets were used for each ornamental plant (4 locations replicated three times), to make it 24 plastic buckets for both ornamental plants (*C. variegatum* and *B. alba*). The planting material was obtained from

LUCADO horticultural garden located in Akure (less than 5 km to the experimental site). The seeds of *B. alba* were planted while the seedlings of *C. variegatum* were purchased from the horticultural garden and it was repotted. Watering was done daily and weeds were hand-pulled.

Initial soil analysis was conducted on all the soils to determine heavy metal concentration. The heavy metal tested on soil and plant samples were, Cu, Cd, Ni, Pb, and Zn using Atomic Absorption Spectrometer [9]. The plant growth traits were recording, including; plant height (cm), stem girth (cm), leaf length (cm) and leaf numbers. At 12 weeks after planting (WAP), soil analysis was done to determine the remaining heavy metal concentration in the soil in order to determine the percentage of contamination reduction. In the final week of the experiment (12 weeks after planting), soil and plant (root and shoot) samples were again analyzed to determine the heavy metal concentration. WHO [10] permissible limit for heavy metal concentration in the soil and plant were used as standard and as a rating for each plant phytoremediation potential. The data were subjected to analysis of variance (ANOVA) using Statistical Package for Social Sciences (Version 17). Significant means were from each other using Tukey Test at 5% level of probability.

4. Results and discussion

4.1 Effect of heavy metal on plant growth parameters

Significant differences were recorded across the treatments (soils from different locations) for the growth parameters of *C. variegatum* (**Table 1**). There was no consistency in the growth performance as severe impact was felt on the number of leaves, stem girth, and leaf length of *C. variegatum* planted in soils from the effluent site as they had the least mean value. *C. variegatum* planted in soils collected from dumpsite had the highest mean value for number of leaves, stem girth, and leaf length. The severe impact felt on *C. variegatum* planted on the effluent site could be attributed to excess levels of metals which may have inhibits physiologically active enzymes as earlier speculated by Gadd [11]. Significant differences were recorded across the treatments for the growth parameters of *B. alba* (**Table 1**). The results

Treatments	Plant height	Number of leaves	Stem girth	Leaf length
C. variegatum				
MS	11.88a	10.00ab	0.79a	23.72c
ES	15.10c	8.00a	0.75a	19.27a
DS	14.36b	17.00c	0.96b	24.02d
FS	13.83b	10.00ab	0.89ab	21.68b
B. alba				
MS	21.96a	10.45a	0.51a	10.14a
ES	30.78b	15.17c	0.62a	10.08a
DS	60.19c	15.00c	0.72a	12.89b
FS	65.55d	14.00bc	0.66a	11.23ab

Means with the same letter in the same column are not significantly different from one another at p < 0.05 based Duncan test.
MS—soils from mechanic workshop; ES—effluent site; DS—dumpsite; FS—forest topsoil.

Table 1.
Effect of soil from different sites on growth parameters of the ornamental plants.

revealed that *B. alba* planted in soils from dumpsite and forest topsoil gave the highest mean value for plant height, number of leaves, and leaf length. Plants on the two soils appeared healthy because the forest topsoils served as the control. The good performance of *B. alba* planted on the dumpsite soils could be a result of a high level of organic matter content.

4.2 Initial and final metal concentrations in plant tissues and in soils

The result presented in **Table 2** shows the initial concentration of heavy metals in the root and shoot of *C. variegatum* and *B. alba*. The concentration of the heavy metals present in the plant was within the permissible value recommended by the WHO (**Table 3**) for except for the Zn concentration (in both plants), and Cd (*B. alba* only). The initial Zn concentration present in root part of *B. alba* was above the minimum plant permissible limit (3.056 mg/kg) while the concentration present at the shoot part was below the permissible limit (0.421 mg/kg).

The result presented in **Table 4** and **Figure 1** shows the initial and final heavy metal concentration of soils from the four sources. The results revealed that the initial and final heavy metal concentrations in all the soils were below the target value recommended by WHO for soils. However, soils from the mechanic workshop site show a considerable decrease in the heavy metal concentration present at the end of the experiment. The initial Pb concentration for the soil was 0.215 but was reduced to 0.093 in the pot where *B. alba* was planted while it was absorbed below the detective limit by *C. variegatum.* A similar trend was also observed for soils collected from an effluent and dumpsite site for all the heavy metals measured. However, there was a slight change in this trend for soils collected from forest topsoil, as there was a slight increase in the final heavy metal concentration recorded for metals such as Cu (initial 0.751; final 0.892 *B. alba,* 1.073 *C. variegatum*), Cd (**initial** 0.072; **final** 0.097, *B. alba*), and Zn (**initial** 27.525; **final** 27.095 *B. alba* 28.1 *C. variegatum*).

Result presented in **Table 5** shows the final heavy metal concentration present in the plant parts for all the soils. For soils collected from the mechanic workshop, the Cd (0.06 for *B. alba* and *C. variegatum*) and Zn concentration present in both plants were above the WHO permissible limit, while the remaining metals were within the permissible limit. Similar trends or results were also recorded for soils collected from the effluent site, dumpsite, and forest topsoil. The growth of both plants were affected variably by the stress of heavy metals such as Zn and Cd. High concentrations of Zn and Cd resulted in stunted growth, reduced biomass production and produced characteristic visible effects similar to those described by other workers in different plant species [12, 13]. These observations are substantiated by a significant concentration in the level of Zn and Cd present in the plant tissue of both

Heavy metals	*B. alba*		*C. variegatum*	
	Root	Shoot	Root	Shoot
Cu	0.455	0.193	0.572	0.49
Cd	0.054	0.01	0.01	BDL
Ni	0.082	0.027	0.09	0.01
Pb	0.034	BDL	0.032	BDL
Zn	3.056	0.421	3.25	2.081

*BDL = Below Instrument Detection Limit (<0.001 ppm) *1 mg/kg = 1 ppm.*
MS—soils from mechanic workshop; ES—effluent site; DS—dumpsite; FS—forest topsoil.

Table 2.
Initial analysis to determine heavy metal conc. in plant root and shoot (ppm).

Heavy metals	Target value of soil (mg/kg)	Permissible value of Plant (mg/kg)
Cu	36	10
Cd	0.8	0.02
Ni	35	10
Pb	85	2
Zn	50	0.60

Target values are specified to indicate desirable maximum levels of elements in unpolluted soils.
Source: WHO [10].

Table 3.
WHO permissible limit of Cu, Cd, Ni, Pb and Zn in soil and plant by WHO [10].

Soil source	Initial	Final Conc.	
		B. alba	*C. variegatum*
Mechanic			
Cu	1.567	0.836	1.484
Cd	0.11	0.088	0.085
Ni	0.89	0.26	0.314
Pb	0.215	0.093	BDL
Zn	37.17	24.9	31.274
Effluent			
Cu	2.122	1.75	1.823
Cd	0.153	0.11	0.142
Ni	1.27	0.494	0.829
Pb	0.262	0.21	0.069
Zn	42.57	26.35	32.923
Dumpsite			
Cu	2.014	0.962	1.216
Cd	0.289	0.068	0.092
Ni	1.276	0.398	0.483
Pb	0.312	BDL	0.077
Zn	32.036	28.719	30.136
Forest topsoil			
Cu	0.751	0.892	1.073
Cd	0.072	0.097	0.047
Ni	0.558	0.085	0.048
Pb	0.134	0.145	0.066
Zn	27.525	27.095	28.1

*BDL = Below Instrument Detection Limit (< 0.001 ppm) *1 mg/kg = 1 ppm.*
MS—soils from mechanic workshop; ES—effluent site; DS—dumpsite; FS—forest topsoil.

Table 4.
Soil heavy metal concentration (mg/kg).

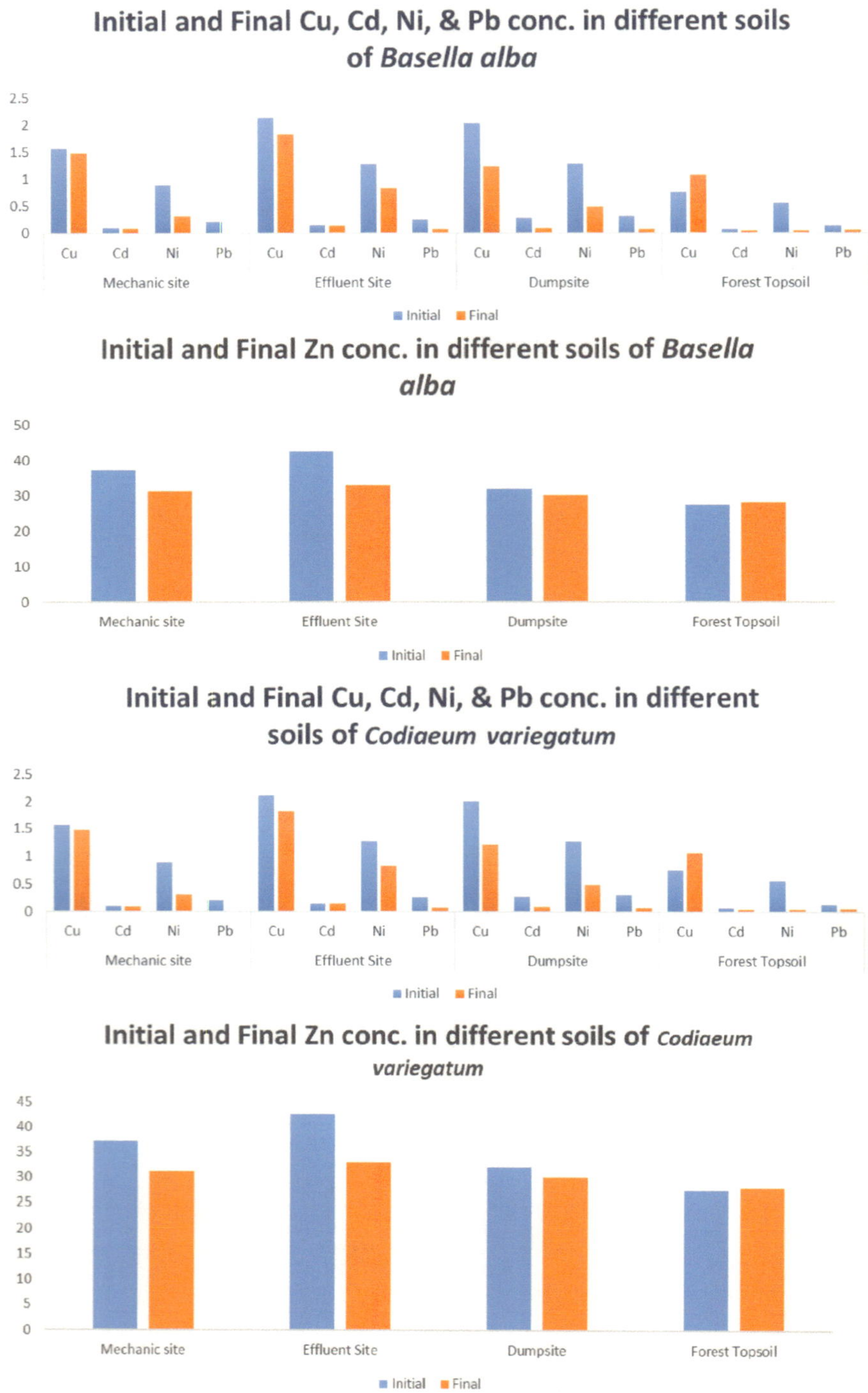

Figure 1.
Initial and final heavy metal concentration (mg/kg) for C. variegatum *and* B. alba.

ornamental plants. The decrease in the mean value of growth parameters of *B. alba* and *C. variegatum* planted on soils from effluents and mechanic site may be attributed to the significantly high concentration of Cd and Zn value which is higher than the permissible limit. These findings agree with Pandey and Pathak [14]. Metal stress in plants leads to a decrease in growth parameters and dry matter of plants [14, 15].

	B. alba		*C. variegatum*	
	Root	**Shoot**	**Root**	**Shoot**
Mechanic				
Cu	0.627	0.038	0.915	0.085
Cd	0.049	0.012	0.049	0.011
Ni	0.078	0.031	0.079	0.011
Pb	0.011	BDL	BDL	BDL
Zn	3.038	0.32	3.063	0.475
Effluent				
Cu	0.484	0.041	0.915	0.059
Cd	0.037	0.01	0.051	BDL
Ni	0.087	0.03	0.065	0.02
Pb	BDL	BDL	0.017	BDL
Zn	2.737	0.299	2.873	0.628
Dumpsite				
Cu	1.096	0.12	0.838	0.514
Cd	0.052	0.013	0.024	0.01
Ni	0.07	0.025	0.085	0.026
Pb	0.015	0.01	0.02	0.01
Zn	3.173	0.125	2.955	0.315
Forest topsoil				
Cu	0.537	0.3	0.563	0.05
Cd	0.057	0.021	0.023	BDL
Ni	0.092	0.04	0.06	BDL
Pb	0.009	0.003	BDL	BDL
Zn	2.859	0.538	3.425	0.211

*BDL = Below Instrument Detection Limit (<0.001 ppm)*1 mg/kg = 1 ppm.*
MS—soils from mechanic workshop; ES—effluent site; DS—dumpsite; FS—forest topsoil.

Table 5.
Concentration of heavy metals in plants part soil source.

5. Conclusion

This study was conducted to determine the phytoremediation potential of two ornamental plants (*B. alba* and *C. variegatum*). The study reveals the ability of both plants in removing heavy metals (hyperaccumulators), but most heavy concentration was accumulated in the roots more than shoots. However, the accumulation of Cd and Zn at the end of the study was higher than the permissible limit. However, the use of *B. alba* to remediate the soil may not be advisable because of its less phytoremediation potential compare to *C. variegatum*. Also the former is edible and could pose a serious threat to health when consumed. Finally, additional studies are needed to investigate the phytoremediation performance of more indigenous ornamental plants in Nigeria.

Author details

Adeyela Ibironke Okunlola, Dotun Nathaniel Arije*
and Katherine Olayinka Olajugbagbe
Department of Crop, Soil, and Pest Management, The Federal University of Technology, Akure, Ondo State, Nigeria

*Address all correspondence to: dotunarije@gmail.com

References

[1] Woranan N, Orapan M, Prasad M. Chapter 9: Potential of ornamental plants for phytoremediation of heavy metals and income generation. In: Prasad MNV, editor. Bioremediation and Bioeconomy. UK: Elsevier; 2016. pp. 177-217. DOI: 10.1016/B978-0-12-802830-8.00009-5

[2] Alaboudi K, Ahmed B, Brodie G. Phytoremediation of Pb and Cd contaminated soils by using sunflower (*Helianthus annuus*) plant. Annals of Agricultural Science. 2018;**63**(1):123-127. DOI: 10.1016/j.aoas.2018.05.007. ISSN: 0570-1783

[3] Nica DV, Bura M, Gergen I, Harmanescu M, Bordean D-M. Bioaccumulative and conchological assessment of heavy metal transfer in a soil-plant-snail food chain. Chemistry Central Journal. 2012;**6**(1):55

[4] Orisakwe OE, Nduka JK, Amadi CN, Dike DO, Bede O. Heavy metals health risk assessment for population via consumption of food crops and fruits in Owerri, South Eastern, Nigeria. Chemistry Central Journal. 2012;**6**(1):77

[5] Liu H, Zhao H, Wu L, Liu A, Zhao F, Xu W. Heavy metal ATPase 3 (HMA3) confers cadmium hypertolerance on the cadmium/zinc hyperaccumulator *Sedum plumbizincicola*. The New Phytologist. 2017;**215**(2):687-698

[6] Shuang C, Tingan Z, Shanlin Z, Ping L, Qixing Z, Qianru Z, et al. Evaluation of three ornamental plants for phytoremediation of Pb-contaminated soil. International Journal of Phytoremediation. 2013;**15**(4):299-306. DOI: 10.1080/15226514.2012.694502

[7] Miao Q, Yan J. Comparison of three ornamental plants for phytoextraction potential of chromium removal from tannery sludge. Journal of Material Cycles and Waste Management. 2013;**15**(1):98-105

[8] Wang X, Zhou Q. Ecotoxicological effects of cadmium on three ornamental plants. Chemosphere. 2005;**60**(1):16-21

[9] Shaltout A, Ibrahim M. Detection limit enhancement of Cd, Ni, Pb and Zn determined by flame atomic absorption spectroscopy. Canadian Journal of Analytical Sciences and Spectroscopy. 2007;**52**:5

[10] World Health Organization. Permissible Limits of Heavy Metals in Soil and Plants. Geneva, Switzerland: WHO; 1996

[11] Gadd GM. Geomycology: Biogeochemical transformations of rocks, minerals, metals and radionuclides by fungi, bioweathering and bioremediation. Mycological Research. 2007;**2007**(111):3-49

[12] Zhou W, Qiu B. Effects of cadmium hyperaccumulation on physiological characteristics of *Sedum alfredii* Hance (Crassulaceae). Plant Science. 2005;**169**:737-745

[13] Gajewska E, Sklodowska M. Relations between tocopherol, chlorophyll and lipid peroxides contents in shots of Ni-treated wheat. Plant Physiology. 2007;**164**:364-366

[14] Pandey N, Pathak GC. Nickel alters antioxidative defence and water status in greengram. Indian Journal of Plant Physiology. 2006;**11**:113-118

[15] Ryser P, Sauder WR. Effects of heavy-metal-contaminated soil on growth, phenology and biomass turnover of *Hieracium piloselloides*. Environmental Pollutution. 2006;**140**:52-61

Chapter 9

HCH-Contaminated Soils and Remediation Technologies

Aurora Santos López, Carmen M. Dominguez Torre and David Lorenzo Fernández

Abstract

The production of lindane (gamma isomer of hexachlorocyclohexane, γ-HCH) in the second half of the twentieth century was an inefficient process that generated vast amounts of residues of other HCH isomers, without any pesticide activity. These residues were often dumped in an uncontrolled way or unsecured landfills in points near the production sites, causing hot spots of soil and groundwater contamination on all continents. Given the persistence and hydrophobicity of these pollutants, the problem generated decades ago is still a challenge to be solved. Several technologies, based on physical, chemical, and biological treatments, have been proposed for the remediation of HCH-contaminated soils. This chapter reviews the particularities of this kind of contamination and critically examines the bases and results of the technologies applied, paying special attention to the physicochemical remediation processes.

Keywords: lindane, HCHs, polluted soils, remediation technologies

1. Introduction

The intensive use of organochlorine pesticides (OCPs) during the last decades around the world and the inadequate management of the wastes generated during the production of these compounds represents a huge environmental problem. That is the case of lindane production, the gamma isomer of hexachlorocyclohexane (γ-HCH), whose production during the last century has generated large amounts of solid wastes, consisting of a mixture of other HCH isomers, that has caused hot points of soil and groundwater contamination [1].

Lindane was synthesized for the first time in 1825 by Michael Faraday [2] and deeply used as a broad-spectrum organochlorine insecticide since the 1940s [3, 4]. Among the eight isomers of HCH, lindane is the only one with insecticidal properties. Unfortunately, the lindane production, schematically summarized in **Figure 1**, is an inefficient process, generating large volumes of the other HCH isomers (mainly α-, β- and δ-HCH). The mixture of HCH isomers obtained in the chlorination of benzene is called technical-HCH, and it was usually subjected to a purification process to separate the γ-HCH isomer. After this step, about 10 kg of HCH wastes were obtained per kg of purified lindane. The solid HCH wastes (consisting of a white powder of HCH isomers) were inappropriately dumped during decades in the production sites nearby, resulting in environmental contamination with global dimension [5–14].

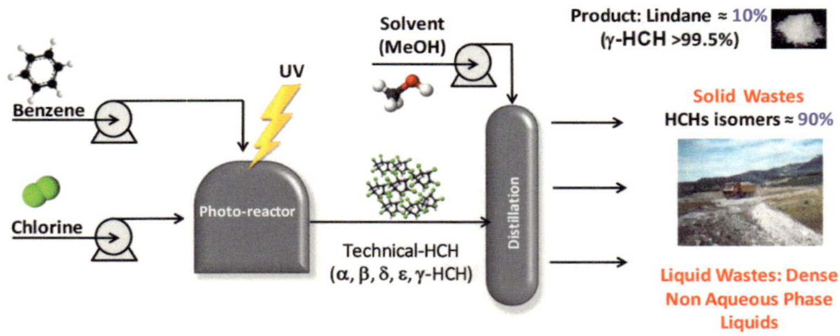

Figure 1.
Scheme of lindane production and purification processes.

Figure 2.
Location of sites polluted with HCH wastes around the world, modified from [10].

It is estimated that approximately 450,000 tons of lindane were used worldwide between 1950 and 2000. Approximately 63% of the lindane produced was consumed in Europe, 17% in Asia, and about 4.2% in the United States, resulting in the ubiquitous presence of HCH wastes, as is shown in **Figure 2**.

Due to the high refractoriness and adverse effects of HCHs on the ecosystem and human beings [3, 15, 16], several HCH isomers are considered persistent organic pollutants (POPs) by the Stockholm Convention [10] and classified as neurotoxic, carcinogen, and teratogen by the Environmental Protection Agency (EPA) and the World Health Organization (WHO) [17, 18]. The structure and main chemical properties of HCH isomers are given in **Table 1**.

Due to its toxicity, the production and use of lindane have been banned in most countries, including Europe and the United States [24, 25], but many landfills and the surroundings of the lindane production sites remain polluted nowadays, with soil and groundwater contaminated by these compounds [5–14]. The low tolerance limits allowed for HCHs in water and soils have prompted a growing interest of the scientific community to develop simple, cost-effective, and fast methods for the degradation of these pollutants. Conventional methods commonly used include the excavation of polluted soil and its further containment in secure landfills. The traditional groundwater treatment consists of pump-and-treat, with adsorption in activated carbon as a common treatment. However, these options are very expensive and are not a definitive solution since the destruction of the pollutants is not achieved. Therefore, they are considered neither sustainable nor definitive remediation methods [1].

Some studies have focused on the remediation of HCHs in the aqueous phase, dealing with groundwater treatments applied in situ [26] or on-site [16, 27, 28].

Isomer	CAS	Solubility (mg/L)	MW (g/mol)	Structure
α-HCH*	319-84-6	1.2-2 [19–22]	290.83	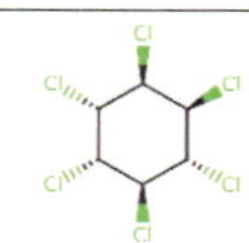
β-HCH*	319-85-7	0.15-0.7 [20–22]	290.83	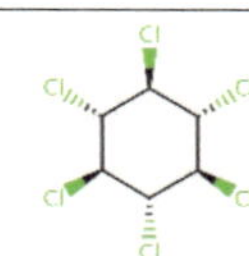
γ-HCH	319–86-8	8.6–31 [19, 21–23]	290.83	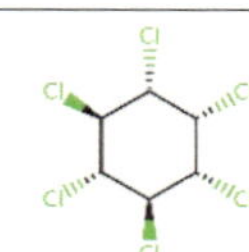
δ-HCH*	58-89-9	2.1-15.3 [19–21]	290.83	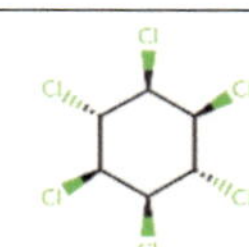
ε-HCH	6108-10-7	7.8**	290.83	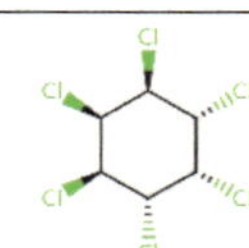

**HCH isomers included in the Stockholm Convention.*
***https://comptox.epa.gov/dashboard/dsstoxdb/results?search=DTXSID0024135.*

Table 1.
Name, CAS, water-solubility, molecular weight (MW), and chemical structure of the main HCH isomers.

However, only a few works are found in the literature concerning the remediation of soils contaminated by HCH wastes. The objective of these treatments is the chemical or biological degradation of HCHs. They were carried out to the remediation of soils artificially spiked with HCH isomers and soils with real HCH contamination. The chemical technologies used for the remediation of soils polluted with high HCH concentration are analyzed and discussed in the following sections. Biological treatment of these highly contaminated soils requires long times [29, 30] and are not treated here. Moreover, to better understand the particularities and characteristics of the problem of HCH polluted soils, a description of the main sites contaminated by HCH wastes is also included.

2. Soils with historical HCH contamination

Soils contaminated by HCH isomers show a wide range of concentration values. In general, these soils come from areas where residues of the lindane manufacture or other chlorinated pesticides have been dumped and uncontrollably accumulated. The contamination is present in the form of particulate matter (white particles of HCH wastes in soil distinguishable with the naked eye) and/or adsorbed into the soil. When a particulate matter of HCH wastes is not reported, the concentration of β-HCH in the soil is usually higher than that of α-HCH [31], indicating that HCH isomers are adsorbed into the soil. Real soils polluted with adsorbed HCH isomers have been reported in the following works:

- Riparian area of the Mulde river (Germany). In this case, two highly contaminated sites were characterized by Keller (Kel) and Spittel (Spi) [32]. The concentration of **β**-HCH with the depth of the soil is shown in **Figure 3**. It can be seen that the HCH concentration decreases with this variable.

- Teltow Canal (Berlin) [33], where the concentrations of HCH isomers in the soils were analyzed. In general, the isomer with the highest concentration in the different sediments studied was **β**-HCH, as shown in **Table 2**. The lack of particulate matter of HCH wastes in these sediments could explain the higher concentration of β-HCH detected in the soil.

- A contaminated area in Bitterfield (Germany) was described by Wycisk et al. [13], including an old landfill used for the discharge of HCHs and other chlorinated pesticides. The concentration of β-HCH in the soil of the old landfill was higher than the concentration α-HCH, as shown in **Table 3**.

- A gravel pit located in the northeast of France [34], contaminated by HCH wastes dumped by the PCUK company. This company stopped the manufacture of lindane in 1974. In this place, the lindane wastes were encapsulated, but

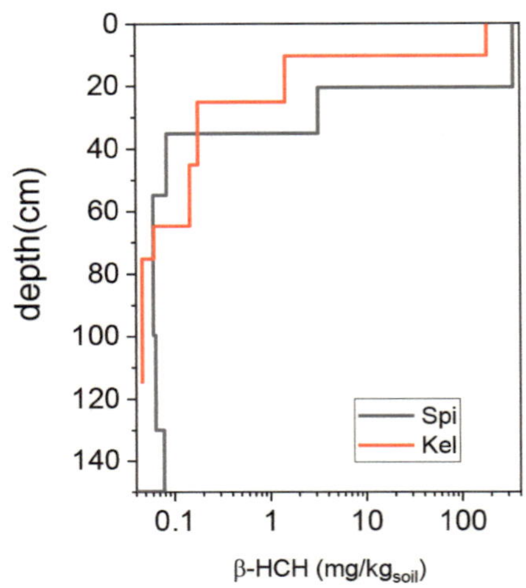

Figure 3.
Soil contamination by β-HCH as a function of depth [32].

Depth, cm	α-HCH	β-HCH	γ-HCH	δ-HCH	ε-HCH	ΣHCH
00–10	17	120	41	64	b.d.l.	242
10–15	48	110	29	44	b.d.l.	231
15–20	61	140	42	68	b.d.l.	311
25–30	79	170	50	97	b.d.l.	396
55–60	120	65	47	130	b.d.l.	362
60–65	28	20	12	40	b.d.l.	100
65–70	18	13	7.6	25	b.d.l.	64
95–100	0.3	0.3	0.2	2.1	b.d.l.	3

b.d.l = below detection limit

Table 2.
Concentration (μg/kg) of HCH isomers in Teltow Canal sediments, Berlin [33].

β-HCH (45 mg/kg) and γ-HCH (25 mg/kg) isomers remained in the soil. The presence of α-HCH was not described in this case.

- Agricultural soils in India [29], where chlorinated pesticides were probably stored in the past. The concentration of HCH isomers in these agricultural soils varied from 4.9 to 7.4 mg/kg soil, with a concentration of α-HCH lower than that one of β-HCH.

- A contaminated industrial site in Beijing, China. In this case, Liang et al. [35] collected and analyzed soil samples from an old organochlorine pesticide plant located in Beijing. The soil, which was sieved (2 mm) and homogenized, was characterized, obtaining the following parameters: pH 7.8, total organic carbon 6.2 g/kg, total nitrogen 0.82 g/kg, and moisture 2.8%. The soil contained α-HCH, β-HCH, p,p′-DDT, or p′-DDT, p,p′-DDE, and p,p′-DDD with concentrations of 2.7, 10.8, 12.9, 3.1, 2.0, and 2.6 mg/kg, respectively.

- Farm soils in Kazakhstan, where obsolete pesticides were stored during decades [36]. As can be seen in **Table 4**, the concentration of β-HCH was higher than the concentration of α-HCH.

On the other hand, other works dealing with soils contaminated by HCH wastes reported the presence of white granules of particulate HCHs noticed with the naked eye. In these studies, the average concentration of α-HCH measured in the soil was higher than that of β-HCH, which agrees with the composition of technical-HCH. These studies are listed below:

- Contaminated soils in Galicia (Spain) affected by the industrial activity of lindane production [37]. The soils were analyzed at different points and depths, and the concentration values of HCH isomers found ranged from 5 to around 80,000 mg/kg, with very different concentrations depending on the depth analyzed. The highest values of HCHs (81,035 mg/kg) probably corresponds to the presence of HCH isomers in the form of particulate matter. Fragments and

Depth, cm	α-HCH	β-HCH	γ-HCH	δ-HCH
00–10	440.0	702.5	23.3	10.3
10–20	535.6	574.3	9.5	b.d.l.
20–30	109.0	60.1	0.4	b.d.l.
30–60	0.5	6.4	0.3	b.d.l.
60–80	0.4	2.4	0.1	b.d.l.
80–100	0.2	7.0	b.d.l.	b.d.l.

b.d.l = below detection limit

Table 3.
HCH content (mg/kg) in vertical floor profiles on Spitelwasser [13].

Hot points	α-HCH	β-HCH	γ-HCH
MAC	0	100	100
Point 1	67.1 ± 9.1	176.0 ± 23.3	22.2 ± 3.2
Point 2	15.3 ± 7.3	83.2 ± 5.5	13.0 ± 4.2
Point 3	b.d.l.	b.d.l.	b.d.l.

b.d.l = below detection limit

Table 4.
Concentration values of HCH isomers (μg/kg) in various soils analyzed [36].

dust of this white substance were present in the studied area. This material was also analyzed to determine the presence of technical-HCH wastes. The technical HCH produced in that fabric presented the following composition: 77% of α-HCH, 16% of β-HCH, 5% of γ-HCH, and 2% of δ-HCH. The composition of some soil samples analyzed in this work is shown in **Figure 4**. It was noticed that samples with the highest proportion of β-HCH correspond to those with the lowest total concentration of HCHs, whereas samples with the highest percentage of α-HCH correspond to soils with higher total HCH concentration (probably as grains of technical-HCH wastes). The presence of HCH in the form of particulate matter could add difficulties in the remediation of these sites.

- Sabiñánigo (Huesca, Spain), with two landfills contaminated with HCH wastes dumped by INQUINOSA, a lindane factory which operated from 1975 to 1988 [11]. High concentrations of HCHs in the soil were measured (**Table 5**), and a higher concentration of α-HCH isomer than that of β-HCH was reported, which is in agreement with the presence of HCHs as particulate matter detected in that soil.

- City of Meninos, Brazil, where contaminated soils were found near to a former lindane factory, which operated between 1950 and 1962 [12]. Although the distribution of HCH isomers in the soil was not reported, the high concentration of these pollutants measured (several thousand mg/kg), indicates the presence of HCH-wastes as particulate matter.

- Santo André', Sao Paulo (Brazil) with HCH-contaminated land [38]. In this study, there is no explicit indication of the presence of HCH particulate matter, but the high proportion of α-HCH, shown in **Figure 5**, seems to confirm this hypothesis.

- Soil contaminated by HCH wastes in the Midwest (USA). Phillips et al. [30] studied three areas (A, B, and C) with a high concentration of HCHs, mainly due to the presence of HCHs granules. The total concentration of HCHs along

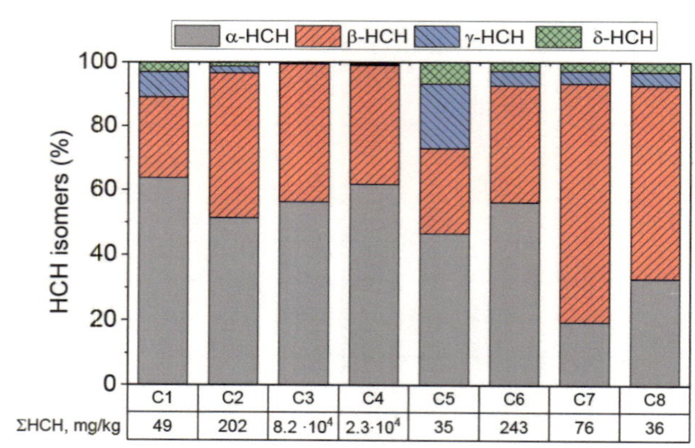

Figure 4.
Percentages of HCH isomers found in soil samples [37].

	α-HCH	β-HCH	γ-HCH	δ-HCH	ε-HCH	ΣHCH
Maximum	57,000	5600	9700	2200	2700	74,730
Mean	2303.2	245.5	406.8	105.7	138.9	3200.2

Table 5.
HCH concentration (mg/kg) in soils located at Sabiñánigo landfills [11].

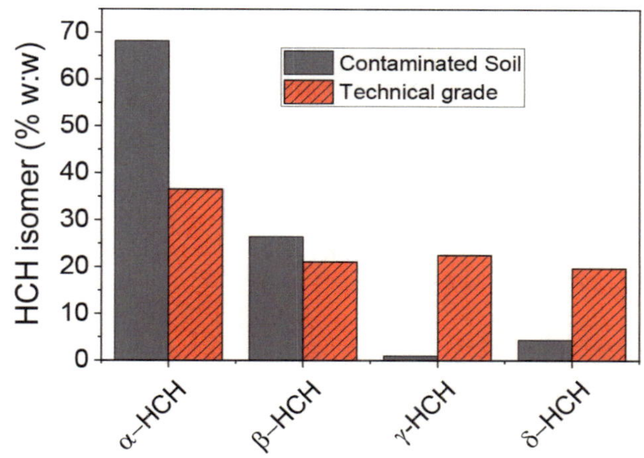

Figure 5.
Distribution of HCH isomers in soils located at Santo André and Sao Paulo (Brazil) [38] and in technical HCH.

a west to east gradient ranged from 22,430 to 1069 mg/kg in the A zone and from 21,100 to 730 mg/kg in the B zone, whereas in zone C, the concentration ranged from 52 to 1427 mg/kg. The composition of HCH wastes was rich in the isomer α (α-HCH 20,000 mg/kg and β-HCH 2000 mg/kg).

3. Physicochemical remediation of HCH-contaminated soils

Physicochemical treatments have been studied in the literature applied to real or spiked contaminated soils. The last ones obtained by contacting the soil with concentrated solutions of HCHs solved in different organic solvents and the subsequent evaporation of the solvents. In the first case (real soils), the contaminants can be absorbed into the soil (a higher concentration of the isomer β-HCH is noticed) or present as HCH granules (with a higher concentration of the isomer α-HCH). The proportion of HCH isomers found also depends on the composition of the dumped HCH wastes or the spiking procedure (in the case of spiked soils).

In general, β-HCH is always the most recalcitrant isomer regardless of the treatment tested (biological or chemical oxidation and biological or chemical reduction). The following sections summarize the works found in the bibliography related to physicochemical remediation treatments.

3.1 Thermal treatments

Thermal treatments have been traditionally applied to the remediation of soils contaminated with persistent organic pollutants, as HCH wastes. However, the use of high temperatures has major drawbacks, such as the low-cost effectivity of the process and the generation of compounds even more toxic than the starting ones when chlorine is in the structure of the organic pollutant, such as dioxins and furans. The main thermal treatments found in the literature for the remediation of HCH-polluted soils are described below:

3.1.1 Thermal desorption

This thermal process was applied to the remediation of real soils located in Sao Paulo (the distribution of HCH isomers suggests the presence of HCHs in the form of particulate matter) [38]. The excavated soils were subjected to high temperatures (up to 450°C), as is indicated in the scheme of the heat treatment plant shown in **Figure 6**. The results obtained for the abatement of the different HCH isomers are shown in **Figure 7**, as a function of the reaction time and the temperature of the treatment. An important degradation of HCHs is achieved in only a few hours of reaction, although temperatures above 250°C are required. It should be noted that β-HCH is also the most recalcitrant isomer.

3.1.2 Deep oxidation by using basic catalysts

Rozdyalovskaya and Chekryshkin [39] studied the destruction of pure lindane at high temperature by using basic catalysts. The deep oxidation of lindane on a catalyst can be represented by the following reaction (Eq. (1)):

$$C_6H_6Cl_6 + 6O_2 + 6Na(K)OH \rightleftarrows 6CO_2 + 6Na(K)Cl + 6H_2O \qquad (1)$$

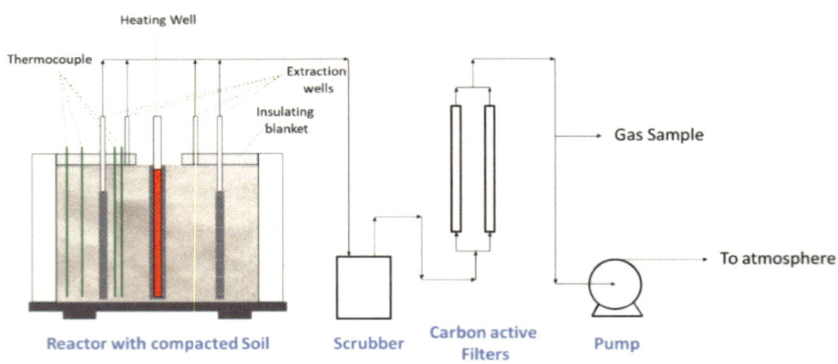

Figure 6.
Heat treatment plant for HCH-contaminated soils [38].

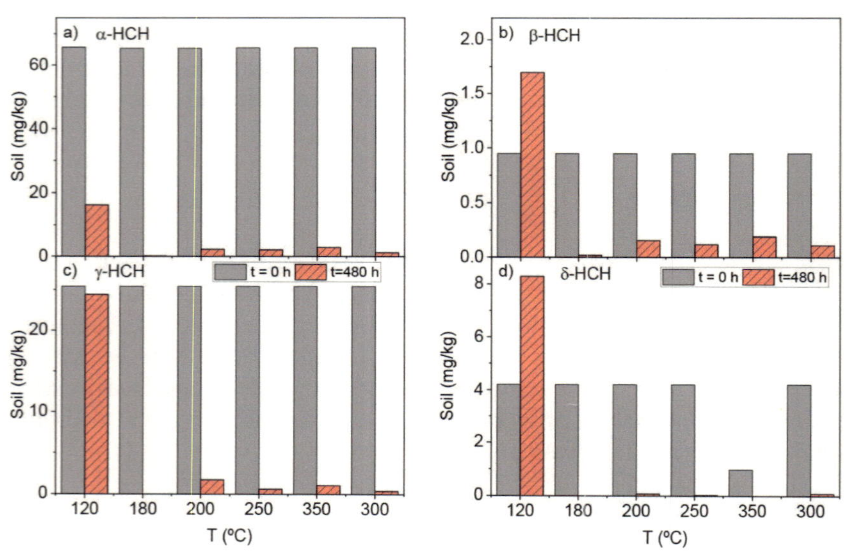

Figure 7.
Removal of HCHs from the soil under different heat treatment conditions [38].

The oxidation was performed in the temperature range of 400–750°C in the presence of some fused catalysts. The catalyst samples were prepared by dehydration, weighing, and mixing of the components and liquid phase synthesis (melting).

The highest activity in the reaction of deep oxidation of lindane was obtained using a molten catalyst based on a eutectic mixture of carbonates of alkali metals with 10 wt % of V_2O_5 and CuO. Moreover, simultaneously with the reaction of deep oxidation of lindane, its dehydrochlorination in a melt of sodium and potassium hydroxides was also noticed. The temperature required for lindane destruction in this process was higher than 450°C. Although there is no mention about dioxins and furans in the work, when working at these temperatures, these compounds are usually generated.

3.1.3 Dehalogenation

This treatment consists of the addition of specific reagents to the soil contaminated with halogenated organic compounds under strong temperature conditions. The process of dehalogenation is achieved by replacing halogen atoms or by the decomposition and partial volatilization of the contaminants [40]. Among dehalogenation processes, base-catalyzed decomposition (BCD) and alkali glycol/polyethylene glycol (APEG) processes [41–43] can be considered. These treatments have been successfully applied to remediate soils and sediments contaminated with chlorinated organic compounds, especially PCBs, dioxins, and furans, but high temperatures are required (150–330°C).

3.2 Chemical oxidation

Different oxidants and activators have been tested in the treatment of soils with real or simulated HCH contamination. However, no studies have been found to date on soils contaminated with HCHs in the form of particulate matter. The presence of this kind of pollution (particulate matter) could pose an additional limitation since the prior solubilization of these granules would be necessary. This phenomenon would be controlled by the interfacial surface between water and the solid phase. The contact between the two phases will increase (i) as the particle size of the HCH granules decreases and (ii) the agitation of the slurry soil-aqueous phase increases.

The main results obtained in the remediation of HCH-contaminated soils with oxidation technologies are described below:

3.2.1 Oxidation with activated persulfate, Fenton reagent, and permanganate

Peng et al. [44] tested the thermal activation of persulfate (PS), at 20 and 40°C, in the treatment of soils artificially contaminated with 800 mg/kg of lindane (γ-HCH was the only HCH isomer studied in this work). The water/soil mass ratio selected was 4:1, and the concentration of PS in the aqueous phase was 0, 5 and 50 g/L. At 20°C, there was no reaction noticed, whereas at 40°C, lindane was eliminated with 50 g/L of PS in 15 days reaction time. When a lower concentration of PS was used (5 g/L) at the same temperature (40°C), the reaction extent was small, as shown in **Figure 8**.

These authors also used the alkaline activation of PS. When this treatment was applied, it was observed that lindane was converted into trichlorobenzenes [44]. Regrettably, there is no information about how these compounds (trichlorobenzenes, TCBs), or other reaction by-products, disappear once formed. The authors did not study the abatement of other HCH isomers than γ-HCH. Recently, Dominguez et al. [45] studied the oxidation of real soils polluted with α-HCH

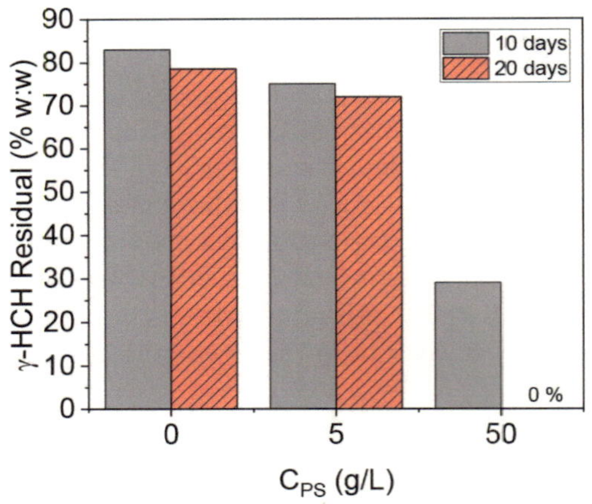

Figure 8.
Removal of lindane from soil after 10 and 20 days with temperature-activated PS at 40°C and using different concentrations of PS [44].

(120 mg/kg) and β-HCH (35 mg/kg) isomers by persulfate activated by alkali finding that the hydrolysis of the β-HCH was the limiting step and that the oxidation rate of TCBs increases notably when the reaction temperature rises from 20 to 40°C.

Usman et al. [34] used artificially contaminated soils (100 mg/kg of each HCH isomer: α, β, γ, δ in sand) and real contaminated soils (concentration of β-HCH = 45 mg/kg and γ-HCH = 25 mg/kg). The high concentration values of β-HCH in the real soil indicate that no particulate matter was present in that soil. The oxidation treatments tested by the authors were persulfate activated by temperature, Fenton reagent (H_2O_2 + Fe), and permanganate. The carbonate content in the real soil was relatively high (195 g CO_3Ca/kg soil), which is relevant for the potential application of H_2O_2 as an oxidant, since it would lead to high unproductive consumption of the oxidant. Moreover, the pH of the soil was slightly alkaline (8.05), which also hinders the application of iron as an activator due to its precipitation at this pH.

A water/soil mass ratio = 20:1 and a large excess of oxidant (17 g/L H_2O_2 and 71 g/L PS) were used, with molar ratios of Fe/H_2O_2 = 1/10 and Fe/PS = 1:2. The higher proportion of Fe used in the activation of PS than in the Fenton process is due to the fact that in the first case, iron is a reagent that is consumed with the progress of the reaction (Eq. (2)), whereas in the case of Fenton reagent, iron is a catalyst, which is continuously regenerated during the radical species production. Fe(II) reacts with hydrogen peroxide to give hydroxyl radicals and Fe(III) (Eq. (3)), which is after regenerated to Fe(II) reacting with another molecule of hydrogen peroxide (Eq. (4)). The reaction of HCHs with the radical species generated by both processes, $OH\cdot$ or $SO_4^{-}\cdot$, yields oxidized by-products or the complete mineralization of the pollutant (oxidation to carbon dioxide, water and salts) (Eq. (6) and Eq. (7)).

$$S_2O_8^{2-} + Fe^{2+} \rightarrow SO_4^{2-} + SO_4^{\cdot-} + Fe^{3+} \quad (2)$$

$$Fe^{2+} + H_2O_2 \rightarrow Fe^{3+} + OH^{\cdot} + OH^{-} \quad (3)$$

$$Fe^{3+} + H_2O_2 \rightarrow Fe^{2+} + OOH^{\cdot} + H^{+} \quad (4)$$

$$C_6H_6Cl_6\left(HCH\right) + 24\,OH\cdot \rightarrow 6\,CO_2 + 12\,H_2O + 6\,HCl \quad (5)$$

$$C_6H_6Cl_6\left(HCH\right) + 12\,H_2O + 24\,SO_4^{\cdot-} \rightarrow 6\,CO_2 + 6\,HCl + 24\left(HSO_4\right)^{-} \quad (6)$$

In the aforementioned work, the authors compare the results obtained with the following treatments: H_2O_2 only, Fenton reagent (H_2O_2 + Fe (II)), PS only, PS activated with Fe(II) and potassium permanganate after 24 hours [34]. When iron is used, it is necessary to carry out the reaction at acid pH to avoid iron precipitation, which results unaffordable in the case of soils with high carbonate content. The results and specific conditions obtained in each treatment for both spiked (a) and real contaminated (b) soils are shown in **Figure 9**. The most recalcitrant HCH isomer was β-HCH regardless of the treatment tested, and the best results were obtained with Fe-activated PS (it should be noted that it was necessary to bring the pH to the acidic zone 2–3). Data about the consumption of the different oxidants are not supplied in the article, but it is expected that H_2O_2 reacted unproductively when this oxidant is applied to the remediation of the real polluted soils (Eq. (7)), being the reason for the cause of the lower HCH conversion obtained with this treatment.

$$H_2O_2 + soil \rightarrow H_2O + \frac{1}{2}O_2 \qquad (7)$$

García-Cervilla et al. have recently studied the remediation of a soil located at 14 m below the ground level in an alluvial of an old landfill contaminated with liquid wastes of lindane production [46] at Sabiñanigo (Spain). A high carbonate concentration was also found in this soil (>45%), and the alkaline activation of PS was selected as a remediation technology. The organic and inorganic composition of the soil sieved at two particle sizes: F (dp < 0.25 mm) and G (0.25–2 mm) is summarized in **Table 6**. This soil presented high HCH concentration in some points (up to 9000 mg/kg) due to the adsorption of DNAPL (dense non aqueous phase liquid) that percolated through the soil and reached the alluvial. The absence of the isomer β-HCH in the DNAPL is the reason of the lack of this HCH isomer in the soil studied.

As previously commented, persulfate activated by alkali was applied for the remediation of this soil. This method follows a free radical mechanism [47–49], summarized in Eqs. (8) and (9).

$$2S_2O_8^{2-} + 2H_2O \xrightarrow{OH^-} 3SO_4^{2-} + SO_4^{\cdot-} + O_2^{\cdot-} + 4H^+ \qquad (8)$$

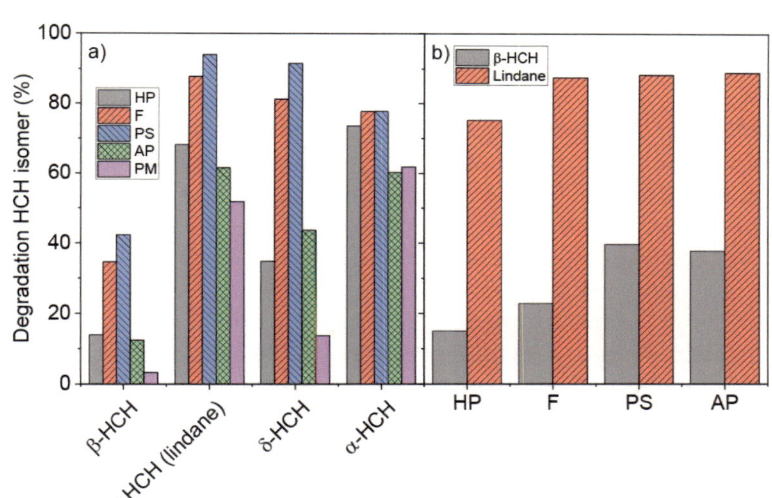

Figure 9.
Degradation of HCH isomers in (a) spiked sand with HCHs and (b) real contaminated soil [34].

	F, dp < 0.25 mm	G 0.25 < dp < 2 mm
TOC, mg/kg	2820	840
TC, mg/kg	54,660	54,840
Carbonates (as $CaCO_3$) (%w)	43.2	45.0
Fe, mg/kg	33,078	31,662
ΣHCH, mg/kg	6597.3	1735.5
ΣHeptachlorocyclohexanes, mg/kg	1997.0	690.7
Total mg/kg	10,109	3346

Table 6.
Inorganic and organic composition of polluted soil (14 mg g l) [46].

$$SO_4^{\cdot-} + OH^- \rightarrow SO_4^{2-} + OH^{\cdot} \quad (9)$$

The addition of an alkali provoked that HCH and heptachlorocyclohexane isomers adsorbed into the soil as a residual phase were converted to trichlorobenzenes and tetrachlorobenzenes, respectively, in less than 48 h. The dehydrochlorination reactions at alkaline conditions, shown in **Figure 10**, were previously described elsewhere [50–52].

At pH above 12, it has been noted that hydroxyl radicals ($OH\cdot$, E^0 = 2.7 V) are predominant against sulfate radicals ($SO_4\cdot^-$, E^0 = 2.6 V) [53]. In addition to hydroxyl radical, superoxide radical is also produced in the alkaline activation of persulfate, as can be seen in Eq. (8). These species are capable of producing a nucleophilic substitution when reacting with halides, as described in **Figures 11** and **12**, where trichlorobenzene is mineralized by the attack of both superoxide [54] and hydroxyl [44] radicals.

The main results obtained in this work are shown in **Figure 13** [46]. The molar ratio NaOH/PS ratio was 2:1, the mass ratio water/soil was 10:1, and the concentration of PS varied between 25 and 100 g/L. As can be seen, more than 1 month was required for the remediation of the soil with a particle size <0.25 mm, due to the high pollutant concentration and the strong adsorption of the pollutants to the soil (higher presence of clays than the other fraction) found in the fraction "F" .

3.2.2 Electrochemical oxidation of HCHs in washing solutions

Muñoz Morales et al. [55] studied the remediation of a soil artificially contaminated with lindane (100 mg/kg). An anionic surfactant, SDS, was used to extract the pollutant from the soil (0.1 g SDS per g soil) using a liquid to solid phase mass ratio of 10. Subsequently, lindane extracted from the soil and solved in the aqueous phase was removed by electrooxidation. Therefore, this was a treatment train consisting, firstly, in the solubilization of the pollutant and secondly, in the selective oxidation of the pollutant, in aqueous emulsion. A diamond electrode was used to in-situ generate hydrogen peroxide by injecting air. Remediation times of 400 min were needed, and the surfactant was recycled for further washing cycles.

3.3 Chemical reduction

Zero-valent iron (ZVI) has attracted the interest of the scientific community over the past decade for its potential to remediate a wide variety of environmental contaminants both in superficial and groundwater [56]. The use of ZVI

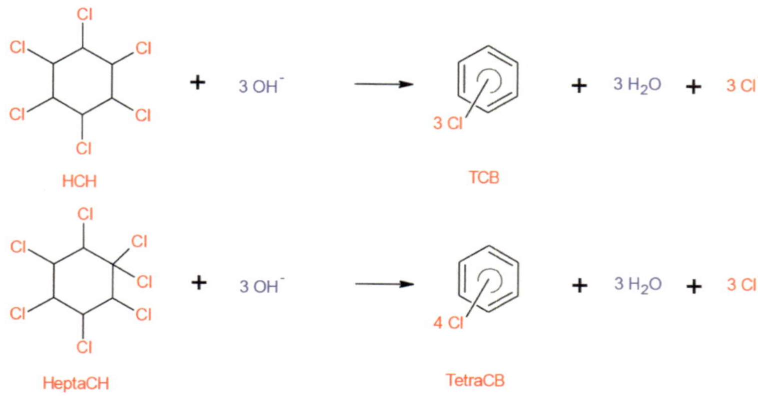

Figure 10.
Dehydrochlorination reactions at alkaline pH [52].

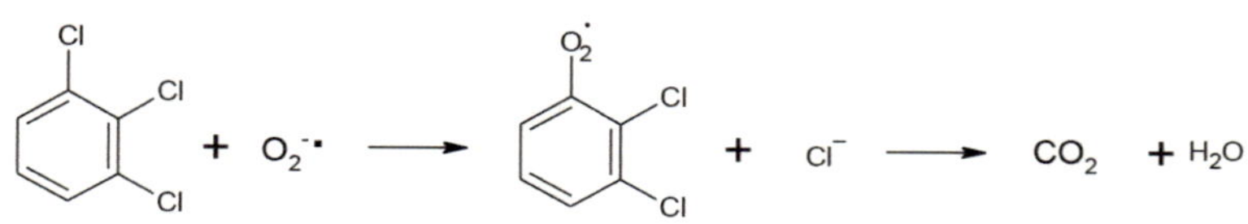

Figure 11.
Nucleophilic substitution of the superoxide radical in the reaction with trichlorobenzene as an example of oxidation reaction. Tetrachlorobenzene isomers follow the same reaction mechanism [54].

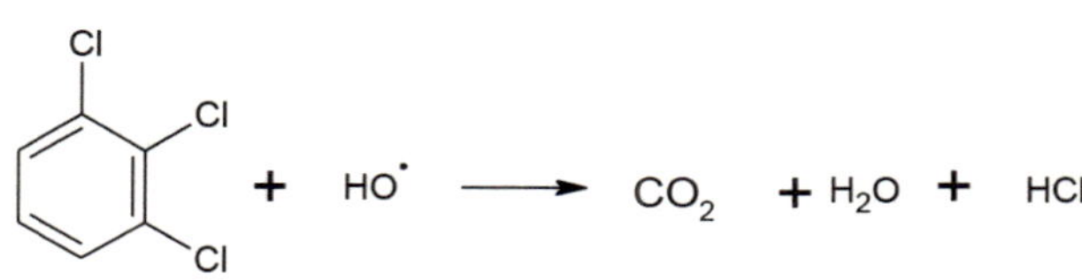

Figure 12.
Attack of hydroxyl radicals on trichlorobenzene as an example of oxidation reaction. Tetrachlorobenzene isomers follow the same reaction mechanism [44].

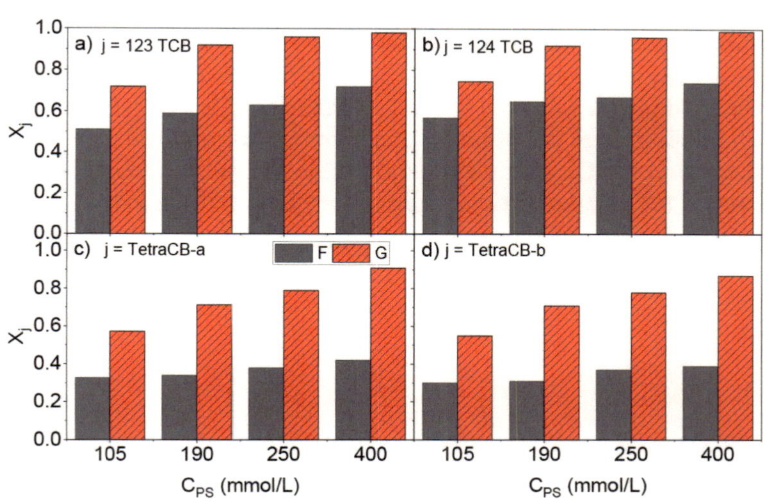

Figure 13.
Conversion of isomers: (a) 1,2,3 TCB, (b) 1,2,4 TCB, (c) TetraCBs-a in soil F, and (d) TetraCBs-b in soil F and G after 509 h. C_{PS} / C_{NaOH} = 1. Soil F: Diameter lower than 0.25 mm, soil G: Diameter between 0.25 and 2 mm [46].

over other metals is a preferred choice due to its high abundance, low cost, and benign environmental impact [25, 56, 57]. Among other pollutants, ZVI showed high efficiency in the treatment of chlorinated organic compounds such as HCH isomers [2, 4, 17, 25, 26, 57–65].

Most of the reported works are focused on the degradation of lindane and the use of ZVI nanoparticles [2–4, 17, 25, 58, 60–63] or the combination of ZVI with other metals, Pd being the most studied [17, 25, 60]. In the presence of ZVI nanoparticles, lindane can be eliminated in 24 hours reaction time when this pollutant is dissolved in water [2] or present in spiked soils [4]. In the case of using bimetallic Pd-Fe nanoparticles [17, 25, 60] or more complex systems, like carbon-supported Cu-ZVI nanoparticles [3] or carboxymethylcellulose Fe/Ni nanoparticles [63], the reaction times for lindane dichlorination can be even decreased. It has been reported that anaerobic conditions favor lindane degradation in the presence of stabilized iron nanoparticles [17], and the temperature has a beneficial effect on the pollutant degradation rate [60], whereas lindane degradation decreases with pH increasing, initial lindane concentration, and in the presence of cations [60]. Several degradation pathways for lindane degradation have been proposed based on the detection of certain reaction intermediates during lindane dichlorination reactions in the presence of ZVI nanoparticles [3, 25, 62, 63].

Although encouraging results in HCH treatment in the presence of these materials have been achieved, the low stability of iron nanoparticles due to aggregation [3, 17, 25, 62] and the unaffordable cost of noble metals like Pd [57] has encouraged the use of ZVI in the form of microparticles during the last years, with lower cost and higher stability [26, 64, 65].

The predominant mechanism for the degradation of lindane using ZVI is the reductive dehalogenation of the pollutant, owing to the electron exchange between the HCH molecule and zero-valent iron [3, 58, 60, 62, 65]. Benzene is obtained as the final product of lindane reduction (along with chlorides), as is shown in **Figure 14**.

As occurred with chemical oxidation, **β**-HCH presents high recalcitrance towards chemical reduction, in both aqueous and soil phases [2, 65, 66] due to the chlorine's position and the low water solubility of this HCH isomer.

Even though promising results have been obtained with ZVI in the degradation of HCHs in the aqueous phase, the use of this material for soil remediation entails additional problems. The application of ZVI in the form of microparticles on contaminated soils would yield low HCH conversion due to the hindered contact between the solid phases (soil and ZVI microparticles). Using ZVI nanoparticles for soil remediation is limited by problems of agglomeration and the high cost associated. Furthermore, if HCH granules are present in the soil, a remarkable decrease in the efficiency of the dechlorination treatment is expected, due to the expected poor contact between the two solid phases.

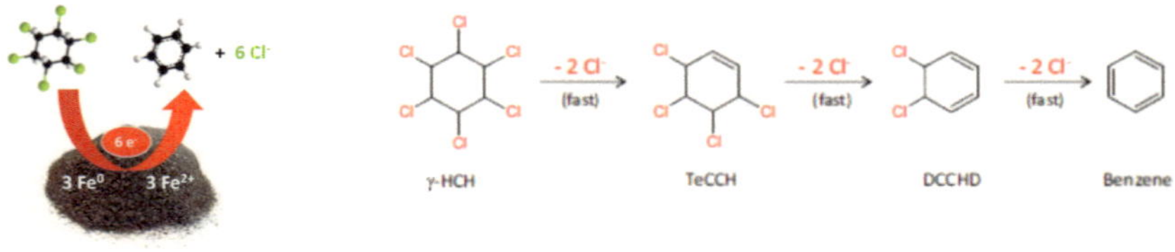

Figure 14.
Dechlorination pathway of lindane over zero-valent iron microparticles [65].

3.4 Soil washing

In this technology, surfactants are used to solubilize the contaminants absorbed into the soil in the aqueous phase. The resulting solution requires a second stage in which the objective is the selective oxidation of the contaminant from the emulsion and the surfactant recover for a next use [67, 68]. However, the solubilization of pollutants from solid phases is hindered by the pollutant transport from the soil to the aqueous phase.

There are few papers in the literature using this technology for the treatment of HCH-contaminated soils, none dealing with the presence of particulate matter, and only spiked soils with HCHs or DNAPL were used.

Muñoz-Morales et al. [55] used soil washing as a first stage to remediate a soil spiked with lindane. For that purpose, the authors used an anionic surfactant, SDS. This surfactant was selected because the next step of the remediation treatment was the electrochemical oxidation of the pollutant in the emulsion, and high conductivity of the solution is required in this oxidation treatment. Using a surfactant concentration of 10 g/L in the aqueous phase, the concentration of lindane found in the aqueous emulsion was 10 mg/L, which was further oxidized by the electrochemical treatment.

Dominguez et al. combined soil flushing (with a nonionic surfactant) and Fenton oxidation [67]. A nonionic commercial surfactant (E-Mulse 3®) was used to extract most of the residual DNAPL in the soil at column conditions. The resulting surfactant flushing solution showed a high concentration of chlorinated organic compounds (COCs = 3693 mg/L, 40% of this amount corresponded to HCH isomers, although β-HCH was not in the mixture). This emulsion was treated by the Fenton process using different concentrations of hydrogen peroxide (200%, 100%, and 50% of the theoretical stoichiometric amount for the complete mineralization of the COCs) and a molar ratio of H_2O_2:Fe = 32. A degradation of COCs >80% was obtained using a concentration of $H_2O_2 \geq 100\%$ of the stoichiometric amount. HCHs (and other nonaromatic COCs) were less prone to oxidation by hydroxyl radicals than chlorobenzenes. The surfactant was recovered at the end of the treatment for further flushing steps.

Regrettably, there are no studies in the literature dealing with soil washing of soils polluted with β-HCH, the least soluble and the most stable HCH isomer against oxidation and reduction. Moreover, if the contamination of the soil by HCHs involves also the presence of particulate matter, transport resistances will be more limiting, and the step of soil washing will slow down. In this case, a good agitation or ultrasound application will be required to improve the contact between the phases and, therefore, to improve the efficiency of the process.

4. Conclusions

Soil contamination by the solid residues generated from the manufacture of lindane, a chlorinated organic pesticide whose use and production has been prohibited, is a great environmental problem, ubiquitous and persistent, given the high toxicity and low biodegradability of these residues in the environment. These soils contain a mixture of HCH isomers, mainly α and β, isomer β being the most recalcitrant to both chemical and biological treatments, due to its lower water solubility and higher chemical stability. This kind of contamination appears as solid HCH particles mixed with soil (usually with a higher concentration of the isomer α-HCH) or adsorbed onto the soil (with a higher concentration of the isomerβ HCH) reaching values up to several hundreds of mg HCH/kg soil. It represents a serious problem due to the large volume of wastes to be treated. Among the physicochemical treatments used, thermal processes are the traditional ones but the less sustainable because the requirement of high temperatures and, therefore, the associated costs are prohibitive for treating large amounts of wastes. In the last decade, chemical treatments have shown promising results. Amon them, oxidation with Fenton reagent or activated persulfate seems to be more suitable than reduction using zero-valent iron particles, because of the greater limitations for the contact between phases in the last one treatment. The selection of the most suitable oxidation method will depend on the type of soil (presence of carbonates and pH). On the other side, the time and method of contact will also be strongly influenced by how the contamination is present (in the form of particulate or adsorbed matter).

Acknowledgements

The authors acknowledge the financial support from SARGA, a public company of the Government of Aragon (project ref. 5507001-182 funded), from the Regional Government of Madrid, through the CARESOIL project (S2018/EMT-4317), and by the Spanish Ministry of Science (project CTM2016-77151-C2-1-R). The authors thank SARGA and the Department of Climate Change and Environmental Education, Government of Aragon, for their support during this work.

Author details

Aurora Santos López*, Carmen M. Dominguez Torre and David Lorenzo Fernández
Chemical and Materials Engineering Department, Complutense University of Madrid, Madrid, Spain

*Address all correspondence to: aursan@ucm.es

References

[1] Vega M, Romano D, Utila E. Lindane (persistent organic pollutant) in the EU. Brussels: European Parliament. [Visited: 07/06/2020]. Study for the PETI Committee (PE 571.398). European Parliament. B-1047 Brussels 2016. Available from: http://www.europarl.europa.eu/RegData/etudes/STUD/2016/571398/IPOL_STU(2016)571398_EN.pdf

[2] Elliott DW, Lien H-L, Zhang W-X. Zerovalent iron nanoparticles for treatment of ground water contaminated by hexachlorocyclohexanes. Journal of Environmental Quality. 2008;**37**:2192-2201

[3] Chang C, Lian F, Zhu L. Simultaneous adsorption and degradation of gamma-HCH by nZVI/Cu bimetallic nanoparticles with activated carbon support. Environmental Pollution. 2011;**159**:2507-2514

[4] Singh R, Singh A, Misra V, Singh RP. Degradation of lindane contaminated soil using zero-valent iron nanoparticles. Journal of Biomedical Nanotechnology. 2011;**7**:175-176

[5] Götz R, Sokollek V, Weber R. The dioxin/POPs legacy of pesticide production in Hamburg: Part 2—Waste deposits and remediation of Georgswerder landfill. Environmental Science and Pollution Research. 2013;**20**:1925-1936

[6] Li Y-F, Scholtz MT, Van Heyst BJ. Global gridded emission inventories of β-hexachlorocyclohexane. Environmental Science & Technology. 2003;**37**:3493-3498

[7] Vijgen J. The legacy of lindane HCH isomer production. Main Report IHPA; January. 2006

[8] Weber R, Gaus C, Tysklind M, Johnston P, Forter M, Hollert H, et al. Dioxin-and POP-contaminated sites—Contemporary and future relevance and challenges. Environmental Science and Pollution Research. 2008;**15**:363

[9] Weber R, Varbelow HG. Dioxin/POPs legacy of pesticide production in Hamburg: Part 1—Securing of the production area. Environmental Science and Pollution Research. 2013;**20**:1918-1924

[10] Vijgen J, Abhilash P, Li YF, Lal R, Forter M, Torres J, et al. Hexachlorocyclohexane (HCH) as new Stockholm convention POPs—A global perspective on the management of lindane and its waste isomers. Environmental Science and Pollution Research. 2011;**18**:152-162

[11] Fernandez J, Arjol MA, Cacho C. POP-contaminated sites from HCH production in Sabinanigo, Spain. Environmental Science and Pollution Research. 2013;**20**:1937-1950

[12] Torres JPM, Froes-Asmus CIR, Weber R, Vijgen JMH. HCH contamination from former pesticide production in Brazil—A challenge for the Stockholm convention implementation. Environmental Science and Pollution Research. 2013;**20**:1951-1957

[13] Wycisk P, Stollberg R, Neumann C, Gossel W, Weiss H, Weber R. Integrated methodology for assessing the HCH groundwater pollution at the multi-source contaminated mega-site Bitterfeld/Wolfen. Environmental Science and Pollution Research. 2013;**20**:1907-1917

[14] Vijgen J, de Borst B, Weber R, Stobiecki T, Forter M. HCH and lindane contaminated sites: European and global need for a permanent solution for a long-time neglected issue. Environmental Pollution. 2019;**248**:696-705

[15] Walker K, Vallero DA, Lewis RG. Factors influencing the distribution of lindane and other hexachlorocyclohexanes in the environment. Environmental Science & Technology. 1999;**33**:4373-4378

[16] Wacławek S, Antoš V, Hrabák P, Černík M, Elliott D. Remediation of hexachlorocyclohexanes by electrochemically activated persulfates. Environmental Science and Pollution Research. 2016;**23**:765-773

[17] Joo SH, Zhao D. Destruction of lindane and atrazine using stabilized iron nanoparticles under aerobic and anaerobic conditions: Effects of catalyst and stabilizer. Chemosphere. 2008;**70**:418-425

[18] WHO. Environment Health Criteria 124: Lindane. International Programme on Chemical Safety. WHO EHC; 1991

[19] Health UDo, Services H. Hazardous substances data bank (HSDB, online database). In: National Toxicology Information Program. Bethesda, MD: National Library of Medicine; 1993

[20] Xiao H, Li NQ, Wania F. Compilation, evaluation, and selection of physical-chemical property data for alpha-, beta-, and gamma-hexachlorocyclohexane. Journal of Chemical & Engineering Data. 2004;**49**:173-185

[21] Shiu WY, Ma KC, Mackay D, Seiber JN, Wauchope RD. Solubilities of pesticide chemicals in water. 2. Data compilation. Reviews of Environmental Contamination and Toxicology. 1990;**116**:15-187

[22] Ke R, Wang Z, Huang S, Khan SU. Accurate quantification of freely dissolved organochlorine pesticides in water in the presence of dissolved organic matter using triolein-embedded cellulose acetate membrane. Analytical and Bioanalytical Chemistry. 2007;**387**:2871-2879

[23] Weil L, Dure G, Quentin K. Wasserlöslichkeit von insektiziden chlorierten Kohlenwasserstoffen und polychlorierten Biphenylen im Hinblick auf eine Gewässerbelastung mit diesen Stoffen. Wasser Abwasser Forsch. 1974;**6**:169-175

[24] Zinovyev SS, Shinkova NA, Perosa A, Tundo P. Dechlorination of lindane in the multiphase catalytic reduction system with Pd/C, Pt/C and Raney-Ni. Applied Catalysis B: Environmental. 2004;**47**:27-36

[25] Nagpal V, Bokare AD, Chikate RC, Rode CV, Paknikar KM. Reductive dechlorinationofγ-hexachlorocyclohexane using Fe–Pd bimetallic nanoparticles. Journal of Hazardous Materials. 2010;**175**:680-687

[26] Dominguez CM, Romero A, Fernandez J, Santos A. In situ chemical reduction of chlorinated organic compounds from lindane production wastes by zero valent iron microparticles. Journal of water process engineering. 2018;**26**:146-155

[27] Dominguez CM, Romero A, Lorenzo D, Santos A. Thermally activated persulfate for the chemical oxidation of chlorinated organic compounds in groundwater. Journal of Environmental Management. 2020;**261**:110240

[28] Cruz-González G, Julcour C, Chaumat H, Bourdon V, Ramon-Portugal F, Gaspard S, et al. Degradation of chlordecone and beta-hexachlorocyclohexane by photolysis, (photo-)Fenton oxidation and ozonation. Journal of Environmental Science and Health, Part B. 2018;**53**:121-125

[29] Garg N, Lata P, Jit S, Sangwan N, Singh AK, Dwivedi V, et al. Laboratory and field scale bioremediation of

hexachlorocyclohexane (HCH) contaminated soils by means of bioaugmentation and biostimulation. Biodegradation. 2016;**27**:179-193

[30] Phillips TM, Lee H, Trevors JT, Seech AG. Full-scale in situ bioremediation of hexachlorocyclohexane-contaminated soil. Journal of Chemical Technology & Biotechnology: International Research in Process, Environmental & Clean Technology. 2006;**81**:289-298

[31] Srivastava V, Srivastava T, Kumar MS. Fate of the persistent organic pollutant (POP) Hexachlorocyclohexane (HCH) and remediation challenges. International Biodeterioration and Biodegradation. 2019;**140**:43-56

[32] Kalbitz K, Popp P, Geyer W, Hanschmann G. Beta-HCH mobilization in polluted wetland soils as influenced by dissolved organic matter. The Science of the Total Environment. 1997;**204**:37-48

[33] Ricking M, Schwarzbauer J. HCH residues in point-source contaminated samples of the Teltow Canal in Berlin, Germany. Environmental Chemistry Letters. 2008;**6**:83-89

[34] Usman M, Tascone O, Faure P, Hanna K. Chemical oxidation of hexachlorocyclohexanes (HCHs) in contaminated soils. The Science of the Total Environment. 2014;**476-477**:434-439

[35] Liang Q, Lei M, Chen T, Yang J, Wan X, Yang S. Application of sewage sludge and intermittent aeration strategy to the bioremediation of DDT- and HCH-contaminated soil. Journal of Environmental Sciences (China). 2014;**26**:1673-1680

[36] Nurzhanova A, Kalugin S, Zhambakin K. Obsolete pesticides and application of colonizing plant species for remediation of contaminated soil in Kazakhstan. Environmental Science and Pollution Research. 2013;**20**:2054-2063

[37] Concha-Grana E, Turnes-Carou M, Muniategui-Lorenzo S, Lopez-Mahia P, Prada-Rodriguez D, Fernández-Fernández E. Evaluation of HCH isomers and metabolites in soils, leachates, river water and sediments of a highly contaminated area. Chemosphere. 2006;**64**:588-595

[38] Araujo MM, Ignatius SG, Oliveira AO, Oliveira SS, Fertonani FL, Paste IA. Thermal desorption of HCH. Journal of Thermal Analysis and Calorimetry. 2016;**123**:1019-1029

[39] Rozdyalovskaya TA, Chekryshkin YS. Deep hexachlorocyclohexane oxidation on molten high-basic catalysts. Russian Journal of Applied Chemistry. 2013;**86**:648-653

[40] Felsot AS, Racke KD, Hamilton DJ. Disposal and degradation of pesticide waste. In: Ware GW, Albert LA, Bro-Rasmussen F, Crosby DG, de Voogt P, Frehse H, et al., editors. Reviews of Environmental Contamination and Toxicology: Continuation of Residue Reviews. New York, NY: Springer New York; 2003. pp. 123-200

[41] Rahuman M, Pistone L, Trifirò F, Miertus S, editors. Destruction technologies for polychlorinated biphenyls (PCBs). In: Proceedings of Expert Group Meetings on POPs and Pesticides Contamination. 2000

[42] Tiernan TO, Wagel DJ, Garrett JH, Vanness GF, Solch JG, Harden LA, et al. Laboratory and field-tests to demonstrate the efficacy of KPEG reagent for detoxification of hazardous wastes containing polychlorinated dibenzo-para-dioxins (PCDD) and dibenzofurans (PCDF) and soils contaminated with such chemical wastes. Chemosphere. 1989;**18**:835-841

[43] Tiernan TO, Wagel DJ, Vanness GF, Garrett JH, Solch JG, Rogers C. Dechlorination of PCDD and PCDF sorbed on activated carbon using the KPEG reagent. Chemosphere. 1989;**19**:573-578

[44] Peng L, Deng D, Guan M, Fang X, Zhu Q. Remediation HCHs POPs-contaminated soil by activated persulfate technologies: Feasibility, impact of activation methods and mechanistic implications. Separation and Purification Technology. 2015;**150**:215-222

[45] Domínguez AR CM, Checa-Fernandez A, Santos A. Remediation of HCHs-contaminated sediments by chemical oxidation treatments. Science of the Total Environment. 2020

[46] García-Cervilla R, Santos A, Romero A, Lorenzo D. Remediation of soil contaminated by lindane wastes using alkaline activated persulfate: Kinetic model. Chemical Engineering Journal. 2020;**124646**:1-16

[47] Devi P, Das U, Dalai AK. In-situ chemical oxidation: Principle and applications of peroxide and persulfate treatments in wastewater systems. Science of the Total Environment. 2016;**571**:643-657

[48] Waclawek S, Lutze HV, Grubel K, Padil VVT, Cernik M, Dionysiou DD. Chemistry of persulfates in water and wastewater treatment: A review. Chemical Engineering Journal. 2017;**330**:44-62

[49] Kolthoff IM, Miller IK. The chemistry of persulfate. I. the kinetics and mechanism of the decomposition of the persulfate ion in aqueous medium. Journal of the American Chemical Society. 1951;**73**:3055-3059

[50] Santos A, Fernandez J, Guadaño J, Lorenzo D, Romero A. Chlorinated organic compounds in liquid wastes (DNAPL) from lindane production dumped in landfills in Sabiñanigo (Spain). Environmental Pollution. 2018;**242**:1616-1624

[51] Santos A, Fernandez J, Rodriguez S, Dominguez CM, Lominchar MA, Lorenzo D, et al. Abatement of chlorinated compounds in groundwater contaminated by HCH wastes using ISCO with alkali activated persulfate. The Science of the Total Environment. 2018;**615**:1070-1077

[52] Lorenzo D, García-Cervilla R, Romero A, Santos A. Partitioning of chlorinated organic compounds from dense non-aqueous phase liquids and contaminated soils from lindane production wastes to the aqueous phase. Chemosphere. 2020;**239**:124798

[53] Furman OS, Teel AL, Watts RJ. Mechanism of base activation of persulfate. Environmental Science & Technology. 2010;**44**:6423-6428

[54] Hayyan M, Hashim MA, AlNashef IM. Superoxide ion: Generation and chemical implications. Chemical Reviews. 2016;**116**:3029-3085

[55] Muñoz-Morales M, Braojos M, Sáez C, Cañizares P, Rodrigo MA. Remediation of soils polluted with lindane using surfactant-aided soil washing and electrochemical oxidation. Journal of Hazardous Materials. 2017;**339**:232-238

[56] Fu F, Dionysiou DD, Liu H. The use of zero-valent iron for groundwater remediation and wastewater treatment: A review. Journal of Hazardous Materials. 2014;**267**:194-205

[57] Wang Z, Peng PA, Huang W. Dechlorination of γ-hexachlorocyclohexane by zero-valent metallic iron. Journal of Hazardous Materials. 2009;**166**:992-997

[58] Elliott DW, Lien H-L, Zhang W-X. Degradation of lindane by zero-valent iron nanoparticles. Journal of Environmental Engineering. 2009;**135**:317-324

[59] Yang S-C, Lei M, Chen T-B, Li X-Y, Liang Q, Ma C. Application of zerovalent iron (Fe_0) to enhance degradation of HCHs and DDX in soil from a former organochlorine pesticides manufacturing plant. Chemosphere. 2010;**79**:727-732

[60] Singh R, Misra V, Mudiam MKR, Chauhan LKS, Singh RP. Degradation of gamma-HCH spiked soil using stabilized Pd/Fe-0 bimetallic nanoparticles: Pathways, kinetics and effect of reaction conditions. Journal of Hazardous Materials. 2012;**237**:355-364

[61] Lacinova L, Cernik M, Sourkova H, editors. Degradation of chlorinated cyclohexanes (lindane) by zero-valent iron nanoparticles (NZVI). In: NANOCON 2013, 5th International Conference Brno, CZ, EU. 2013

[62] San Román I, Alonso M, Bartolomé L, Galdames A, Goiti E, Ocejo M, et al. Relevance study of bare and coated zero valent iron nanoparticles for lindane degradation from its by-product monitorization. Chemosphere. 2013;**93**:1324-1332

[63] Yang J, Sun H. Degradation of γ-hexachlorocyclohexane using carboxymethylcellulose-stabilized Fe/Ni nanoparticles. Water, Air, and Soil Pollution. 2015;**226**:280

[64] Dominguez CM, Parchão J, Rodriguez S, Lorenzo D, Romero A, Santos A. Kinetics of lindane dechlorination by zerovalent iron microparticles: Effect of different salts and stability study. Industrial and Engineering Chemistry Research. 2016;**55**:12776-12785

[65] Dominguez CM, Rodriguez S, Lorenzo D, Romero A, Santos A. Degradation of hexachlorocyclohexanes (HCHs) by stable zero valent iron (ZVI) microparticles. Water, Air, and Soil Pollution. 2016;**227**:446

[66] Cong X, Xue N, Wang S, Li K, Li F. Reductive dechlorination of organochlorine pesticides in soils from an abandoned manufacturing facility by zero-valent iron. The Science of the Total Environment. 2010;**408**:3418-3423

[67] Dominguez CM, Romero A, Santos A. Selective removal of chlorinated organic compounds from lindane wastes by combination of nonionic surfactant soil flushing and Fenton oxidation. Chemical Engineering Journal. 2019;**376**:120009

[68] Trellu C, Mousset E, Pechaud Y, Huguenot D, van Hullebusch ED, Esposito G, et al. Removal of hydrophobic organic pollutants from soil washing/flushing solutions: A critical review. Journal of Hazardous Materials. 2016;**306**:149-174

Chapter 10

Hazardous Components of Landfill Leachates and Its Bioremediation

Menaka Salam and Namdol Nilza

Abstract

Landfill leachates contain both dissolved and suspended material and may pose a threat to the environment because of the toxic substances that it carries and contaminates surface water and groundwater. They are composed of several different categories of components out of which many of them are recalcitrant and highly toxic. Major components of landfill leachates are dissolved organic compounds, inorganic macro compounds, heavy metals such as copper, lead, cadmium, chromium, nickel etc. and xenobiotic compounds such as polychlorinated biphenyls. Complex organic compounds which are released from industrial effluents like perfluorooctanoic acid and benzothiazole are also common in many of the landfill leachates. Biological treatment is a low cost effective method for the treatment of landfill leachates which can act as an accelerator for further treatment by either chemical or physical method. Improved strategies have been developed in the biological treatment of leachates which shows the efficiency of the system. But, as leachate characteristics vary depending on the rainfall and other environmental factors, it is important to first thoroughly analyze the physical and chemical properties of the landfill under study. A combined effort involving proper analysis of the leachate components, monitoring leachate flow, risk assessment, and treatment of the leachate before its release is required to efficiently control its impact to the environment.

Keywords: landfill leachates, xenobiotic, bioremediation, microorganisms, bioreactor

1. Introduction

Landfill leachate is a liquid composed of absorbed components which may be soluble solids or any other undesirable components present in the landfill. It is formed when water passes through the landfill waste in the form of rain and seeps through the stockpile consisting of waste materials from different sources such as municipal and industrial wastes. Since the waste material can comprise of various chemicals, organic and inorganic compounds in large volume which also gets decomposed, the leachate that is formed is generally high in toxicity.

Due to generation of huge amount of solid wastes, many of which contain toxic and recalcitrant substances, its control and management has become the utmost need in recent years. Segregation of waste materials and recycling are two major ways of reducing the pollution in a solid waste dumpsite or a landfill. Among the hazardous components of landfill leachates, microplastics in the size range of 100 to 1000 μm is also found in many of the municipal solid waste landfill leachates. This has been developed after a long time period of time, and the major types are polyethylene and polypropylene which are increasingly used throughout the years because of its convenience to use. Other hazardous components which have been of major concern since long time are toxic heavy metals like lead, arsenic, cadmium, mercury etc. and xenobiotic compounds like halogenated organic compounds. Proper and timely monitoring of the landfill and its leachate is required to control the pollution that can occur to the groundwater, soil and air. Gases that get released due to decomposition and microbial activity have caused severe accidental bursts in many landfill areas due to improper management and lack of monitoring. Therefore, risk assessment is an important criterion for the management of landfill areas. A number of treatment methods comprising of physical, chemical and biological methods have been developed, but none of these methods alone can be declared as having the highest efficiency as the landfill leachates also vary in their composition, volume and migration. However, bioremediation or treatment using biological processes is of major interest because of its low-cost, high efficiency and environment friendly. Physical and chemical methods, although very efficient, have high cost and some of the chemical methods pose environmental concerns in the long run. Regarding biological processes, a number of microorganisms are yet to be explored which can be used as a bioremediating agent. Many bacteria and fungi are already proved to have high removal capacity of leachate parameters like BOD, COD, nitrate and sulphate. In recent years, microorganisms able to remediate heavy metals, xenobiotic compounds and even plastics have been discovered. This shows that there is huge potential of microorganisms to help in the treatment of landfill leachates.

2. Components of landfill leachates

Leachates contain both organic and inorganic pollutants, out of which some constituents are highly toxic and pose a threat to the environment. The actual composition of landfill leachate varies according to the type of waste which has been dumped and accumulated. In developing countries waste segregation is not practiced, waste management and treatment system are not well established and landfill systems are also not properly designed to prevent leachate flow. Because of lack of waste segregation, landfill dumpsites may have mixed waste materials. For example, in a municipal solid waste (MSW), food waste, household discards, plastics, paints, mercury containing waste, batteries and other products made up of toxic compounds and heavy metals may be present. However, in general, the dissolved organic matter comprises of acids, alcohols, aldehydes and sugars, and the inorganic components such as calcium, magnesium, sulphate, chloride and ammonia. These constituents are present in much higher concentrations than normal aquifers. Inorganic pollutants found in leachates are ammonium, phosphorous, sulphate and heavy metals. Volatile fatty acids are majorly present in young leachates and tend to decrease gradually in aged landfills. Humic acids and fulvic acids are also commonly present in old landfills and leachate plumes. Among the heavy metals, the common ones are Fe, Pb, Ni, Cd, As, Cr, Cu and Hg. Other toxic pollutants which are present in landfill leachates are the aromatic hydrocarbons (Benzene, Toluene, Ethylbenzene, and Xylene), phenols, pesticides, polyethylene, plasticizers, and halogenated organic compounds like PCBs and dioxins. Landfills are also a shelter for pathogenic microorganisms, mostly coliform bacteria and a few viruses. The pH and temperature changes may, however, inactivate these microorganisms.

DOI: http://dx.doi.org/10.5772/intechopen.94890

Landfill site	TDS (mg/l)	Electrical Conductivity (mS/cm)	pH	Chloride (mg/l)	Nitrate (mg/l)	Sulphates (mg/l)	BOD (mg/l)	COD (mg/l)	Heavy metals	References
Ghazipur Landfill site, Delhi, India	34,560	23,221	8.4	1598	449	ND	9250	20,992	Fe, Zn, Cu, Pb, Ni, Cr, Cd	[1]
Mavallipura landfill site, Bangaluru, India	2027	4120	7.4	660	22.36	ND	1500	10,400	Fe, Ni, Pb, Cu, Cd, Cr	[2]
Borg El-Arab landfill site, Egypt	27,452	40,921	7.8	11,387	1.4	596	11,700	15,629	Fe, Zn, Ni, Cd, Cr	[3]
Sukawinatan landfill, Indonesia	4640	ND	7.45	162.5	93.6	199	145.7	900	Fe, Cu	[4]
Alhend'ın landfill site, Spain	34,470	ND	7.96	7298	ND	166.7	5750	30,994	Fe, Cd, Cu, Ni, Pb, Zn	[5]
Landfill at Bizerte, North Eastern Tunisia	1770	19.6	7.4	3300	1628	ND	5200	26,200	Fe, Cu	[6]

(ND: Not defined)

Table 1.
Comparision of some important parameters of various landfill sites.

A variety of microorganisms degrades the various components present in the landfill waste and produces their metabolic products and other decaying organic matters. Through decomposition of organic constituents by microorganisms dwelling in the landfill site and also through chemical reactions between the components of waste, landfill gases are released in the form of methane and carbon dioxide which are greenhouse gases. The gases release increases gradually over time and poses environmental threat and security issues to the people residing near the landfill sites. **Table 1** shows a comparision of some important pollution parameters of landfill sites at various places. It is observed that the pHs of these landfill sites are mostly in the alkaline range, there is increased level of electrical conductivity, COD, chloride and nitrate concentrations, and the common heavy metals which are present are Fe, Ni, Cu and Cr.

3. Effect of leachates on the soil environment

Landfill leachates cause serious environmental issues mostly in developing countries polluting the groundwater, soil and air. Even in an engineered landfill site with landfill liners, the barriers tend to get damaged or deteriorate with time, therefore, leachate may get leaked and pass through the soil. The consequence is harmful effects to human health and also causes hindrance to economic health and development. Heavy metals such as Pb, As, Cd, Cr and Hg leach out from uncontrolled landfill sites and cause a major threat to human health. Due to rapid urbanization, areas near the landfill sites in many cases are gradually transformed into residential areas and also covered by agricultural fields. Inorganic cations and anions like sodium, calcium, chloride, sulphate etc. seep through the leachate and contaminate groundwater and soil. These inorganic substances are not altered by the soil type and remain a pollutant of the water and soil [7]. This ultimately gives rise to changes in soil composition and fertility. Studies have shown that iron and zinc are the major pollutants of the soil samples in the vicinity of landfill sites. Zinc is present mostly in the upper soil layer and iron has highest dispersion rate [2]. Leachate percolation in soil reduces the hydraulic conductivity resulting in clogging of the soil and these changes the properties of the soil such as water retention, field capacity etc. The soil microbial community gets changed and formation of biofilms with metal precipitation may arise.

4. Role of microorganisms in remediation of landfill leachates

Remediation of landfill leachate using microorganisms is a cost effective approach as compared to conventional treatment processes. However, a lot need to be explored and studied regarding this area. The conventional methods of treatment are chemical and physical methods such as coagulation/flocculation, chemical oxidation, air stripping and membrane filtration [8]. These methods can remove COD and other toxicity levels ranging from 40–90%. However, in many cases, cost of the whole process and also production of non-degradable sludge remains a disadvantage. Use of microorganisms does not have these disadvantages and are also effective in the treatment process as there are a number of microorganisms which produce various extracellular enzymes through which they can degrade toxic compounds to less toxic or non-toxic products. Landfill leachate samples have been studied to investigate the microorganisms dwelling in them and which have the capability of degrading the main pollutants of leachate such as nitrate, phosphate and ammonia. It has been found that certain fungi, actinomycetes, and bacteria

Organism	Source of isolation	Substance degraded	Percent removal	Heavy metal resistance	Reference
Lysinibacillus sp., *Bruvundimonas* sp., *Brevibacterium* sp., *Thermococcus* sp.	Powerstown Landfill, Co. Carlow, Ireland	Nitrate, Phosphate, ammonia	35%, 55%, 88%	As, Cd, Fe, Ni, Cu	[9]
Not characterized, wastewater effluent as microbial source	Weltevreden Sanitary landfill site, Brakpan, South Africa	Ammonia COD	99% 36%	—	[10]
Chlorella vulgaris and *Chlamydomonas reinhardii*	Municipal landfill management, Istanbul	Nitrogen Phosphorous	69.03% 100%	—	[11]
Actinomycetes, Bacillus, Pseudomonas and Burkholderia	Jebel Chekir landfill leachate, Tunisia	TOC	70–80%	As, Fe, Pb	[12]
Trametes versicolor	Nonthaburi landfill site, Thailand	Color, BOD, COD	78%, 68%, 57%	—	[13]
Immobilized microorganisms	—	COD and Ammoniacal Nitrogen	98.3%, 99.9%,	—	[14]

Table 2.
Removal efficiency of leachate pollutants by some organisms.

belonging to *Firmicutes* and *Proteobacteria* have efficient degrading potential. Many of these microorganisms are also tolerant to heavy metals like Arsenic, iron, nickel, cadmium and copper. **Table 2** shows the removal efficiency of some microorganisms as well as microalgae isolated from landfill leachates.

5. Effectiveness of bioremediation

Bioremediation using microorganisms is a method of choice due to its low cost and simplicity in operation mechanism. The various types of biological treatment processes are upflow anaerobic sludge blanket (UASB), activated sludge reactor, membrane bioreactor, rotating biological contactor, batch reactor and moving bed biofilm reactor. All of them are efficient in treatment of young leachates. The UASB and batch reactor processes are also efficient in treatment of middle aged and mature leachates. Anaerobic reactor using seed sludge as inoculants has been found to tolerate high levels of COD and also precipitate heavy metals such as Fe, Zn, Ni, Cd, Pb, Cu and Cr [15]. Phytoremediation is one efficient and inexpensive process for remediation of mature leachates. However, phytoremediation has many limitations like remediation only limited to the surface and depth where the roots can reach, slow growth, and inadequacy in preventing the contaminant from leaching into the groundwater.

Various bioremediation processes have been experimented and their efficiency studied so far, but it has been observed that bioremediation when combined with physical and chemical processes in a monitored manner shows much promise and efficacy in removal of pollutants of leachate plumes.

6. Management of soil contamination caused by leachates

Landfill is the preferred method for solid waste disposal all over the world due to its ease and low-cost operations. However, landfilling requires proper designing of its structure and planning of disposal in order to avoid its pollution effects to the environment. Landfill leachates if not properly controlled can give rise to serious consequences like soil contamination and ultimately damage to crops and vegetation. In some cases, there may be accumulation of heavy metals in vegetation and consequently in fishes dwelling in contaminated water bodies like lakes and streams. The ultimate effect is biomagnification and serious health issues to people inhabiting the nearby areas of the landfill. Most of the components are carcinogenic and genotoxic in the long run. Therefore, the landfill leachates need to be properly monitored and controlled. Due to lack of engineered systems in most of the landfills all over the world, and also due to the practice of merely dumping wastes in the landfill sites of developing countries, underlying soil and groundwater pollution by the toxic leachates have become a major concern in the recent years [16].

Management of leachate has posed a challenging task as it requires various stages such as (i) monitoring of its formation and flow or migration, (ii) assessment of the various parameters of leachates, (iii) investigation of its hazardous components, and (iv) its treatment before finally releasing to the environment.

6.1 Monitoring of leachate formation and flow or migration

Leachate formation and its duration may depend on the components of a particular landfill site and its microbial community composition. The flow rate and the leachate volume may also vary depending on the season and rainfall. Therefore, proper investigation of the leachate formation is required in order to control the flow rate and migration of leachate plume. Various techniques used for monitoring leachate plume migration are hydro-geological techniques, electromagnetic methods, fluorescence methods, stable isotopes labelling, microbial analysis etc. [16]. Microbes play important role in characterizing the important parameters of leachates. Present day advancement in molecular biology techniques shows great promise to delineate the microbial community composition in a short time period. Metagenomics and next generation sequencing technology would greatly help in finding out novel microorganisms with novel pathways of complex compounds degradation. This information would further help in managing the landfill sites through natural attenuation.

6.2 Assessment of various parameters of leachates

The important parameters like pH, alkalinity, BOD, COD, TDS, nitrate, sulphate, chloride, electrical conductivity etc. should be assessed regularly and check whether their values are according to the permissible limits set by various regulators. Multi parameter analyzer can be used to measure different parameters such as pH, electrical conductivity, oxidation reduction potential, salinity, temperature etc. Spectrophotometers are used to check many of the parameters which involve color development [17]. Procedure for measurement of parameters should be according to the standard procedures adopted by United States Environment Protection Agency (USEPA) or similar professional organizations which have set the standard protocols. The values obtained can be compared with the standards of various regulatory limits so as to assess the risk of the leachate to the environment.

6.3 Investigation of hazardous components

Hazardous components of landfill leachates such as toxic heavy metals, halogenated compounds, aromatic compounds, pesticides and other recalcitrant organic compounds, if present even in minute quantities can eventually give rise to contamination of soil and the groundwater. Therefore, the hazardous components need to be analyzed and proper treatment methods such as physico-chemical and biological treatments can be recommended. Assessment of heavy metals in landfill leachate is often an arduous task as we have to understand the metal speciation and its flow [18]. Heavy metals tend to get oxidized and this fact has to be considered in the assessment studies. Most of the heavy metals are present in the form of complexes and the metal speciation is also influenced by biological activities. Risk assessment i.e. the effect of hazardous components to the environment needs a systematic approach because of the uncertainty of the soil environment and groundwater flow. Various predication models and softwares are available for risk assessment of landfill leachates [19]. The implication of biological toxicity tests of leachate samples is also commonly applied by many investigators. Phytotoxicity tests of leachate samples have been reported using *Sinapis alba*, *Lemna minor*, *Vicia faba*, *Zea mays*, *Hordeum vulgare*, *Lepidium sativum*, *Lycopersicon esculentum*, *Helianthus annus*, *Medicago sativa* etc. and seed germination and root elongation have been mostly studied [20, 21].

6.4 Treatment of landfill leachate before its disposal

The various reliable treatment methods of landfill leachates are (i) Biological reactors, (ii) Physico-chemical treatment, and (iii) natural attenuation. Biological treatment methods have been found to be more efficient in case of young leachates which are easily biodegradable whereas in case of old leachates, physic-chemical treatment methods are more suitable. Combined treatment methods where biological, physical and chemical treatment methods are applied have been shown to have highest efficiency [16]. Natural attenuation, which involves the merging of physical chemical and biological processes occurring in nature, has been found to be very useful for remediation of leachate plume. Natural processes in constructed wetlands and aerated lagoons have shown high efficiency (60–99%) in removing BOD, COD, ammoniacal nitrogen, Chloride, phosphate, iron and phenols [22–24]. Biological reactors with aerobic and anaerobic digestion processes have shown high removal capacity of BOD and COD as well as heavy metals in some cases. Among the advanced oxidation processes, ozonation has potential in removing color and organic acids i.e. humic acid and fulvic acid [25]. Coagulation using Ferric chloride followed by filteration and reverse osmosis has been shown to completely remove organic pollutants like Di-(2-ethylhexyl) phthalate (DEHP) and Bisphenol A [26]. Among the latest treatment technologies, leachate treatment using magnetic adsorbents and nanomaterials are of special interests as they have high efficiency in removal of organic acids and heavy metals [27, 28]. But the use of these techniques has environmental impacts and further research need to be done with green and environmental friendly processes.

7. Conclusions

Landfill leachates pose serious threats to the environment. In many cases, groundwater contamination has occurred due to lack of monitoring and efficient treatment system. The general parameters which are measured in a landfill site are pH, alkalinity, electrical conductivity, TDS, BOD, COD, chloride, ammoniacal nitrogen, nitrate, sulphate etc. Analysis of hazardous components like toxic heavy metals, halogenated compounds, polymers etc. are also necessary. Treatment methods using microorganisms is an attractive way for removing or remediating the pollutants in the leachates as it is cost effective and environment friendly. There is a great potential in studying the efficiency of microorganisms in leachate treatment, but a lot need to be explored in this field. Novel microorganisms having potential use in leachate treatment can be explored using modern techniques like metagenomics. Biological treatment of leachates from young landfill sites has already been proved efficient. The treatment efficiency for leachates from old landfill sites can be improved by integrating biological method with physical and chemical methods. Natural attenuation through aerated lagoons and phytoremediation is also a means for eliminating pollutants in the vicinity of landfill sites. In the present scenario, reducing solid waste and hazardous waste components by reuse and recycling should be strictly considered. The landfill sites should be properly designed to avoid leachate contaminants to soil and groundwater. Proper monitoring, risk assessment, and leachate treatment with advanced technologies are very much necessary to avoid any kind of serious environmental impact.

Author details

Menaka Salam* and Namdol Nilza
Amity Institute of Microbial Technology, Amity University Uttar Pradesh, Noida, India

*Address all correspondence to: salammenaka@gmail.com

References

[1] Kumari P, Kaur A, Gupta NC. Extent of Groundwater Contamination Due to Leachate Migration Adjacent to Unlined Landfill Site of Delhi, Environmental Claims Journal. 2018. DOI: 10.1080/10406026.2018.1543825

[2] Naveen BP, Sumalatha J, Malik RK. A study on contamination of ground and surface water bodies by leachate leakage from a landfill in Bangalore, India. Geo-Engineering. 2018; 9, 27 DOI: 10.1186/s40703-018-0095-x

[3] Abd El-Salam MM, Abu-Zuid GI. Impact of Landfill Leachate on the Groundwater Quality: a Case Study in Egypt. Journal of Advanced Research. 2014;6(4): 579-586. DOI:10.1016/j.jare.2014.02.003

[4] Eka Sri Y, Dedi S, Ridwan, Marsi, Faizal. Characteristics of Leachate at Sukawinatan Landfill, Palembang, Indonesia. 2013 IOP Publishing Journal of Physics: Conference Series 423 012048. DOI:10.1088/1742-6596/423/1/012048

[5] Gálvez ML, Rodríguez M, Zamorano, Ramos-Ridao AF. Evaluation of the quality and treatability of leachate produced at a landfill connected to an urban waste composting and recovery plant at Alhendín (Granada, Spain). Journal of Environmental Science and Health Part A. 2010;45(5):612-621. DOI: 10.1080/10934521003595688

[6] Turki N, Bouzid J. Effects of Landfill Leachate application on Crops growth and Properties of a Mediterranean Sandy Soil. Journal of Pollution Effects and Control. 2017;5: 186. DOI: 10.4176/2375-4397.1000186

[7] Mishra S, Tiwary D, Ohri A. Leachate characterisation and evaluation of leachate pollution potential of urban municipal landfill sites. International Journal of Environment and Waste Management. 2018;21(4)

[8] Yanan R, Qiuyan Y. Fungi in Landfill Leachate Treatment Process, Biodegradation and Bioremediation of Polluted Systems - New Advances and Technologies, Rolando Chamy, Francisca Rosenkranz and Lorena Soler, IntechOpen, DOI: 10.5772/60863

[9] Morris S, Garcia-Cabellos G, Enright D, Ryan D, Enright AM. Bioremediation of Landfill Leachate Using Isolated Bacterial Strains. International Journal of Environmental Bioremediation & Biodegradation. 2018;6(1):26-35. DOI: 10.12691/ijebb-6-1-4

[10] Mpenyana L, Coetzee MAA, Schwarzer S. Removal of Ammonia from Landfill Leachate in a Two-Stage Biofiltration Process. Journal of Biological Sciences. 2008;8:368-373. DOI: 10.3923/jbs.2008.368.373

[11] Khanzada ZT. Phosphorus removal from landfill leachate by microalgae. Biotechnology Reports. 2020; 25. DOI:10.1016/j.btre.2020.e00419

[12] Yahmed AB, Saidi N, Trabelsi I, Murano F, Dhaifallah T, Bousselmi L, Ghrabi A. Microbial characterization during aerobic biological treatment of landfill leachate (Tunisia). Desalination. 2009;246:378-388

[13] Saetang J, Babel S. Biodegradation of organics in landfill leachate by immobilized white rot fungi, *Trametes versicolor* BCC 8725. Environmental Technology. 2012; 1-10

[14] Ye Z, Yu H, Wen L, Ni J. Treatment of landfill leachate by immobilized microorganisms. Science in China Series B: Chemistry. 2008;51:1014-1020. DOI:10.1007/s11426-008-0101-9

[15] Kawai M, Purwanti IF, Nagao N, Slamet A, Hermana J, Toda T. Seasonal variation in chemical properties and

degradability by anaerobic digestion of landfill leachate at Benowo in Surabaya, Indonesia. Journal of Environmental Management. 2012; 110:267-275. DOI: 10.1016/j.jenvman.2012.06.022

[16] Mukherjee S, Mukhopadhyay S, Hashim MA, Sen Gupta B. Contemporary Environmental Issues of Landfill Leachate: Assessment and Remedies. Critical Reviews in Environmental Science and Technology. 2015;45(5):472-590, DOI: 10.1080/10643389.2013.876524

[17] Aziz SQ, Aziz HA, Bashir MJK, Mojiri A. Assessment of various tropical municipal landfill leachate characteristics and treatment opportunities. Global NEST Journal. 2015;17(3): 439-450. DOI:10.30955/gnj.001521

[18] Claret F, Tournassat C, Crouzet C, Gaucher EC, Schäfer T. Metal speciation in landfill leachates with a focus on the influence of organic matter. WasteManagement.2011;31(9-10): 20362045. DOI:10.1016/j.wasman. 2011.05.014ff. ffhal-00605888

[19] Slack RJ, Gronow JR, Hall DH, Voulvoulis N. Household hazardous waste disposal to landfill: Using LandSim to model leachate migration. Environmental Pollution. 2007;146: 501-509.

[20] Vaverková MD, Elbl J, Koda E, Adamcová D, Bilgin A, Lukas V, Podlasek A, Kintl A, Wdowska M, Brtnický M, Zloch J. Chemical Composition and Hazardous Effects of Leachate from the Active Municipal Solid Waste Landfill Surrounded by Farmlands. Sustainability. 2020; 12, 4531

[21] Turki N, Bouzid J. Effects of Landfill Leachate application on Crops growth and Properties of a Mediterranean Sandy Soil. Journal of Pollution Effects and Control. 2017; 5: 186. DOI: 10.4176/2375-4397.1000186

[22] Sawaittayothin V, Polprasert C. Nitrogen mass balance and microbial analysis of constructed wetlands treating municipal landfill leachate. Bioresource Technology. 2007; 98: 565-570

[23] Mehmood MK, Adetutu E, Nedwell DB, Ball AS. In situ microbial treatment of landfill leachate using aerated lagoons. Bioresource Technology. 2009; 100: 2741-2744

[24] Justin MZ and Zupancic M. Combined purification and reuse of landfill leachate by constructed wetland and irrigation of grass and willows. Desalination. 2009; 246: 157-168

[25] Poznyak T, Bautista GL, Chaírez I, Córdova RI, Ríos LE. Decomposition of toxic pollutants in landfill leachate by ozone after coagulation treatment. Journal of Hazardous Materials. 2008; 152: 1108-1114

[26] Theepharaksapan S, Chiemchaisri C, Chiemchaisri W, Yamamoto, K. Removal of pollutants and reduction of bio-toxicity in a full scale chemical coagulation and reverse osmosis leachate treatment system. Bioresource Technology. 2011;102: 5381-5388

[27] Reshadi MAM, Bazargan A, McKay G. A review of the application of adsorbents for landfill leachate treatment: Focus on magnetic adsorption. The Science of the Total Environment. 2020;731:138863. DOI: 10.1016/j.scitotenv.2020.138863

[28] de Almeida R, de Souza Couto JM, Gouvea RM, et al. Nanofiltration applied to the landfill leachate treatment and preliminary cost estimation. Waste Management and Research. 2020;38(10):1119-1128. doi:10.1177/0734242X20933333

Chapter 11

Phytoremediation: An Ecological Solution for Decontamination of Polluted Urban Soils

My Dung Jusselme, Noureddine Bousserrhine, Samir Abbad-Andaloussi, Florence Brondeau and Clarisse Balland-Bolou-Bi

Abstract

Urbanization and industrialization are the main causes of increasing contaminated soils in cities all around the world. This leads to numerous abandoned lands, reduction in biodiversity, and thereby posing a serious health risk for urban inhabitants. The development of effective and ecological remediation approaches is necessary. Phytoremediation is well known as an ecological solution with good acceptation for remediation of contaminated soils. Since, urban soils are particularly characterized by their highly disturbed, heterogeneous and low fertility, the application of phytoremediation to rehabilitate contaminated soils in urban areas is until now very limited at the laboratory scale and even less at the field scale. In this context, we have to take into account all these parameters and precautions when it's application. The main objective of this chapter is to discuss how to take phytoremediation approaches from a proven technology to an accepted practice in an urban context. An overview of urban soil types is provided following phytoremediation's application for urban soils with the focus on inorganic and organic pollutants, to provide a frame of reference for the subsequent discussion on better utilization of phytoremediation. At last, we offer suggestion on how to gain greater acceptance for phytoremediation by urban inhabitants.

Keywords: phytoremediation, ecological solution, urban soils, social sciences, ecological garden

1. Introduction

Although occupied only a small (<3%) proportion of the Earth's terrestrial surface, urban soils provide a wide range of ecosystem services to inhabitants of cities [1]. In the current context of population growth and urbanization as well as rapid industrialization, urban soils have largely disappeared and polluted by different types of organic and inorganic pollutants. According to urban scholars, although there is an increase of the cultural levels and diverse with more various cities, urbanization however generally leads to a reduction in biodiversity and ecosystem quality. Over the last decade or more, urban gardening is privileged and growing trend in many cities all around the world. For this development, the inhabitants should be assured of that the land is clean and safe. It is urgent that urban soil remediation projects must be to encourage investments.

Conventional methods of soil decontamination possess disadvantages in forms of environmental cost and financial burden. This truth leads to the search of ecological technologies for restoration of urban soils. One such approach includes phytoremediation. Phytoremediation is a process that uses plant for biological treatment of both organic and inorganic from polluted soils in non-urban and urban areas. Operating costs are very low, ranging from $ 0.02 to 1.00 per m^3 of soil [2]. Phytoremediation is based on the use of plant species to extract, retain, immobilize or degrade pollutants in soils. This technique provides good recovery of soils contaminated with heavy metals, and petroleum hydrocarbons.

In the urban context, there are two challenges in attracting the application of phytoremediation for contaminated soils. First, how do make the application of this approach operate and effective? Second, how do inform and train professionals and also non-professionals of the remediation of the contaminated soils potential offered by phytoremediation approaches. This will encourage the use of an ecologically, viable and socially accepted depollution technique.

In this chapter, we will discuss how to take phytoremediation approaches from a proven technology to an accepted practice in the urban context. An overview of urban soil types is provided following phytoremediation's application for urban soils with the focus on inorganic and organic pollutants, to provide a frame of reference for the subsequent discussion on better utilization of phytoremediation. At last, we offer suggestion on how to gain greater acceptance for phytoremediation by urban inhabitant.

2. An overview of urban soil contaminations

2.1 Urban soil type

"Urban soils" could have several definitions according to scientific or technic domain considered. For World Reference Base for Soil Resources (WRB), urban soils are composed of "any material within two meters of the Earth's surface that is in contact with the atmosphere, excluding living organisms, areas with continuous ice not covered by other material, and water bodies deeper than two meters" [3]. The Morel and Schwartz team's works made it possible to complete the definition by adding that these soils are under strong human influence in the urban and suburban landscape [4–6]. These soils are called Technosols [3]. Their studies begin to be more and more important at the beginning of the 21st century with an exponential increase in the number of publications concerning urban soils (**Figure 1**). Indeed, before the 2000s, the urban soils were considered too disturbed, polluted and poor fertility. Nevertheless, with the ever-increasing population in the city and the growing public concern about environment and human health, the restoration or rehabilitation and remediation of these soils have become a priority. In the urban area, soil is a key issue, subject to very rapid changes in allocation and use (green space, gardens, peri-urban agriculture, urban and industrial activities). Soils provide many essential ecosystems services in urban area, such as carbon and mineral nutrients storage, biota's habitat, role in hydrologic cycle by reducing runoff and promoting infiltration, water supply and reduction of pollutant bioavailability.

The main characteristics of urban soils are strong vertical and horizontal spatial heterogeneity in terms of physical, chemical and biological properties [7]. This strong variability can be explained by differences in occupation and use, such as the soils supporting buildings and infrastructures, landscaping areas.

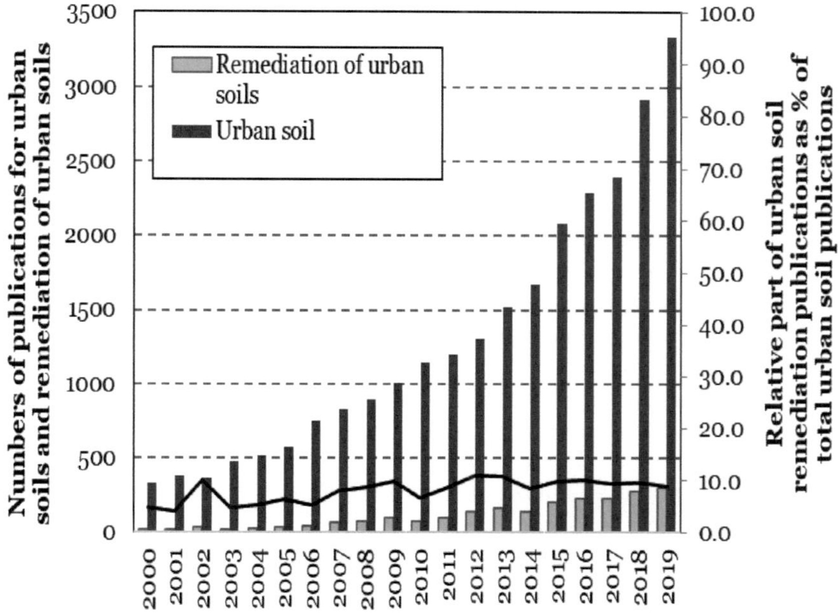

Figure 1.
Evolution of the number of annual scientific publications on soils (dark gray histogram) and urban soils (histogram light gray) in the international scientific literature over the period 2000–2019. Evolution of the relative share of publications on soil remediation urban is represented by the black curve, which is estimated as % of the total number of publications on urban soils. Bibliometrics on the state of scientific and technological knowledge on urban soils has been evaluated with two search engines: Web of Science and Medline, using these keywords "urban soil", "remediation", "restoration", "rehabilitation" with different combinations.

Various anthropogenic factors lead to a modification of the initial state of the soil in urban zones. Moreover, most of urban land are the new soils created through mixing, incorporation, and export of earthy materials, compaction or sealing. Unfortunately, the incorporation of these materials leads to frequent pollution of these soils. In general, urban soils display raised pH values due to addition of calcareous and other waste building materials.

2.2 Pollutant types in urban soil

Due to the human activities, urban soils are contaminated with various organic and inorganic pollutants. Among which, polycyclic aromatic hydrocarbons (PAH), pesticides, biphenyl-polychlorinated (PCB), metals, metalloids and radionuclides are the most abundant. Their presence in soil is undesirable due to their highly toxic and the environmental disturbances they create. Soils contain natural quantity of potentially toxic metals due to constitution of parent rock materials. Trace metals including lead (Pb), cadmium (Cd), mercury (Hg), chromium (Cr), zinc (Zn), copper (Cu), nickel (Ni) and some metalloids such as arsenic (As), Selenium (Se), manganese (Mn) are toxic for living organisms even at low concentration in soils. Whereas some trace metals such as zinc (Zn), copper (Cu), nickel (Ni) are vital elements for living organisms and their physiological properties (enzyme activators, electron transfer system in photosynthesis and respiration). Moreover, the presence of hydrocarbons and metals in soils affects negatively seed germination and plant growth [8], soil microbial community and activities [9], metabolic capacities of plants and microorganisms [10].

Nevertheless, since several decades, the anthropogenic origins of all the urbans pollutants are various and mainly attributed to (i) transport sources (traffic, vehicle emission, brake and tyre wear), (ii) commercial and industrial emissions (energy production, electronics, metallurgical and chemical industries, fuel combustion, incineration), (iii) domestic activities (construction and demolition, waste disposal, wastewater), and (iv) agricultural activities (application of fertilizers and pesticides, wastewater irrigation) [11].

Soil erosion and storm water runoff in urban areas are the main contributor to diffuse pollution according to the United States Environmental Protection Agency [12]. Moreover, the incorporation of calcareous and other waste building materials into urban soils are no negligible and several inorganic pollutants, especially trace metals, are being introduced into these soils leading their use dangerous for human health. Degradation of trace metals is not possible; therefore, they are accumulated and persist in the soil for many years. The concentration of various pollutants in rural and urban areas in France are presented in **Table 1**. We can see that the concentrations of the most of pollutants are superior in urban area than in rural area. These data confirm also the heterogeneity of these urban soils and are coherent with the concentrations of urban soils of other metropolises (USA, Spain, China, Ireland, Finland, Algeria, Nigeria and Iran) [7].

The review of the literature indicates that most scientific articles (>80%) focus on metals and little data are available on traditional or emerging organic pollutants that are now being detected. Many studies still need to be carried out to assess the impact of these pollutants on urban soils and consequently on ecosystem services provided by these soils, and more broadly on human health.

2.3 Ecological methods for restoration of contaminated urban soils

As seen earlier in Section 2.1, urban soils are much polluted. It is therefore necessary to treat them before any other use, be it for parks or gardens. Obviously, depending on the nature of the pollutants (organic or inorganic), their concentrations, and the soil physic-chemical properties, the appropriate technique will differ. Moreover, the reason for which monitoring will also be a criterion for the choice of operational staff. The remediation technics used for the depollution of contaminated site can be in situ or ex situ, on site or off site and biological, physical and chemical. They are often employed in combination with each other in order to optimize the system more efficiently and cost-effectively.

Ecological methods for soil remediation have received considerable interest in the last decade (**Figure 1**) and exhibit almost 10% of the publications on urban soils. This growing interest has several reasons such as potential cost savings compared to conventional non biological techniques and the benefit effects of this techniques on urban soil that are often polluted with a poor fertility. Ecological methods the most used in urban soils are phytoremediation, microbes-assisted-remediation, and amendment incorporation. Phytoremediation can be used in combination with this other technique.

Phytoremediation [10, 11] consists to use of plants to remediate and revegetate contaminated sites. Phytoremediation technique was first developed to clean up heavy metal(loid)s contaminated soils, thus, the first publications on the subject appears at the end of 1980s and beginning of the 2000s for urban soils. Phytoremediation is considered environmentally friendly, esthetically pleasing, non-invasive and cost-effective technology to clean up the sites with low-to-moderate levels of heavy metal(loid)s (see Section 2).

Amendment incorporation in urban soils corresponds mainly to organic amendment such as compost or biochar [12, 13]. In urban soils, this technique is used since 2000s for disturbed soils with poor structure and low levels of OM and fertility in order to improve the physical properties (such as bulk density, infiltration rate,

			Data acquired in mainly rural areas			Data acquired in mainly urban areas		
Family	**Name**	**Unit**	**Min**	**Med**	**Max**	**Min**	**Med**	**Max**
Trace metals	As	mg/kg	1.00	—	25.00	1.00	**8.80**	50.20
	Pb	mg/kg	2.20	**34.10**	91.50	5.30	**57.40**	650.00
	Zn	mg/kg	<5	**80.00**	275.00	13.00	**94.90**	2600.00
	Ni	mg/kg	<2	**31.00**	78.90	4.00	**15.00**	6200.00
	Hg	mg/kg	0.02	—	0.10	0.05	**0.20**	28.00
	Cd	mg/kg	<0.02	**0.16**	6.99	0.05	**0.43**	3.63
	Cr	mg/kg	<2	**66.30**	118.00	0.90	**21.00**	111.30
	Cu	mg/kg	<2	**12.80**	27.20	4.20	**27.00**	190.00

			Data acquired in mainly rural areas			Data acquired in mainly urban areas		
Family	**Name**	**Unit**	**Min**	**Med**	**Max**	**Min**	**Med**	**Max**
HAP	Naphtalene	mg/kg	0.00	**0.00**	1.03	0.01	**0.11**	11.00
	Acenaphtylene	mg/kg	0.00	**0.00**	0.53	0.01	**0.14**	15.00
	Acenaphtene	mg/kg	0.00	**0.00**	0.16	0.02	**0.16**	13.00
	Fluorene	mg/kg	0.00	**0.00**	0.25	0.01	**0.08**	6.40
	Phenanthrene	mg/kg	0.00	**0.01**	3.47	0.01	**0.12**	7.80
	Anthracene	mg/kg	0.00	**0.00**	0.56	0.02	**0.21**	33.00
	Fluoranthene	mg/kg	0.00	**0.01**	6.08	0.01	**0.12**	10.00
	Pyrene	mg/kg	0.00	**0.00**	4.37	0.01	**0.02**	0.64
	Benzo(a)anthracene	mg/kg	0.00	**0.00**	2.18	0.01	**0.05**	1.90
	Chrysene	mg/kg	0.00	**0.00**	4.14	0.02	**0.12**	10.00
	Benzo(b)anthracene	mg/kg	0.00	**0.00**	2.22	0.01	**0.02**	0.60
	Benzo(k)anthracene	mg/kg	0.00	**0.00**	1.46	0.01	**0.08**	16.00
	Benzo(a)pyrene	mg/kg	0.00	**0.00**	1.73	0.02	**0.17**	29.00
	Indeno(1, 2, 3-cd)pyrene	mg/kg	0.00	**0.00**	1.83	0.01	**0.05**	1.20
	Dibenzo(a,h) anthracane	mg/kg	0.00	**0.00**	1.13	0.01	**0.05**	0.70
	Benzo(g, h, i)perylene	mg/kg	0.00	**0.00**	1.53	0.01	**0.02**	12.00
	Σ 16 HAP	**mg/kg**	**0.13**	**0.16**	**31.67**	**0.28**	**1.56**	**167.31**
PCB	**Σ PCB**	**µg/kg**	**0.20**	**0.70**	**17404.20**	—	—	—
Dioxines/furanes	Σ Dioxines/furanes	**ng/kg**	24.75	**28.17**	2095.28	27.58	**162.70**	4678.40
Cyanure	Cyanure	mg/kg	—	—	—	0.10	**1.00**	6.10
Phenol	Indice phenol	mg/kg	—	—	—	0.01	**0.48**	86.00
Hydrocarbures	C_{10}, C_{40}	mg/kg	—	—	—	0.50	**20.00**	260.00

Table 1.
Concentration of organic and inorganic pollutants in rural and urban soils in France (values extracted from Ademe [13]).

hydraulic conductivity, water content, aggregate stability, and porosity) and function (such as water and nutrients available for plants, support for living organisms, etc.). Concerning contaminated urban soils, the studies on biochar has shown its ability to bind metals, decrease their mobility and bioavailability, stimulate microbial activity and promote soil revegetation and recovery (see Section 3.3).

Microbes-assisted-remediation [14] or bioremediation is a method involving the use of microorganisms to breakdown hazardous contaminants/pollutants to nontoxic or harmless forms. This technique was mainly used for organic pollutants. It can be also used for inorganic pollutant to stabilize metals or metalloids into soil or extract them when associated to phytotechnologies. Bioremediation techniques are mainly of two types: in situ (at the site of contamination) and ex-situ. Bioremediation presents several benefits such as economic viability, social acceptability, and eco-friendly (see Section 3.1).

3. Phytoremediation in urban context

3.1 Phytoremediation of inorganic pollutants

Inorganic pollutants which include heavy metals and metalloids are release into the environment due to human activities of industry, transportation and also urban activities. In order to remediate the soils polluted by inorganic pollutants, several conventional chemical and physical techniques have been used for decades; however, they are expensive and often hard to set-up. Recently, phytoremediation is admitted as an appropriate method using plants for the depollution of inorganic pollutants. The number of publications related to phytoremediation has only increased since the early 2000s with an average of 700 articles per over the last 5 years (source: Web of science) with 3–5% focused on urban soil. Moreover, 90% of these publications are related to phytoremediation of soils contaminated by trace metals and metalloids.

Phytoremediation of inorganic pollutants refers to phytoextraction, phytostabilization, phytovolatilization and rhizofiltration [14, 15]. Phytovolatilization (only for mercury and selenium) and rhizofiltration are still techniques with an experimental approach and mostly under controlled conditions unlike phytoextraction and phytostabilization which have been applied in the field, and most used to rehabilitate urban soils.

Phytostabilization consist to cover contaminated soil by plants either by seeding or planting. As a consequence, the biological, physical and chemical properties of the soils will be improved. The presence of vegetal cover, especially dense root system will permit to decrease the dispersion/mobilization of inorganic pollutants by promoting (i) water infiltration rather than runoff, (ii) evapotranspiration which will limit the percolation of water and thus the leaching of contaminants, and (iii) by retaining fine particles. Thus, plants will stabilize inorganic pollutants by accumulating them in the rhizosphere or into roots and will decrease their bioavailability. Phytostabilization, despite these many advantages (improvement of biological, physical and chemical qualities and consequently the increase in soil ecosystem services), is above all more a management strategy for polluted urban soils than a depollution technique since trace metals and metalloids remain in the soil. The application of amendments promotes the heavy metal stabilization in soils. Recently, aided phytostabilization have been used for remediation of urban soils [16–18]. This technique consists in the chemical stabilization of inorganic pollutants with the combined use of a wide range of soil amendments with a selected plant. This soil amendment can be natural mineral (phyllosilicates, zeolites, and oxides), organic substances, industrial or urban wastes and agriculture (manure, straw, and composts). This amendment will increase the soil pH and sorption capacity of soil rhizosphere (see Section 2.3).

Phytoextraction is based on the ability of plants to grow on contaminated soils, absorb inorganic pollutants by their roots and then transfer and accumulate them in significant quantities in their aerial organs (stem, leaves, and reproductive organs). The pollutant presented in soils must therefore be bioavailable for plants. Thus, the phytoremediation process will increase the fraction of metals bioavailable for plants depending on a combination between plant physiology, soil microorganisms (see Section 3.1), soil chemistry and the interaction between plant and microbes. There are many reviews that inventory these hyperaccumulators or high biomass accumulating plants used as a function of the major trace metals or metalloids they accumulate [14, 19, 20].

Moreover, in order to improve the efficiency of plants involved in phytoextraction process, many authors proposed the transfer of the hyperaccumulator phenotype from small and slow growing hyperaccumulator species to fast growing, high biomass-producing non-accumulator plants. Many genes involved in the acquisition, allocation and detoxification of metals come from bacteria and yeasts [21]. For example, some works on bioengineering have used plants capable of removing methyl-mercury from contaminated mining and urban soils [22], a strong neurotoxic agents, is biosynthesized in Hg-contaminated soils. To detoxify this compound, transgenic plants have been engineered to express modified bacterial genes *merB* and *merA*.

In the case of lead (Pb) which is one of the most trace metals presented in urban soils (see Section 2.2), the content of bioavailability lead in the soils is very low and it is difficult for plant to uptake them. Therefore the rehabilitation of soils polluted be lead is often difficulty. To overcome the problem, it is necessary to realize assisted phytoremediation [23]. This technique consists of adding to the soil various chemical compounds that can increase the availability of trace metals or metalloids in the soil solution. The chemical compounds used are generally aminopolycarboxylic acids (APCA), molecules chelating metal cations such as ethylenediamine tetraacetic acid (EDTA), nitrilotriacetic acid (NTA), hydroxyethylenediamine-tetraacetic acid (HEDTA) or diethylenepentaacetic acid (DTPA). Nevertheless, it has been shown that the aminopolycarboxylic acids can be toxic for some plants, microorganisms or nematodes. Meanwhile organic acids such as citric or oxalic acids which are less toxic can be used, but they are less effective in increasing the fraction of trace elements easily assimilated by plants. Moreover, transgenic plants have been engineered too to overproduce recombinant proteins and chelating molecules such as citrate, phytochelatins, metallothioneins, phytosiderophores playing roles in chelation and assimilation of metal.

3.2 Phytoremediation of organic pollutants

Due to increased human activities including urbanization and industrialization, the pollution of organic pollutants in urban areas has been increased over the last decade. Urban and peri-urban soils are often polluted as consequence of human activities. The main sources of the urban organic pollutants are (1) the utilization of the pesticides in the urban environment, (2) the atmospheric deposition of organic pollutants in form gaseous and particulate by transport, (3) the using of urban waste composts as amendments in urban agriculture and (4) the development of urban industry. According to the results of bibliographic research over the last 20 years on website Web of Sciences, phytoremediation of organic pollutants in non-urban and urban soils generally involved several classes of compounds which are mostly polycyclic aromatic hydrocarbons (PAHs) [24, 25], polychlorinated biphenyls (PCBs) [26] and petroleum hydrocarbons (PHCs) [27] and others low molecular weight compounds such as benzene, toluene and xylene [2] (**Table 1**).

Phytoremediation for organic contaminants takes place at two levels: inside and outside of plant cells. Like the mechanisms of phytoextraction (absorption) which is the primary of phytoremediation for inorganic pollutants as described above (see Section 2.1), some low molecular weight organic contaminants can be taken up by root and then to be accumulated and/or degraded *in planta* [28]. However, most of organic contaminants are generally too large and/or hydrophobic therefore they cannot to be absorbed by plants. Two primary *ex planta* mechanisms of phytoremediation for organic contaminants are (1) rhizodegradation *via* the active microbial communities in the rhizosphere, and (2) phytodegradation *via* the plant enzymes. For rhizodegradation, rhizosphere microbial community through by their metabolic process transform the organic pollutants (hydrocarbon) to microbial biomass, bioenergy, carbon dioxide and also water for their development [2, 29]. For phytodegradation, plants used for phytoremediation excrete various extracellular enzymes including laccases, dehalogenases, nitrilase, nitroreductases and peroxidases degrading the organic contaminants [30]. Recently, numerous works have reported that different plant species and varieties are able to be used for phytoremediation of organic contaminants. Most of plant used belong to ornamental woody and herbaceous species [31]. Particularly, the utilization of different plant species of Asteraceae family, potential and suitable candidates, for phytoremediation of organic in urban areas was well quoted in the review presented by [32].

Over recent years, the number of works in phytoremediation for organic contaminants has intensely increased with many encouraging results that have emerged regarding the capacities of several plants to degrade specific organic contaminants. To make phytoremediation for organic compounds successful, it is fundamental to understand (1) the type of soil to be treated, (2) the concentration and the fate of each organic pollutants and (3) the relations between the physical, chemical and biological parameters. Urban soils are known to have particular characteristics that have mentioned above, therefore the application of this technology in urban polluted soils remains a daunting challenge for scientists. An exploratory bibliographic research on the Web of Science from 2000 to 2020 show that a few works use greenery to eliminate the organic pollutants in urban context since its application can be limited by many factors including climate and anthropogenic modifications of the soil (e.g. impacts on soils by urban-rural temperature contrast also known as urban heat islands) [33].

3.3 Challenges and perspectives of phytoremediation's application in urban soils

The urban context is very particular with regard to its location, spatial heterogeneity, pollution and usage. Even if urban soils are not intended to be reclaimed, there is still a risk to the health of the local population. It is for this reason that it is necessary to rehabilitate these soils. Many studies present the evidence results in utilization of different ornamental plant species for phytoremediation (e.g. family Asteraceae) can survive under such adverse urban conditions. In situations where the city budgets are limited and no alternative treatment can be carried out, the use of phytoremediated-plants could be affordable, sufficient, economically and community acceptable. Thus, plants play also a significant role in preservation of green spaces through enforcement of environmentally sustainable city planning. This application presents wealth of opportunities for city designers of urban landscapes and a good compromise to enhance urban diversity using phytoremediation in association with water infrastructures and open space on multiple scales.

Phytoremediation seems to be a promising technique but there are still many challenges, especially in an urban context. Indeed, the use of this technique is long (several decades) and restricted. Phytoremediation is thus limited by the area explored by plant roots and the low growth and low biomass produced. Moreover, this biomass cannot be used as compost because it is considered as contaminated waste. It is therefore necessary to select the right plant, adapted to urban soils, non-invasive in order not to alter the floristic diversity and capable of mobilizing metals even if they are not bioavailable. Thus, for each urban soil, a risk assessment should be carried out to protect local biodiversity before introducing alien species, but also a study should be carried out to better understand the interaction between the factors in the rhizosphere (metals/soil/microorganisms/plant roots).

Urban soils are increasingly being used for urban agriculture, either for private use or for small-scale local production. Thus, one of the big challenges is to cultivate while respecting food security and human health but there is a lack of data. To remediate to its problem, more and more works were focused on the combination of phytoremediation and food production [34]. At present, there are no large-scale studies, and most of this work reports on experiments with crop/phytoremediating plants combinations. There is always the problem of the biomass produced, can it be consumed? Can it be used as compost? Legislation in all countries is very vague or non-existent and needs to be strengthened. Research needs to be further continued to overcome these challenges of establishing food production on urban soils by carrying out studies on the translocation of pollutants in plants and their bioaccumulations, eco-toxicological risk assessment and soil legislation.

4. Improving the efficiency of phytoremediation in urban context

In spite of the fact that phytoremediation has a great of advantages in comparison to other technologies, it has also some limitations. The process of the phytoremediation is very slow from a few months to several years. The most of the plant used for phytoremediation have often small aboveground biomass and slow growth rate, and shallow root system, therefore very limits for their application in large-scale operations. Also, the low concentration of contaminants in form bioavailability in soils cause a low ability of contaminant absorption by plants.

To improve these limitations, one alternative that we will mention in this chapter is the use of (1) specific microorganisms such as fungi and bacteria, (2) earthworms, considered as 'ecosystem engineers' of soil, and (3) amendment such as biochar. All these complementary methods will permit to increase the growth of plants, biotic and abiotic stress tolerance and all the processes associated, such as mineral nutrient absorption, roots exudation and rhizosphere microbial activities, will be improve the process of the phytoremediation.

4.1 Using microorganisms (symbiont: fungi and bacteria)

4.1.1 Using fungi-assisted phytoremediation

A fungus (plural: fungi) belongs to the group of eukaryotic organisms. These organisms forms a kingdom that is separate from the other eukaryotic life kingdoms of plants and animals. Fungi are heterotroph, since they obtain carbon and energy from organic matter. Two major functional categories of fungi are saprophytic and mycorrhizal fungi. Saprophytic fungi decompose nonliving organic matter and they are important agents in soil mineralization processes and carbon cycle. Mycorrhiza are symbiotic species associated with vascular plants. There are eight main types of mycorrhizal symbioses based on their morphology and not on a biological reality [35].

According to pollutant type (organic and inorganic), the mycorrhizal fungi will be different. Whatever the pollutants, the selection of an appropriate host plant with mycorrhizae is of primary importance to improve phytoremediation. For organic pollutants such as polycyclic aromatic hydrocarbons (PAH), endophytic fungi is preferentially used to increase the efficiency of phytoremediation [36, 37]. For example, arbuscular myccorhizal fungi (AMF), belonging to the phylum *Glomeromycota,* form ubiquitous mutualistic interactions with roots of 80–90% of vascular plants species. AMF is widely used to degrade PAH. The hydrocarbons remediating potential of other endophytic fungi have been reported since the last decades. Thus, *Pestaliotopsis microspora* associated to the *Dendrobium* plant species have shown an efficient degradation potential of plastic polyester polyurethane. *Phomopsis liquidambari* degrade efficiently PAH in *Bischofia polycarpa* [36]. These symbiosis between endophytic fungi and vascular plants permit an increase of plant growth and hydrocarbons biodegradation by roots and its microflora associated, an improvement of adsorption and bioaccumulation of hydrocarbons by roots [38, 39].

For inorganic pollutants such as trace metals or metalloïds, some endophytic fungi, especially AMF that can increase the uptake of arsenic or other metals such as zinc, copper or lead [39]. Nevertheless, it has been shown that the most effective fungi in terms of host plant adaptation are ectomycorrhizae and ericoid mycorrhizae [35, 40, 41]. Indeed, the great development of the extraracinar mycelium allows it to explore a large volume of soil but also to store more metals and transform them into a less toxic form thanks to a wide range of enzymatic activities.

The interaction mycorrhizae-plant symbiosis and inorganic pollutants has three advantages. First, fungi can tolerate a high level of metal toxicity. Second, they are able to remove inorganic pollutants from soil and water. Finally, they promote plant growth even in polluted soils.

4.1.2 Using bacteria-assisted phytoremediation

In healthy soil, bacteria represents billons of unicellular organism and thousands of different species. Bacteria play a crucial role in ecosystem service of soil such as decomposers. As a consequent, bacteria release nutrients that other organisms could not access. Nevertheless, environmental and structural characteristics of urban soil greatly influence soil microbes. Indeed, anthropogenic impacts such as organic and inorganic pollutants in technosols and in urban runoff can shift the abundance and diversity of bacterial communities [42]. For example, it has been shown that in urban soils the main phyla identified are Acidobacteria, Actinobacteria and Proteobacteria.

In the rhizosphere zones, bacteria interact with plant root in form of commensalism or mutualism. These root associated beneficial bacteria that plays an important role in acquisition for nutrient, tolerance to abiotic stress and also defense against pests are referred to as the plant-growth-promoting rhizobacteria (PGPR) [43]. Therefore, PGPR have been mainly considered to use in phytoremediation in order to increase the efficiency of the phytoremediation. Recently, another bacterial type called plant growth-promoting endophytic bacteria (PGPE) which have been shown to act as PGPR are widely used in phytoremediation [44].

In the phytoremediation context, the microbial mechanisms direct and indirect that can improve the efficiency of phytoremediation are differ depending the pollutant types including organic or inorganic. Generally, root assisted-bacteria are used in order to improve the adaptation of hyperaccumulator plants to suboptimal urban soil conditions (see Section 2.1, 2.2 and 2.3) and ameliorate the efficiency of phytoremediation. For inorganic pollutants including trace metals, the mechanisms employed for enhance the phytoremediation involve improvement of plant growth by increasing mineral contents, plant metal tolerance by phytohormones products, and capacity of absorption and accumulation by producing organic acid and metal-specific ligands (e.g. siderophores) [45]. We can here cite some research works on the phytoremediation of metals facilitated by soil bacteria. The bacterial species *Bacillus* sp. MN3-4 which is a lead-resistant bacterium enhanced phytoremediation potential of plant *Alnus firma* by reducing the phytotoxic effects of metals [46]. A nickel-resistant PGPB *Pseudomonas* sp. A3R3 increased the capacity of Ni-accumulation of *Alyssum serpyllifolium* plant by production of ACC deaminase and IAA, siderophore synthesis and polymer hydrolyzing enzyme [47]. Besides, many works show that the use of plant growth-promoting rhizobacteria (PGPRs) as complementary process for metal phytoremediation leads to (i) higher plant growth by improving soil properties and biological activities under toxic metal stress, (ii) decrease phytotoxicity, and (iii) decrease oxidative damage to plant tissues that are exposed to high metal trace content by increasing antioxidant enzymatic systems [48, 49].

Unlike inorganic pollutants, for organic pollutants whose molecules contain principally carbon, the principal bacterial mechanisms when phytoremediation's applied is related to pollutant co-metabolism and/or degradation pathways [50]. In fact, exogenous as well as endogenous bacteria have a system of co-metabolism of the organic pollutants as the sole carbon source with amino acid, lipid, fatty acids and organic acids. Alternatively, these bacteria come to colonize in the rhizosphere and benefit the production of root exudates, consisting of sugar, fatty-acid, organic acids, amino acids and other carbon-containing compounds for growth and degrade these organic pollutants [51].

Although a lot of research points out many advantages this alternative technology, to our knowledge, no work on phytoremediation of pollutants facilitated by soil bacteria in urban areas has been carried out. To apply this technique in urban context, we must take into account all the parameters, consisting of bacterium, plant species, soil composition and nutrient (see Section 2), pollutant type and concentration as well as the competition with other organisms that can limit the use of phytoremediation in the field.

4.2 Using soil fauna: in case of earthworms

Earthworms act as soil ecosystem engineers because of their crucial role in building galleries and in the decomposition of organic matter; therefore they play an important role in agriculture production [52, 53]. In polluted soils, various species of earthworms including *Eisenia fetida*, *Lumbricus terrestris*, *Lumbricus rubellus* and *Aporrectodea caliginosa* can survive in soils polluted with metals and even accumulate heavy metals including Cd, Pb, Cu and Zn [54]. This leads to the ideas of earthworm's application for phytoremediation. On the one hand, earthworms can improve the soil physical and chemical properties and increase the soil fertility through an amelioration of the microbial activities. On the other hand, through their activity, earthworms increase the bioavailability of heavy metals in soils which is a primordial factor controlling the success of heavy metal phytoextraction [54–56]. In the case of mercury, for example, mercury changed from the stable crystalline iron oxide state to the mobile amorphous oxide state by earthworm's activities [57]. In spite of their important role in the bioavailability of heavy metals allowing the improvement of phytoremediation, the majority of studies using earthworms for phytoremediation has been developed to improve the capacity of microorganisms inoculated in soils (call bioaugmentation) to establish, survive and colonize the rhizosphere. Earthworms are known to help (1) settlement of inoculated microorganism, (2) enhancement of microbial survival (e.g. by supplying nutrients) and (3) distribution of microorganisms in soil, earthworms insuring transport.

A summary of the mechanisms direct and indirect of earthworm's effect on soil microorganisms and plants was presented in **Figure 2**.

Despite a large body of literature on the benefit for soil and plants by earthworm actions, the research on earthworms-assisted phytoremediation has just started on a laboratory scale with some encouraging results [55, 56]. The attention of this research topic is expanding by the time with an increasing the sum of times cited per year according to the citation report from Web of Science Core Collection between 2010 and 2020 (**Figure 3**). Outdoor experiments up to fields scale need to be investigated and documented.

4.3 Using soil amendment (biochar)

Urban soils are often nutrient poor and polluted. They are degrading more and more quickly with the loss of organic matter and soil permeability that cause the negative impacts on soil structure with increasing in soil density due to soil compaction and other factors. To overcome these deficiencies, the addition of natural organic matter including compost has been recognized to increase the bio-physico-chemical qualities of these urban soils [58–60]. Among the different composts, the application of biochar, which is a carbonaceous solid material, is used preferentially for urban soils. Biochar is derived from the pyrolysis of biomass. All cellulose, lignin and other non-carbonic materials gasify and are burned. Only pure carbon remains with approximately 40% of the carbon originally contained in biomass.

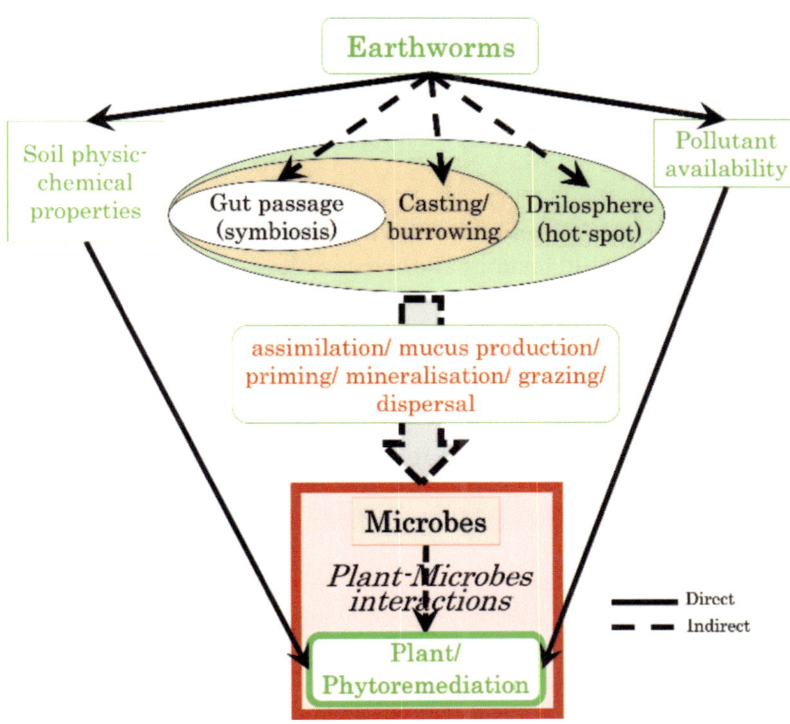

Figure 2.
Mechanisms direct and indirect of earthworm's effect on plant and microorganisms in the phytoremediation context.

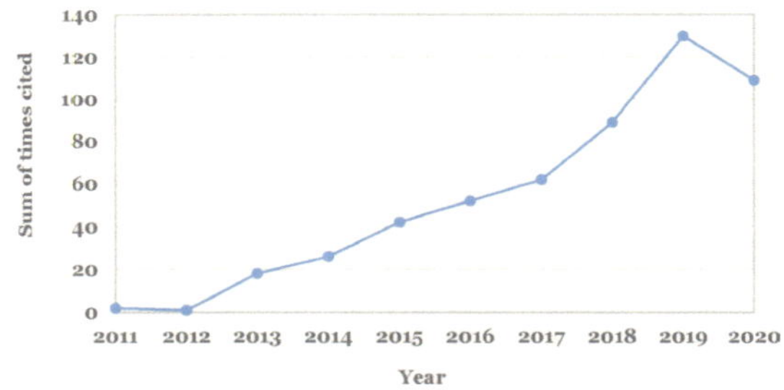

Figure 3.
Citation report of the sum of times cited per year on the topic "earthworms" and "phytoremediation" from web of sciences. This report reflects citations to source items indexes within web of science Core collection. Perform a cited reference search to include citations to items not indexed within web of science core collection.

Rather than an amendment (because it is very poor in nutrients), biochar would behave as a soil structure and perhaps as a catalyst, via mechanisms of action that are still poorly understood. The incorporation of biochar decreases the mobility and bioavailability of metals, thus decreasing their translocation in plants while improving the soil characteristics such as infiltration rate, hydraulic conductivity, porosity and therefore the water content. The growing of plants and water cycle is also improved.

Biochar, as a carbon-rich, stable and sustainable product, also acts as a carbon sink, which explains why it is attracting growing interest in the context of concerns about human-induced global warming. It could be one of the immediate solutions to the overall negative impact of urban and agricultural activities with the use of fossil carbon in the form of fuels, greenhouse gas emissions and tillage that degrades the carbon sink that humus constitutes.

Nevertheless, the application of biochar presents possible negative effects. Biochar may contain toxic elements naturally present in its composition and which may lead to an increase in pollution when incorporated. This can affect living organisms and the functioning of the soil. Moreover, because of the dust formed during their application, it present a risk for human health. There is still little data on its negative impacts.

To date, most of the studies has focused on the impact of compost on soil characteristics in agricultural area and relatively little data has been carried out in urban area. Future research should focus on the optimization of compost rates (quantity, depth...) in order to standardize the use of biochar on soil to minimize the bioaccessibility of pollutants and maximize soil/water relations and plants reestablishment [59].

5. Social aspects linked to phytoremediation in urban context

5.1 Perceptions and social acceptability of phytoremediation method: some elements for thought

The use of the words acceptability, social acceptance or social reception gives rise to terminological debates [61]. Acceptability is indeed a term vague enough to be used frequently [61]. We can nevertheless consider the social acceptability of a project as a process of social construction born from the confrontation of the arguments of the different actors and which results in an identification of the population concerned with the values carried by the said project. Some stress the fact that this dialog often comes down to the implementation of a communication strategy intended to convince the target audience as part of a top-down conception of a project [62]. The acceptance term is sometimes preferred but can imply a form of resignation of the inhabitants compared to a project conceived in a non-concerted way ([61], according to [63]). Some therefore prefer to use the term "acceptance" [61] after [64], others prefer the term "social reception". In fact, we can speak of acceptance of a project when it is appropriate by a population that identifies with the objectives pursued and the methods mobilized by it. This appropriation is conditioned by the perception of the project.

For psychology, perception is the function that allows the body to receive, process and interpret information received which comes from the surroundings through the senses. This construction is obviously specific to the type of information, to the individual or group who receives it and to the context in which it is disseminated. Thus a project will be perceived and therefore appropriate differently according to the economic, social, historical context, according to the modalities of diffusion of the information and the nature of this one, and obviously according to the type of actors diffusing and receiving the information and their expectations.

If we particularly consider phytoremediation projects, the perception by the population concerned is influenced by multiple factors: first of all, the identification of the risk associated with soil pollution and the potential benefits expected from phytoremediation [65]. This identification is closely linked to knowledge of the health risks involved. It was highlighted in a Quebec mining site, that the knowledge by all of a strong soil pollution whose effects on the health of populations are clearly highlighted, facilitates the acceptance of phytoremediation projects. In this case, the benefit is clearly identifiable and the populations are extremely favorable to a method of depollution considered as ecological.

However, if the populations of mining sites are alerted to the health risks linked to these forms of pollution [66] which is not necessarily the case in urban areas where pollution is old and associated with activities considered to be less polluting. Thus, the spreading of Parisian mud on the fields of farmers located in the immediate suburbs of Paris in the 19th century was not initially considered as a polluting activity [67]. In addition, the renewal of the population in a good number of urban regions leads to a lack of knowledge of the history of soils and associated pollution.

In most cases, the esthetic and landscaping criteria has an essential role in the reception that can be given to this type of project [68]. The revegetation of soils in neighborhoods that the image is devalued by an industrial or mining past and the presence of brownfields, constitutes a benefit clearly identifiable by the population who have been living there for a long time or more recently. Revegetation is often equated with an embellishment and an improvement of the living environment from an ecological point of view.

The different phytoremediation methods used, can, however, raise questions about the choice of species (sometimes non-native and poorly accepted by local residents), the fate of pollutants and the time required to obtain results [65]. Phytoextraction raises the question, for example, of the fate of plants that have absorbed a certain amount of pollutants, including trace metals, and their treatment [69].

Good reception of the project can be facilitated by working upstream with the inhabitants in order to make them aware of the characteristics of the different phytoremediation methods and their effects. Consultation on the landscapes desired by local residents would make it possible to consider the choice of species that can be used appreciated [61]. This work obviously requires a time of information and consultation that is added to the time necessary to obtain the first effects of the different phytoremediation methods.

It is also difficult to envisage social acceptability without considering the potential economic benefits. In terms of costs, phytoremediation is a much less expensive technique than conventional techniques, however it still seems to be little applied [70]. In this regard, it should be emphasized that local communities such as companies specializing in soil remediation are often ill-informed and poorly trained or little trained in this type of alternative techniques and prefer to apply better known and better controlled methods such as excavation and backfilling of polluted areas. It seems that phytoremediation is struggling to get out of the purely scientific and experimental sphere. The time required to obtain significant results is a constraint both for development companies, local authorities and for the population. In the process of acceptability of phytotechnologies, an articulation between these different temporalities constitutes an issue to be taken up.

In addition, the techniques of economic valuation of the biomass resulting from phytoremediation by the production of energy are still often experimental and little diffused and/or applied. Its transformation into energy, whether by thermodynamic processes (combustion, pyrolyse, roasting) or by biological processes (methanization), poses the problem of becoming pollutants and in particular of the trace metals contained in the biomass, in particular in the case phyto-extraction (ash after combustion, digestate after production of biogas). The acceptability of soil remediation projects through phytoremediation depends on the benefits known to society (population and decision-makers) and the value attributed to them.

5.2 Potential social benefits of phytoremediation

The social benefits attributed to phytoremediation can therefore be considered through the prism of ecosystem services. This concept, first imagined by ecologists, has been mobilized and widely publicized since the Millennium Ecosystem Assessment (2015); the objective sought was to promote the protection of ecosystems by assigning economic and social value to the services provided by them [71]. Ecosystem services can therefore be defined as the benefits provided by ecosystems to human societies. A general distinction is made between production (or supply) services, regulation services and cultural services. Despite the reservations which are made by ecologists and sociologists among others with regard to this concept and the reflections as to a "commodification of nature", this can be useful here to consider the potential economic and social benefits of phytoremediation operations [71, 72]. These are a few lines of inquiry and not an exhaustive analysis. The purpose of phytoremediation is to reconstitute an ecosystem allowing depollution of the soil or stabilization of pollutants in the soil.

The most directly perceptible benefit for the population is undoubtedly landscaped and esthetic. The revegetation of polluted sites, often fallow land can on the one hand radically modify the urban landscape and the image of districts or cities sometimes stigmatized by their industrial or mining past, and thus procure an embellishment to which the local populations are sensitive [61]. On the other hand, this revegetation can in certain conditions and ultimately provide spaces for relaxation and leisure. In this sense, these are the benefits associated with cultural services that can be highlighted.

The benefit most directly sought by this type of project is obviously soil remediation. It can be clearly identified by the population, particularly in regions where health risks are known. Beyond the management of this pollution, it is also the structure and fertility of the soils that will be improved if not restored: the greater permeability of these soils is an asset to limit runoff and potential flooding in certain cases and a restoration of the water cycle more generally, including filtering and purification functions provided by vegetation [13].

We should add that in the context of sustainable city projects, revegetation via phytoremediation can contribute to the objectives of reducing greenhouse gases and improving air quality, plants storing carbon in their tissues via photosynthesis. The plants introduced into phytoremediation operations, whether local or not, participate in the maintenance or dissemination of a certain diversity of flora and therefore fauna and can be integrated into larger projects for the maintenance or development of urban biodiversity. The areas benefiting from these projects can thus be associated with the construction of ecological corridors within the framework of the green and blue frames promoted in recent years at different territorial scales. Phytoremediation can therefore help to provide regulatory services for the restoration of these ecosystems in urban areas.

The valorization of the biomass produced within the framework of these revegetation operations, can in certain cases and in the long term, be envisaged of different forms. Burning and pyrolizing wood products produces gas. Oil from pyrolysis can also be used in the composition of certain fuels, while ash and biochar (vegetable charcoal) can be reincorporated into the soil as fertilizers. The roasting of this woody biomass provides fuel. Non-woody plant waste subjected to anaerobic digestion allows for the production not only of gas but also of digestates; these can also be reintroduced into the soil [13]. These are therefore production or supply services which can be highlighted and fairly easily economically quantifiable.

The assessment of these social and environmental amenities provided by phytoremediation projects are, however, for the most part complex to assess and account for economically, in particular regulation and cultural services. The monetary calculation of the direct or indirect services rendered could however minimize the real costs of soil rehabilitation projects and facilitate their wider implementation.

5.3 Potential eco-garden with plant used in phytoremediation

Phytoremediation is a plant-based technology that make us think about the potential eco-garden whom urban residents can profit the green and beautiful landscapes and easily accept it. Ecological gardens can be viewed in two ways depending on the target audience. For city managers, these gardens are installed in a sustainable way to cover polluted soils and thus limit the risks to the population. The plants that will be used are, in general, ornamental plants that will require little maintenance and will be durable over time. A list of ornamental plant species provided (see more in [31]) belonging to different plant groups: trees, shrub, and herbaceous which have a good potential phytoremediation for heavy metal are already used for remediate the polluted soils. For this purpose, the exploitation of ornamental plants could be an additional option. At the top, we raise the points that we need to take care when application of phytoremediation. We propose also that phytoremediation could be successfully exploited in urban territories; in these contexts, many herbaceous and others are suitable for planting because of their ornamental features and adaptability to inhabited areas.

For the surrounding population, these ecological gardens have several roles, first of all a food production role, an educational role by promoting social cohesion. Thus, one of the big challenges is to cultivate while respecting food security and human health. Research needs to be further continued to overcome these challenges of establishing food production in combination with phytoremediation in urban areas by carrying out studies on eco-toxicological risk assessment.

Phytoremediation consist of different process and mechanisms such as absorption and accumulation of pollutant in plant as well as degradation. In the case of the contaminants are absorbed and accumulated in plant, risks in allotments are higher because of transfer of pollutants to the food chain [73]. Phytoremediation with degradation process maybe more suitable. In all cases, it is recommended to take precautions when you want to install eco-gardens on the polluted soils with hyper accumulator plants. High precautions has to be paid to parks, playgrounds, kindergartens and urban zones where residents come into close contact with soils. There are various species of ornamental plants in the literature, the choice of plant species depends on the climate, the tastes and traditions of each country.

6. Conclusions

From what we can see, phytoremediation is indeed an ecological and economical technology, acceptable and efficient to remediate the polluted soils. However, this technology is not actually widely applied in the urban context but it has many advantages regardless of the technique chosen or the pollutants present. Thus, the redevelopment of urban land in cities has become a priority. Since the implementation in 2006 of the draft European Directive on soil protection, which gives priority to soil diagnosis and remediation, the general objective of the European strategy has been to protect soil and guarantee its sustainable use by preventing its degradation, preserving its functions and restoring degraded soils. Despite these many improvements, legislation on these soils is either non-existent or very vague. Moreover, we have very little experience with trials of remediation of urban soils by the technique of phytoremediation. Nevertheless, the first results are promising with a stabilization of pollution, a decrease in erosion, a decrease in heat islands, and an increase in biodiversity with the implementation of ecological corridors in urban soil management. Research needs to be further continued to overcome these gaps on urban soils.

Author details

My Dung Jusselme[1*], Noureddine Bousserrhine[1], Samir Abbad-Andaloussi[1], Florence Brondeau[2] and Clarisse Balland-Bolou-Bi[1]

1 Laboratoire Eau Environnement et Systèmes Urbains (Leesu), Université Paris Est-Créteil, Créteil, France

2 Laboratoire Espaces, Nature et Culture (ER ENeC), Université Paris-Sorbonne, Paris, France

*Address all correspondence to: thi-my-dung.jusselme@u-pec.fr

References

[1] Colding J, Barthel S. The potential of 'urban green commons' in the resilience building of cities. Ecological Economics. 2013;**86**:156-166

[2] Gerhardt KE, Gerwing PD, Greenberg BM. Opinion: Taking phytoremediation from proven technology to accepted practice. Plant Science. 2017;**256**:170-185

[3] IUSS Working Group. International soil classification system for naming soils and creating legends for soil maps. In: World Soil Resour Reports. Rome: FAO; 2015. p. 106

[4] Morel JL, Schwartz C, Florentin L, de Kimpe C. In: Hillel E, editor. Urban Soils. Oxford: Elsevier; 2005. pp. 202-208

[5] Blanchart A, Séré G, Cherel J, Warot G, Stas M, Consalès J-N, et al. Towards an operational methodology to optimize ecosystem services provided by urban soils. Landscape and Urban Planning. 2018;**176**:1-9

[6] Schwartz C, Séré G, Stas M, Blanchart A, Morel J-L, Consalès J-N. Quelle ressource Sol dans les villes pour quels services et quels aménagements? Results, Innovations, Transfer - INRA. 2015;**45**:1-11

[7] Pouyat RV, Szlavecz K, Yesilonis I, Schwarz P, Groffman K. Chemical, physical and biological characteristics of urban soils. In: Jacqueline A-P, Astrid V, editors. Urban Ecosystem Ecology. Agronomy Monograph 55. Madison, WI: American Society of Agronomy, Crop Science Society of America, Soil Science Society of America; 2010:119-152

[8] Smith MJ, Flowers TH, Duncan HJ, Alder J. Effects of polycyclic aromatic hydrocarbons on germination and subsequent growth of grasses and legumes in freshly contaminated soil and soil with aged PAHs residues. Environmental Pollution. 2006;**141**(6):519-525

[9] Alrumman SA, Standing DB, Paton GI. Effects of hydrocarbon contamination on soil microbial community and enzyme activity. Journal of King Saud University. 2015;**27**(1):31-41

[10] Nie M, Wang Y, Yu J, Xiao M, Jiang L, Yang J, et al. Understanding plant-microbe interactions for phytoremediation of petroleum-polluted soil. PLoS One. 2011;**6**(3):1-8

[11] Modabberi S, Tashakor M, Sharifi Soltani N, Hursthouse AS. Potentially toxic elements in urban soils: Source apportionment and contamination assessment. Environmental Monitoring and Assessment. 2018;**190**(12):1-8

[12] U.S. Environmental Protection Agency. EPA's Report on the Environment (2003 Draft). Washington, DC: U.S. Environmental Protection Agency; 2008. Available from: U.S. Environmental Protection Agency.pdf

[13] ADEME. Méthodologie de détermination des valeurs de fonds dans les sols: Echelle d'un site Groupe de travail sur les valeurs de fonds; 2018. p. 107. Available from: www.ademe.fr/mediatheque

[14] Mahar A, Wang P, Ali A, Awasthi MK, Lahori AH, Wang Q, et al. Challenges and opportunities in the phytoremediation of heavy metals contaminated soils: A review. Ecotoxicology and Environmental Safety. 2016;**126**:111-121

[15] Ashraf S, Ali Q, Zahir ZA, Ashraf S, Asghar HN.

Phytoremediation: Environmentally sustainable way for reclamation of heavy metal polluted soils. Ecotoxicology and Environmental Safety. 2019;**174**:714-727

[16] Radziemska M, Bęś A, Gusiatin ZM, Cerdà A, Jeznach J, Mazur Z, et al. Assisted phytostabilization of soil from a former military area with mineral amendments. Ecotoxicology and Environmental Safety. 2020;**188**:109934

[17] Zhang M, Pu J. Mineral materials as feasible amendments to stabilize heavy metals in polluted urban soils. Journal of Environmental Sciences. 2011;**23**(4):607-615

[18] Rajapaksha AU, Ahmad M, Vithanage M, Kim K-R, Chang JY, Lee SS, et al. The role of biochar, natural iron oxides, and nanomaterials as soil amendments for immobilizing metals in shooting range soil. Environmental Geochemistry and Health. 2015;**37**(6):931-942

[19] Shah V, Daverey A. Phytoremediation: A multidisciplinary approach to clean up heavy metal contaminated soil. Environmental Technology and Innovation. 2020;**18**:100774

[20] Patra DK, Pradhan C, Patra HK. Toxic metal decontamination by phytoremediation approach: Concept, challenges, opportunities and future perspectives. Environmental Technology and Innovation. 2020;**18**:100672

[21] Ehrlich HL. Microbes and metals. Applied Microbiology and Biotechnology. 1997;**48**:687-692

[22] Pilon-Smits E, Pilon M. Breeding mercury-breathing plants for environmental cleanup. Trends in Plant Science. 2000;**5**(6):235-236

[23] Aderholt M, Vogelien DL, Koether M, Greipsson S. Phytoextraction of contaminated urban soils by Panicum virgatum L. enhanced with application of a plant growth regulator (BAP) and citric acid. Chemosphere. 2017;**175**:85-96

[24] Ouvrard S, Barnier C, Bauda P, Beguiristain T, Biache C, Bonnard M, et al. In situ assessment of phytotechnologies for multicontaminated soil management. Multicontaminated Soil Management. 2011;**6514**:245-263

[25] Sun M, Fu D, Teng Y, et al. In Situ Phytoremediation of PAH-Contaminated Soil by Intercropping Alfalfa (*Medicago sativa* L.) with Tall Fescue (*Festuca arundinacea* Schreb.) and Associated Soil Microbial Activity. Journal of Soils Sediments 2011;**11**:980-989

[26] Tu C, Teng Y, Luo Y, et al. PCB Removal, Soil Enzyme Activities, and Microbial Community Structures during the Phytoremediation by Alfalfa in Field Soils. Journal of Soils Sediments. 2011;**11**:649-656

[27] KJ G, Kiwanuka S, Ryan D, Dowling DN. Ecopiling: A combined phytoremediation and passive biopiling system for remediating hydrocarbon impacted soils at field scale. Frontiers in Plant Science. 2015;**5**(January):1-6

[28] Jong MY, Van Aken B, Schnoor JL. Leaching of contaminated leaves following uptake and phytoremediation of RDX, HMX, and TNT by poplar. International Journal of Phytoremediation. 2006;**8**(1):81-94

[29] Schwitzguébel JP. Phytoremediation of soils contaminated by organic compounds: Hype, hope and facts. Journal of Soils and Sediments. 2017;**17**(5):1492-1502

[30] Lee JH, Lee JH. An overview of phytoremediation as a potentially promising technology for environmental pollution control.

Biotechnology and Bioprocess Engineering. 2013;**439**:431-439

[31] Capuana M. A review of the performance of woody and herbaceous ornamental plants for phytoremediation in urban areas. iForest. 2020;**13**(2):139-151

[32] Nikolić M, Stevović S. Family Asteraceae as a sustainable planning tool in phytoremediation and its relevance in urban areas. Urban Forestry & Urban Greening. 2015;**14**(4):782-789

[33] Pavao-Zuckerman MA. The nature of urban soils and their role in ecological restoration in cities. Restoration Ecology. 2008;**16**(4):642-649

[34] Haller H, Jonsson A. Growing food in polluted soils: A review of risks and opportunities associated with combined phytoremediation and food production (CPFP). Chemosphere. 2020;**254**:126826

[35] Garbaye J. La symbose mycorhizienne. Éditions Quæ. Quærendo. 2013:251. ISBN: 978-2-7592-1964-3, ISSN: 1777-4624

[36] Nandy S, Das T, Tudu CK, Pandey DK, Dey A, Ray P. Fungal endophytes: Futuristic tool in recent research area of phytoremediation. South African Journal of Botany. 2020:1-11. DOI: 10.1016/j.sajb.2020.02.015

[37] Deng Z, Cao L. Fungal endophytes and their interactions with plants in phytoremediation: A review. Chemosphere. 2017;**168**:1100-1106

[38] Rajtor M, Piotrowska-Seget Z. Prospects for arbuscular mycorrhizal fungi (AMF) to assist in phytoremediation of soil hydrocarbon contaminants. Chemosphere. 2016;**162**:105-116

[39] Leyval C, Joner EJ, del Val C, Haselwandter K. Potential of arbuscular mycorrhizal fungi for bioremediationNo title. In: Gianinazzi S, Schüepp H, Barea JMHK, editors. Mycorrhizal Technology in Agriculture. Basel: Springer V. Birkhäuser; 2002. pp. 175-186

[40] Martino E, Turnau K, Girlanda M, Bonfante P, Perotto S. Ericoid mycorrhizal fungi from heavy metal polluted soils: Their identification and growth in the presence of zinc ions. Mycological Research. 2000;**104**(3):338-344

[41] Liu B, Wang S, Wang J, Zhang X, Shen Z, Shi L, et al. The great potential for phytoremediation of abandoned tailings pond using ectomycorrhizal Pinus sylvestris. Science of the Total Environment. 2020;**719**:137475

[42] Joyner JL, Kerwin J, Deeb M, Lozefski G, Prithiviraj B, Paltseva A, et al. Green infrastructure design influences communities of urban soil bacteria. Frontiers in Microbiology. 2019;**10**(MAY):1-13

[43] Vacheron J, Desbrosses G, Bouffaud ML, Touraine B, Moënne-Loccoz Y, Muller D, et al. Plant growth-promoting rhizobacteria and root system functioning. Frontiers in Plant Science. 2013;**4**:1-19, Article ID: 356

[44] Kong Z, Glick BR. The Role of Plant Growth - Promoting Bacteria in Metal Phytoremediation. Advances in Microbial Physiology. 2018;**71**:98-124

[45] Kong Z, Glick BR. The role of bacteria in phytoremediation. In: Yoshida T, editor. Applied Bioengineering. 2017. p. 20. Chapter 11

[46] Shin MN, Shim J, You Y, Myung H, Bang KS, Cho M, et al. Characterization of lead resistant endophytic bacillus sp. MN3-4 and its potential for

promoting lead accumulation in metal hyperaccumulator Alnus firma. Journal of Hazardous Materials. 2012;**199-200**:314-320

[47] Ma Y, Rajkumar M, Luo YM, Freitas H. Inoculation of endophytic bacteria on host and non-host plants-effects on plant growth and Ni uptake. Journal of Hazardous Materials. 2011;**195**:230-237

[48] Radziemska M, Bęś A, Gusiatin ZM, Cerdà A, Jeznach J, Mazur Z, et al. Assisted phytostabilization of soil from a former military area with mineral amendments. Ecotoxicology and Environmental Safety. 2020;**188**(October 2019):1-9

[49] Das M, Adholeya A. Role of microorganisms in remediation of contaminated soil. Microorganisms in Environmental Management: Microbes and Environment. 2012;**9789400722**:81-111

[50] Li Y, Lian J, Wu B, Zou H, Tan SK. Phytoremediation of pharmaceutical-contaminated wastewater: Insights into rhizobacterial dynamics related to pollutant degradation mechanisms during plant life cycle. Chemosphere. 2020;**253**:126681

[51] Kotoky R, Rajkumari J, Pandey P. The rhizosphere microbiome: Significance in rhizoremediation of polyaromatic hydrocarbon contaminated soil. Journal of Environmental Management. 2018;**217**:858-870

[52] Lavelle P. Functional domains in soils. Ecological Research. 2002;**17**:441-450

[53] Scheu S, Schlitt N, Tiunov AV, Newington JE, Jones TH. Effects of the presence and community composition of earthworms on microbial community functioning. Oecologia. 2002;**133**:254-260

[54] Morgan JE, Morgan AJ. The accumulation of metals (Cd, Cu, Pb, Zn and Ca) by two ecologically contrasting earthworm species (Lumbricus rubellus and Aporrectodea caliginosa): Implications for ecotoxicological testing. Applied Soil Ecology. 1999;**13**(1):9-20

[55] Jusselme MD, Miambi E, Mora P, Diouf M, Rouland-Lefèvre C. Increased lead availability and enzyme activities in root-adhering soil of Lantana camara during phytoextraction in the presence of earthworms. Science of the Total Environment. 2013;**445-446**:101-109

[56] Jusselme MD, Poly F, Miambi E, Mora P, Blouin M, Pando A, et al. Effect of earthworms on plant Lantana camara Pb-uptake and on bacterial communities in root-adhering soil. Science of the Total Environment. 2012;**416**:200-207

[57] Da Silva E, Nahmani J, Lapied E, Alphonse V, Garnier-Zarli E, Bousserrhine N. Toxicity of mercury to the earthworm Pontoscolex corethrurus in a tropical soil of French Guiana. Applied Soil Ecology. 2016;**104**:79-84

[58] Kranz CN, McLaughlin RA, Johnson A, Miller G, Heitman JL. The effects of compost incorporation on soil physical properties in urban soils – A concise review. Journal of Environmental Management. 2020;**261**:110209

[59] He L, Zhong H, Liu G, Dai Z, Brookes PC, Xu J. Remediation of heavy metal contaminated soils by biochar: Mechanisms, potential risks and applications in China. Environmental Pollution. 2019;**252**:846-855

[60] Cogger CG. Potential compost benefits for restoration of soils disturbed by urban development. Compost Science & Utilization. 2005;**13**:243-251

[61] Amalric M, Cirelli C, Larrue C. Quelle réception sociale pour l'ingénierie écologique industrielle? L_insertion socio-territoriale des zones humides artificielles. VertigO - la revue électronique en sciences de l'environnement; 2015;**15**(3):1-13

[62] Gendron C. Thinking social acceptability: Beyond interest, the values. Communication and Public Engagement. 2014;**11**:1-15

[63] Batellier P. Revoir les processus de décision publique: de l'acceptation sociale à l'acceptabilité sociale. Montréal: GaiaPresse; 2012

[64] Depraz S. Géographie des espaces naturels protégés: Genèse, principes et enjeux territoriaux. Paris: Arman Colin Collect 'U' Série Géographie; 2008

[65] Weir E, Doty S. Social acceptability of phytoremediation: The role of risk and values. International Journal of Phytoremediation. 2016;**18**(10):1029-1036

[66] Vodouhe FG, Khasa DP. Local community perceptions of mine site restoration using phytoremediation in Abitibi-Temiscamingue (Quebec). International Journal of Phytoremediation. 2015;**17**:962-972

[67] Chaline C, Barles S. La ville délétère, médecins et ingénieurs dans l'espace urbain. XVIIIe- XIXe Siècle. 2018;**608**:436

[68] Origo N, Wicherek S, Hotyat M. Réhabilitation des sites pollués par phytoremédiation. VertigO. 2012;**12**(2). Available from: http://journals.openedition.org/vertigo/12633 DOI: 10.4000/vertigo.12633

[69] ADEME, INERIS, ISA-Lille, Mines Saint-Etienne. Les phytotechnologies appliquées aux sites et sols pollués (nouveaux résultats de recherche et démonstration). ADEME, Angers; 2017. p. 68

[70] Montpetit É, Lachapelle E. Information, values and expert decision-making: The case of soil decontamination. Policy Sciences. 2016;**49**(2):155-171

[71] Méral P. Les services écosystémiques. Repenser les relations nature et sociétés. Quae, Paris; 2016. p. 298

[72] Maris V. Nature à vendre: Les limites des services écosystémiques. Quae, Paris; 2014. p. 92

[73] Khalid S, Shahid M, Khan N, Murtaza B, Bibi I, Dumat C. A comparison of technologies for remediation of heavy metal contaminated soils. Journal of Geochemical Exploration. 2017;**182**:247-268

Chapter 12

Phytoremediation: A Synergistic Interaction between Plants and Microbes for Removal of Petroleum Hydrocarbons

Govind Kumar, Pankaj Bhatt and Shatrohan Lal

Abstract

Rapid industrialization leads to the deterioration of quality of life and the environment. Petroleum hydrocarbon pollution is one of the contributing factors to that. Petroleum hydrocarbons (PHCs) are natural products, and under high temperature and pressure, they are produced by the anaerobic conversion of biomass. Excessive use of PHCs leads to pollution in the agriculturally important soils and the ultimate source of potability of water, that is, groundwater which is gaining significant attention throughout the world. The fortuitous release of PHCs such as gasoline, diesel, and heating oil are common sources of groundwater contamination. The PHC concentrations in groundwater are often above drinking water standards and bioremediation actions have to be taken. Due to their organic nature, PHCs are difficult to degrade as unavailable for microbial action. Due to this, PHCs are the most widespread environmental contaminants. Plant-microbe synergistic association for remediation of PHCs is comprehensive and it is an effective tool for reclamation of soil and environment from these kinds of undesirable materials. In addition to providing plant growth promotion, microbes can degrade PHCs effectively.

Keywords: petroleum hydrocarbon, biosurfactants, PGPR, biodegradation

1. Introduction

Different types of petroleum hydrocarbons exist, which include saturates, aromatics, asphaltenes, and resins (i.e., pyridines, quinolines, carbazoles, sulfoxides, and amides). Microbes degrade these PHCs to different extents due to their different bioavailability to microbial action. Soils are complex, highly dynamic systems that are the product results from interactions between abiotic and biotic processes that have taken place over billions of years. The result is a spatially complex environment that leads to the spatial and temporal heterogeneity of microbial activity and their diversity. The role of plants and the soil-living microbes remains, to a large extent, unexplored. However, the action of microbes to degrade organic contaminants into harmless compounds has been explored to treat contaminated environments. This approach is referred to as phytoremediation. Phytoremediation is a term that describes the application of plants to reduce the contaminant and its mobility or toxicity in soil, groundwater, or other media [1]. Phytoremediation has been increasingly considered as an appropriate strategy to restore hydrocarbon-polluted soils in

 IntechOpen

ecologically protected areas and agricultural fields. In addition to this, the bioaugmented bacterial species in ecologically protected areas, even if for bioremediation purposes, remains technically questionable because soils usually contain indigenous microbime capable of metabolizing hydrocarbons [2, 3]. In addition to phytoremediation, the use of hydrocarbon-utilizing microorganisms reduces contaminant toxicity by excretion of variety of biosurfactants which are biodegradable and consequently environmentally safe for the reclamation of polluted environments. Different kinds of microorganisms have been found to produce this surfactant, including plant growth-promoting rhizobacteria (PGPR). These PGPRs which are established bio-enhancers and biocontrol agents, due to the possession of properties to solubilize phosphate, produce IAA (indole acetic acid), and sequester iron under stress conditions through production of siderophores, will be best if used for remediation of oil-contaminated sites as this will serve two purposes. On the one hand, they will promote plant growth, and second, by producing biosurfactants that will enhance the process of remediation of oil-contaminated sites.

2. Petroleum hydrocarbon phytotoxicity

Soil physicochemical and biological properties are majorly deteriorated by petroleum hydrocarbon pollution resulting in deleterious effects to plant health and the environment.

The availability of different nutrients like exchangeable iron, phosphorous, sulfate, soil water, and soil air and such changes affect plants adversely.

Hydrostatic anaerobic conditions interfere with the soil plant water relations and seriously harm to plants [4, 5].

The effects of different contaminants on soil plant and microbes depend on the extent and the type of contamination [6].

Contamination of soil results in deterioration of soil properties leads to the damage of crop and the soil may remain not suitable for plant health for several months or years. The soil microflora and its fertility are drastically reduced by undesirable contamination [7].

Extensive damages of soil due to contamination may be for long term. Diesel fuel kills plants cells on contact but it is not a systemic killer. During taking up water and other nutrients diesel fuels reached to the plant roots and damage the roots, and this restrict the plant from uptake of essential nutrients. It can also create imbalance in soil, plant and water relationship [8].

3. Remediation approach

According to Langbehn and Steinhart [9], various approaches significantly treat the problem soil including thermal treatment (physico-chemical techniques), the extraction of gases or liquid matter, soil washing, solidification, stabilization *etc.* However, these techniques require very heavy equipment, require huge amount of energy and are very expensive. According to Rahman et al. [10], efficient removal of petroleum HCs contamination in the soil remains a challenge.

3.1 Phytoremediation

Bio-phytoremediation or the synergistic association between plants and their rhizosphere microbiome for removal of contaminants from the environment has recently become an area of huge possibilities and the intense experimentation [11].

The environmental contaminants mainly caused threat to the plant development germination and root elongation, and these are the two critical stages that are sensitive for contaminant [12].

Some plants which are tolerant to the contaminants show successful germination and root elongation.

For the removal of oil from contaminated soil plants, including grasses and legumes, proved with higher potential as compared to other plants [13, 14] due to the higher root surface area, root elongation, and better soil compaction [13].

According to Wiltse et al. [15], crude oil contamination reduces in the rhizosphere by 33–56% compared to control by using various strains of alfalfa (*Medico sativa* L.), whereas after 8 weeks, 80% of diesel fuel degraded with an alfalfa treatment (Komisar and Park [16]), and 46% of crude oil was removed in 12 weeks of interval with broad bean (*Vicia faba*) as compared to 33% without plants.

Reilley et al. [17] evaluated switchgrass as an independent species for PHCs remediation.

According to Jordahl et al. [18], for bengene, toluene, xylenes (BTX) phytoremediation, the hybrid poplar trees (*Populus deltoides x nigra*) rhizosphere showed the potential results.

Wild and Jones [19] investigated that carrot peels were accumulated with PAHs to a maximum value of 200 μg total PAHs kg^{-1} dry weight in laboratory condition.

4. Mechanism of action by plants

4.1 Phytodegradation/transformation

The degradation of contaminants by using plant-produced enzymes release of into the soil or through metabolic processes can be divided into components including absorption, translocation, and contaminants metabolism by the plant and the root exudates-mediated degradation of contaminant.

4.1.1 Fate of contaminants by the action of plants

The ability of a plant metabolize organic pollutant is generally dependent on the bioavailability of the pollutant, and it is reflected by the octanol-water partition coefficient, Kow, of the pollutant [20, 21].

For the plant action for the contaminant, the type of plants and contaminant also affect the bioavailability of contaminant [22].

Cunningham and Berti [20] explained depending factors of plant to absorb, translocate, and metabolize the contaminant (Kow values in log scale) as described below.

i. With log Kow ≤ 1, plants are able to absorb, translocate, and metabolize hydrophilic contaminants.

   As these contaminants are with high bioavailability, their absorption is controlled by water influx into the plant and they may cause groundwater contamination.

ii. With log Kow values between 1 and 4, plants are able to absorb, translocate, and metabolize the contaminants. According to Briggs et al. [23], the highest contaminant concentration translocated at shoots part with log Kow of 1.8, with declining concentrations at higher and lower values of log Kow.

iii. With log Kow values larger than 4 plants are generally unable to absorb, translocate, and metobilize contaminants due to high hydrophobicity and the contaminant adsorbs to lipids on the root surface of the plant [20–22].

The contaminant absorbed by the plant may be translocated to different plant parts where it is metabolize partially or completely or incorporated into cellular constituents and volatilized [22, 24]. This whole process that includes absorption, translocation, and volatilization called as phytovolatilization.

4.1.2 Significance of root exudates

Role of root exudates may aid remediation direct degradation of contaminants. Root exudates increase the access of the pollutant, soil lubrication, and acting as co-metabolites with PHCs.

According to Schnoor et al. [22], few plant-based enzymes are able to remediate 2,4,6-trinitrotoluene (TNT) and trichloroethylene (TCE).

The bioavailability of the contaminant is the extent to which a pollutant is accessible for to microbial activity [25].

Root exudates (organic acids in nature) may enhance the pollutant bioavailability by competing with the original pollutant for absorption/adsorption sites in the soil due to structural similarity.

Roots are also release lipids and sterols that have been found to increase the bioavailability of contaminants and making them available for microbial degradation. The root passages of plants are also facilitating by lipids and sterols.

Root exudates may act as co-metabolites with the pollutant as root exudates are structurally similar with PHCs [26]. Root exudates and pollutant that are structurally similar showed in **Figure 1**.

4.2 Phytovolatilization

According to Farrell and Germida [25], there is pollutant movement from the soil or groundwater and into the plants and then to the atmosphere, which is called phytovolatilization. This happened when the pollutant is absorbed by the roots, translocated to the plant, and volatilized into the atmosphere.

4.3 Phytostabilization

According to Farrell and Germida [25], plants use to restrict the pollutant in the soil or groundwater called "phytostabilization" and it can be explained by three mechanisms: (1) absorption, (2) root accumulation, and (3) surfaces adsorption by root and entering into humic matter in the plant rhizosphere. All three mechanisms are dependent on the value of Kow as explained above.

4.3.1 Absorption by root and accumulation

In absorption and accumulation, contaminant remains restricted in the roots of the plant. The translocation of the contaminants is not possible into the rest of the plant and are therefore not degraded, incorporated into the cell structure, or volatilized. This may happen for contaminants with a log Kow value between 1 and 4 [20].

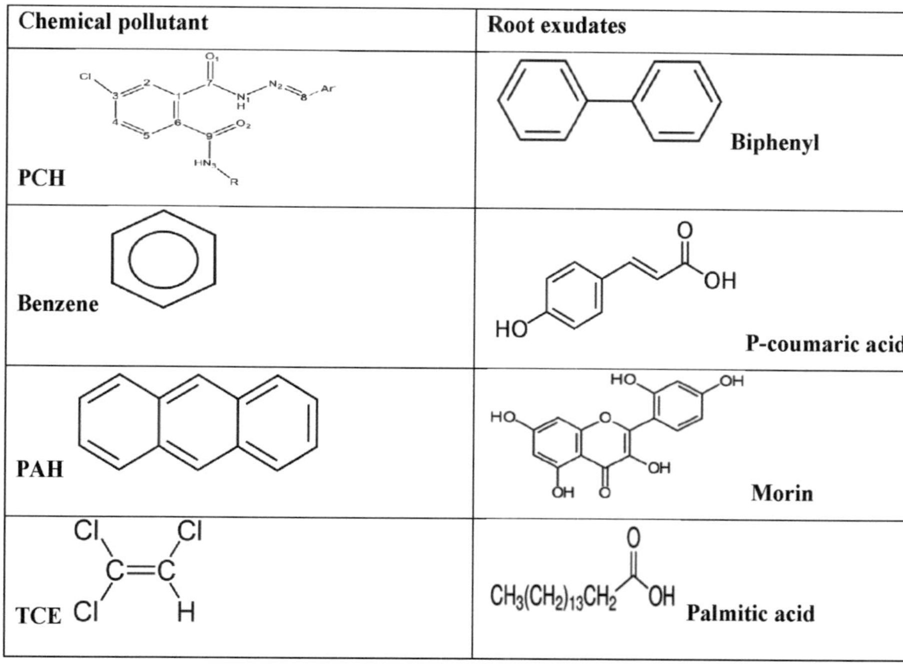

Figure 1.
Similar structures of root exudates and chemical pollutants.

4.3.2 Root adsorption

In these mechanisms, contaminants are immobilized by adsorption to lipids on the surface of the roots. This may happen for contaminants with a log Kow value above 4 [20–22].

5. Degradation of organic contaminants by microorganisms

The halogenated products of petrochemicals and mineral oil constituents are the most important classes of organic contaminants in the environment. Microbial degradation of organic contaminants normally occurs as a result of microorganisms act on the contaminant for their own metabolisms and production cellular constituents or reproduction. Organic contaminants used by microorganisms as a source of carbon and electrons that the organisms use to obtain energy [26].

According to Committee on *In Situ* Bioremediation in the year 1993, the microbial metabolism of contaminants involves aerobic respiration. Anaerobic respiration, co-metabolism, fermentation, reductive dehalogenation, and the use of inorganic compounds as electron donors: these are the variations in metabolism by microorganisms.

Interestingly, bacteria are capable of rapidly distributing genetic information to each other, thus allowing them to adapt rapidly to adverse environment, such as exposure to new pollutants [27].

The *dioxygenase* enzyme plays a significant role in aerobic biodegradation of PAHs by involving the incorporation of two atoms of molecular oxygen into the contaminant and production of less toxic compounds such as acids, alcohols, carbon dioxide, and water [28–30].

In contrast, biodegradation by eukaryotic fungi which is similar to the biodegradation mechanism found in mammals initially it involves the incorporation of only one atom of oxygen into the PAHs (polycyclic aromatic hydrocarbons) [29, 31–33].

Although most of the time fungal transformations result in compounds that are less toxic than the parent PAHs, while some of the minor metabolites are produced, they are more toxic compounds than the PAHs [33].

The microbial breakdown and removal of contaminants are interrelated processes that occur in the soil.

According to Lyman et al. [34], biodegradation is the microbial-mediated chemical transformation of organic compounds, while microbial uptake is the direct removal of the contaminant by adsorbing compounds to the membrane surface or by absorbing compounds through the membrane.

These two processes are interrelated in that the contaminant taken up may be the original contaminant or a biotransformation product.

That microbes are able to degrade and take up pollutants has been well studied, and it is the conceptual basis for other remediation techniques like air sparging, land farming, composting, bioreactors, intrinsic remediation, and others [35].

Depending upon the microorganisms, a number of different microorganisms are able to degrade a number of different PHCs, and the specific catabolic pathway used is dependent on the microbe and pollutants.

In general, microbes degrade PHCs by adsorbing the contaminant to the membrane surface or absorbing the contaminant through the membrane and by using oxygenase enzymes, incorporating oxygen into, and cleaving the structure of the hydrocarbon.

Finally, oxidation of subsequent end products and incorporation into the Krebs cycle may result in the final degradation step and the release of CO_2, H_2O, and energy [35, 36] but the complete degradation does not always happen. Sometimes end products may be directly degraded by microbes and not degraded further or may be degraded to smaller, simpler, more stable intermediaries and then incorporated into the soil as humus or soluble acids, ketones, and alcohols [34].

The biodegradation of the contaminants are depends on the size of the contaminant and the types and geometry of its bonds. PHCs have bonds that microbes have difficulty breaking or are not able to break due to their ring structure and hydrophobic nature. For example, linear alkanes were found to be more readily degradable than branched alkanes or ring structure [35].

Diverse group of microbes able to degrade different contaminant depending on the sites/ locations of contamination. Not all microbes are able to directly take up all contaminants. This results in variations of PHCs and microbial population composition over time and space with the most rapidly degradable HCs and associated microbes being replaced by less degradable hydrocarbons and associated microbes. According to Riser-Roberts [35], one kind of microorganism is very rarely able to fully degrade any other specific contaminant. The diverse microbial populations are able to do effective remediation of the contaminants.

6. Plants and microbes synergy for PHCs biodegradation

Plants have been shown to facilitate organic contaminant degradation principally by providing optimal conditions for microbial proliferation in the root zone. The degradation processes are influenced not only by rhizospheric microorganisms but also by specific properties of the host plant [37]. If plants can be successfully acclimatized on polluted soils, then the plant-microbial interaction in the root zone (rhizosphere) may provide an economical method for enhancing microbial degradation of complex PHCs (**Figure 2**).

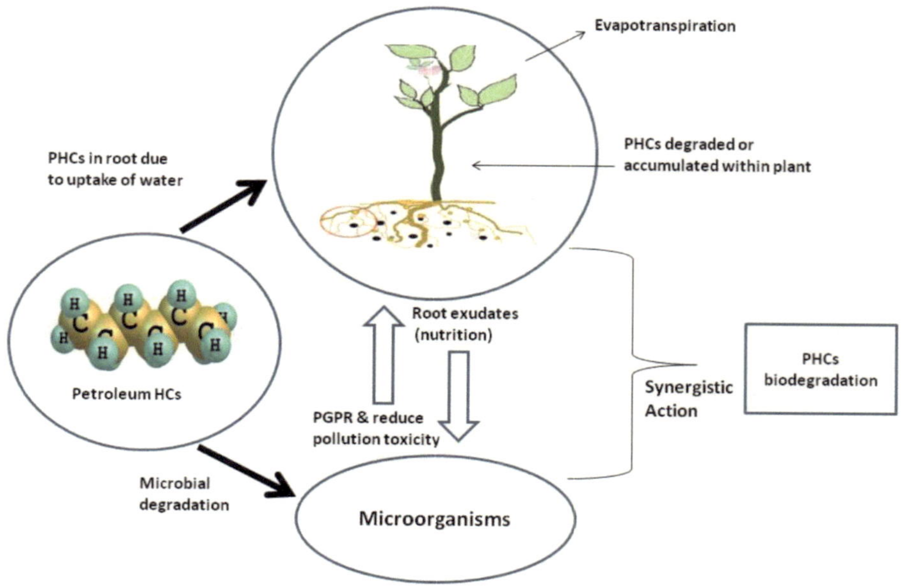

Figure 2.
Mechanism of plant microbe interaction for petroleum HCs pollutants remediation.

7. Enhanced rhizosphere degradation

The enhanced rhizosphere degradation is the breakdown of pollutant in the soil as a result of microbial activity that is enhanced in the presence of the rhizosphere [25] and should include another process in plant-mediated remediation, that is, phytoremediation, the removal of contaminants from the soil. Enhanced microbial activity in the rhizosphere provides the health benefits to the plant and the entire phytoremediation system. Komisar and Park [38] showed that vegetated (alfalfa) soils with diesel contamination observed more microbial count and rapid removal of contaminant. Banks et al. [39] observed that viable counts in soil spiked with polyaromatic and aliphatic hydrocarbons were higher in the presence of alfalfa than in soil without plants. Root elongation and growth opens deeper soil to better water infiltration and oxygen diffusion [40]. Root surfaces provide adhesive zone for soil microorganisms, and roots can disrupt soil aggregates and increase biodegradation of entrapped hydrophobic contaminants [41]. These studies showed how root growth increases microbial activity. Gunther et al. [14] showed that rhizosphere microbial community was mainly responsible for enhanced hydrocarbon disappearance as compared to root free soil. Hou et al. [42] found that due to higher rye grass root intensity, an increase in the degradation of contaminant. Banks et al. [41] observed that the huge reduction in total petroleum hydrocarbon (TPH) concentrations occurred in period with the greatest root growth, but did not evaluate concurrent microbial activity. Gunther et al. [14] found that soil planted with ryegrass reduced a greater amount of a mixture of hydrocarbons than soil without plant. The mixture of hydrocarbon includes *n*-alkanes (C_{10}, C_{14} to C_{18}, C_{22}, and C_{24}), also pristane, hexadecane, phenanthrene, anthracene, fluoranthene, and pyrene. After 22 weeks, the initial hydrocarbon concentration of 4330 mg total hydrocarbon per kg soil decreased to nearly 120 mg per kg soil (97% reduction) in planted soils as compared to 790 mg per kg soil (82% reduction) in soil without plant.

8. Role of microorganisms in reducing phytotoxicity to plants

Other significant role played by microbes involves their ability to reduce the phytotoxicity of contaminants to the plants system and can facilitate plant to grow in adverse soil environment, thereby stimulating the degradation of phytotoxic and non-phytotoxic contaminants [26].

According to Walton et al. [37], the defenses of plants to contaminants may be supplemented rhizopheric microbial activity for degradation of contaminants. The plants and microbes have synergistically work together for dealing with phytotoxicity, where microorganisms benefit from the root exudates while the plants benefit from the ability of microorganisms to break down toxic chemicals and PGPR properties. Rasolomanana and Balandreau [43] observed that rice growth was improved in soil oil residues had been applied.

The authors hypothesized that the increased growth resulted from the co-metabolic action of bacterial species (genus *Bacillus*) by using root exudates in the rhizosphere polluted with the oil residues.

8.1 PGPR

Plant growth-promoting rhizobacteria (PGPR) were first described by Kloepper and Schroth [44] as the soil bacteria that colonize the roots zone of plants by inoculation onto seed and that enhance plant growth. The bacteria inhabiting plant roots and facilitate the plant growth by the mechanisms are referred to as plant growth-promoting rhizobacteria (PGPR).

The plant growth-promoting rhizobacteria (PGPR) facilitate plant growth either directly or indirectly. The direct benefits to the plants are provided by the production of plant growth regulators such as auxine and cytokinines and by increasing the plant uptake of some micro and macro elements in the rhizosphere [45] and indirectly, through the action of biological control of plant pathogens or induction of host defense mechanisms [46–48].

The synergistic action that exists between plants and microbes in the rhizosphere plays significant roles in enhancing the efficacy of phytoremediation [49–54]. Root exudates can stimulate the growth of PGPR, which in turn can alleviate plant stress by lowering stress ethylene, facilitating the nutrient uptake and/or by degrading/sequestering soil contaminants [55].

These microorganisms are nourished and carried through the soil by plant roots [50]. The soil with large volumes of roots results the microbial population can reach to the concentrations upto ~10^{12} microbes per gram of soil [56].

This leads to increase the microbial population to ~500 kg ha^{-1} [57] in the rhizosphere zone. With the PGPR association, plant root growth enhanced and potentially used volume of soil and accelerating salt remediation.

According to Glick [58, 59], among many PGPR properties, the ACC (1-aminocyclopropane-1-carboxylic acid) deaminase activity is the key characteristic because PGPR use the ACC as an N source. Ethylene synthesis significantly reduced by using ACC, the precursor in plants to ethylene. [58, 60, 61] showed that by inhibiting ethylene synthesis, tolerance to stress has been observed. Some PGPR also observed to produce auxin to facilitate root growth.

Many strains of genus *Pseudomonas* observed as potential PGPR due to poses PGPR properties like 1-aminocyclopropane-1-carboxylate (ACC) deaminase activity, indole acetic acid (IAA) and siderophore, P, Zn, K solubilization, etc. Due to *ACC deaminase* activity, ethylene level could not harm root growth and development. *Pseudomonas* spp. provides better root elongation, seedling survival,

Author details

Govind Kumar[1]*, Pankaj Bhatt[2] and Shatrohan Lal[1]

1 ICAR-Central Institute for Subtropical Horticulture (ICAR-CISH), Lucknow, India

2 Integrative Microbiology Research Centre, South China Agricultural University, Guangzhou, China

*Address all correspondence to: govind34093@gmail.com

References

[1] McCutcheon SC, Schnoor JL. Overview of phytotransformation and control of wastes. In: McCutcheon S, Schnoor J, editors. Phytoremediation: Transformation and Control of Contaminants. Hoboken, NJ: John Wiley & Sons, Inc.; 2003

[2] Bento FM, Camargo FAO, Okeke BC, Frankenberger WT. Comparative bioremediation of soils contaminated with diesel oil by natural attenuation, biostimulation and bioaugmentation. Bioresource Technology. 2005;**96**:1049-1055

[3] Macnaughton SJ, Stephen JR, Venosa AD, Davis GA, Chang YJ, White DC. Microbial population changes during bioremediation of an experimental oil spill. Applied and Environmental Microbiology. 1999;**65**:3566-3574

[4] Adams RS, Ellis R. Some physical and chemical changes in soil brought about by saturation with natural gas. Soil Science Society of America Proceedings. 1960;**24**:41-44

[5] Udo EJ, Fayemi AA. The effect of oil pollution on germination, growth and nutrient uptake of corn. Journal of Environmental Quality. 1975;**4**:5377-5540

[6] Boethling RS, Alexander M. Effect of concentration of organic chemicals on their biodegradation by natural microbial communities. Applied and Environmental Microbiology. 1979;**37**:1211-1216

[7] Torstenssen L, Mikaelpell O, Bostenberg C. Need of a strategy for evaluation of arable soil quality. Environmental Pollution. 1998;**27**:4-7

[8] McCown BH, Deneke FJ, Richard WE, Tierzen L. The response of Alaskan terrestrial plant communities to the presence of petroleum. Environmental Pollution. 1972;**1**:34-43

[9] Langbehn A, Steinhart H. Biodegradation studies of hydrocarbons in soils by analyzing metabolites formed. Chemosphere. 1995;**30**:855-868

[10] Rahman KSM, Banat IM, Thahira J, Thayumanavan T, Lakshmanaperumalsamy P. Bioremediation of gasoline contaminated soil by a bacterial consortium amended with poultry litter, coir pith and rhamnolipid biosurfactant. Bioresource Technology. 2002;**81**(1):25-32

[11] Cunningham SD, Anderson TA, Schwab AP, Hsu FC. Phytoremediation of soils contaminated with organic pollutants. Advances in Agronomy. 1996;**56**:55-114

[12] Baud-Grasset F, Baud-Grasset S, Saffernan SI. Evaluation of the bioremediation of a contaminated soil with phytotoxicity tests. Chemosphere. 1993;**26**:1365-1374

[13] Aprill W, Sims RC. Evaluation of the use of prairie grasses for stimulating polycyclic aromatic hydrocarbon treatment in soil. Chemosphere. 1990;**20**:253-265

[14] Gunther T, Dornberger U, Fritsche W. Effects of ryegrass on biodegradation of hydrocarbons in soil. Chemosphere. 1996;**33**(2):203-216

[15] Wiltse CC, Rooney WL, Chen Z, Schwab AP, Banks MK. Greenhouse evaluation of agronomic and crude oil-phytoremediation potential among alfalfa genotypes. Journal of Environmental Quality. 1998;**27**:169-173

[16] Komisar SJ, Park J. Phytoremediation of diesel contaminated soil using alfalfa. In:

Fourth International In Situ and On-Site Bioremediation Symposium; 28 April-1 May 1997; New Orleans, LA. Vol. 3. 1997. pp. 331-336

[17] Reilley KA, Banks MK, Schwab AP. Dissipation of polycyclic aromatic hydrocarbons in the rhizosphere. Journal of Environmental Quality. 1996;**25**:212-219

[18] Jordahl JL, Foster L, Schnoor JL, Alvarez PJJ. Effect of hybrid poplar trees on microbial populations important to hazardous waste bioremediation. Environmental Toxicology and Chemistry. 1997;**16**:1318

[19] Wild SR, Jones KC. Polynuclear aromatic hydrocarbons uptake by carrots grown in sludge-amended soil. Journal of Environmental Quality. 1992;**21**:217-225

[20] Cunningham SD, Berti WR. Remediation of contaminated soils with green plants: An overview. In Vitro Cellular & Developmental Biology-Plant. 1993;**29**:207-212

[21] Alkorta I, Garbisu C. Phytoremediation of organic contaminants in soils. Bioresource Technology. 2001;**79**:273-276

[22] Schnoor JL, Light LA, McCutcheon SC, Wolfe NL, Carreia LH. Phytoremediation of organic and nutrient contaminants. Environmental Science & Technology. 1995;**29**:318A-323A

[23] Briggs GG, Bromilow RH, Evans AA. Relationship between lipophilicity and root uptake and translocation of non-ionized chemicals by barley. Pesticide Science. 1982;**13**:495-504

[24] Salt DE, Smith RD, Raskin I. Phytoremediation. Annual Review of Plant Physiology and Plant Molecular Biology. 1998;**49**:643-668

[25] Farrell RE, Germida JJ. Phytotechnologies: Plant Based System for the Remediation of Oil Impacted Soils. Saskatoon, SK, Canada: Department of Soil Science, University of Saskatchewan; 2000

[26] Siciliano SD, Germida JJ. Mechanisms of phytoremediation: Biochemical and ecological interactions between plants and bacteria. Environmental Reviews. 1998;**6**:65-79

[27] Bollag JM, Mertz T, Otjen L. Role of microorganisms in soil remediation. In: Anderson TA, Coats JR, editors. Bioremediation through Rhizosphere Technology. ACS Symposium Series 563. American Chemical Society. York, PA: Maple Press; 1994. pp. 2-10

[28] Gibson DT, Subramanian V. Microbial degradation of aromatic hydrocarbons. In: Gibson DT, editor. Microbial Degradation of Organic Compounds. New York: Marcel Dekker; 1984. pp. 181-252

[29] Pothuluri JV, Cerniglia CE. Microbial metabolism of polycyclic aromatic hydrocarbons. In: Chaudry GR, editor. Biological Degradation and Remediation of Toxic Chemicals. Porland, Oregon: Dioscorides Press; 1994. pp. 92-124

[30] Ewies JB, Erges SJ, Chang DPY, Schroeder ED. Bioremediation Principles. Toronto: McGraw-Hill, Inc.; 1998

[31] Cerniglia CE, Gibson DT, Van Baden C. Algal oxidation of aromatic hydrocarbons: Formation of 1-napthol from naphthalene from *Agmenellum quadruplicatum*, strain PR-6. Biochemical and Biophysical Research Communications. 1979;**88**:50-58

[32] Cerniglia, C.E., Kelly, D.W., Freeman, J.P., Miller, D.W. 1986.

Microbial metabolism of pyrene. Chemical ± biological interactions 57, 203-216.

[33] Sutherland JB. Detoxification of polycyclic aromatic hydrocarbons by fungi. Journal of Industrial Microbiology. 1992;**9**:53-62

[34] Lyman WJ, Reidy PJ, Levy B. Mobility and Degradation of Organic Contaminants in Subsurface Environments. Beltsville, Maryland and Washington, D.C., United States: C. K. Somley, Inc; 1992. p. 395

[35] Riser-Roberts E. Remediation of Petroleum Contaminated Soils: Biological Physical and Chemical Processes. Boca Raton, FL: Lewis Publishers; 1998. p. 542

[36] Chapelle FH. Ground Water Microbiology and Geochemistry. U.S. Geological Survey in Columbia, South Carolina: John Willey and Sons, Inc; 1992. p. 424

[37] Walton BT, Holyman AM, Perez MM, Anderson TA, Johnson TR, Christman RF. Rhizosphere microbial communities as a plant defense against toxic substances in soils. In: Anderson TA, Coats JR, editors. Bioremidiation through Rhizosphere Technology. Washington D.C.: American Chemical Society; 1994. pp. 42-92

[38] Komsar SJ, Park J. Phytoremediation of Diesel-Contaminated Soil Using Alfalfa. Germany: TIB, Leibniz Information Centre for Science and Technology; 1997. pp. 331-336. (ISBN: 1574770314, 1574770284)

[39] Banks MK, Lee E, Schwab AP. Evaluation of dissipation mechanisms for benzo[a]pyrene in the rhizosphere of tall fescue. Journal of Environmental Quality. 1999;**28**:294-298

[40] Singer AC, Crowley DE, Thompson IP. Secondary plant metabolites in phytoremediation and biotransformation. Trends in Biotechnology. 2003;**21**(3):123-130

[41] Banks MK, Mallede H, Rathbone K. Rhizosphere microbial characterization in petroleum-contaminated soil. Soil and Sediment Contamination. 2003;**12**(3):371-385

[42] Hou FS, Milke MW, Leung DW, MacPherson DJ. Variations in phytoremediation performance with diesel-contaminated soil. Environmental Technology. 2001;**22**:215-222

[43] Rasolomanana JL, Balandreau J. Role de la rhizosphere dans la biodegradation decomposes recalcitrants: cas d'une riziere polluee par des residus petroliers. Revue D' Ecologie et de Biologie du sol. 1987;**24**(3):443-457

[44] Kloepper JW, Schroth MN. Proceedings of the 4th International Conference on Plant Pathogenic Bacteria. Vol. 2. Station de Pathoglogie Vegetale et Phytobacteriologie; 1978. pp. 879-882

[45] Glick BR. The enhancement of plant growth by free living bacteria. Canadian Journal of Microbiology. 1995;**41**:109-117

[46] O'Sullivan DJ, O'Gara F. Traits of fluorescent *Pseudomonas* spp. involved in suppression of plant pathogens. Microbiological Reviews. 1992:662-676

[47] Thomashow LS, Weller DM. Current concepts in the use of introduced bacteria for biological disease control: Mechanisms and antifungal metabolites. In: Stacey G, Keen NT, editors. Plant-Microbe Interactions. Vol. 1. New York: Chapman and Hall; 1996. pp. 187-227

[48] Van Loon LC, Bakker PAHM, Pieterse CMJ. Systemic resistance induced by rhizosphere bacteria.

Annual Review of Phytopathology. 1998;**36**:453-485

[49] Glick BR. Phytoremediation: Synergistic use of plants and bacteria to clean up the environment. Biotechnology Advances. 2003;**21**:383-393

[50] Kuiper I, Lagendijk EL, Bloemberg GV, Lugtenberg BJJ. Rhizoremediation: A beneficial plant microbe interaction. Molecular Plant-Microbe Interactions. 2004;**17**:6-15

[51] Huang XD, El-Alawi Y, Penrose DM, Glick BR, Greenberg BM. A multi-process phytoremediation system for removal of polycyclic aromatic hydrocarbons from contaminated soils. Environmental Pollution. 2004;**130**:465-476

[52] Pilon-Smith WAH. Phytoremediation. Annual Reviews in Plant Biology. 2005;**56**:15-39

[53] Choudhary NK, Fletcher RH, Somurel SB. Systematic review: The relationship between clinical experience and quality of health care. Annals of International Medical and Dental Research. 2005;**142**:260-273

[54] Sinha S, Mukherjee SK. Cadmium-induced siderophore production by a high Cd-resistant bacterial strain relieved Cd toxicity in plants through root colonization. Current Microbiology. 2008;**56**:55-60

[55] Hontzeas N. Expression and characterization of 1-ACC deaminase from the rhizobacterium *Pseudomonas pudta* UW4: A key enzyme in bacterial plant growth promotion. Biochimica et Biophysica Acta. 2004;**1703**:11-19

[56] Whipps JM. Carbon economy. The Rhizosphere. 1990:59-97

[57] Metting J. Soil Microbial Ecology: Applications in Agricultural and Environmental Management. England, U.K.: Taylor and Francis; 1992

[58] Glick BR. Bacterial ACC deaminase and the alleviation of the plant stress. Advances in Applied Microbiology. 2004;**56**:291-312

[59] Glick BR. Using soil bacteria to facilitate phytoremediation. Biotechnology Advances. 2010;**28**(3): 367-374

[60] Mayak S, Tirosh T, Glick BR. Plant growth-promoting bacteria confer resistance in tomato plants to salt stress. Plant Physiology and Biochemistry. 2004;**42**:565-572

[61] Sergeeva E, Shah S, Glick BR. Tolerance of transgenic canola expressing a bacterial ACC deaminase gene to high concentration of salt. World Journal of Microbiology and Biotechnology. 2006;**22**:277-282

[62] Department of Energy (DOE) U.S. A Framework for Net Environmental Benefit Analysis for Remediation or Restoration of Petroleum-Contaminated Sites. ORNL/TM-2003/17. National Petroleum Technology Office; 2003

[63] Environmental Protection Agency. The Quality of Our Nations Waters: A Summary of the National Water Quality Inventory: 1998 Report to Congress. EPA 84100001. Office of Water, U. S. Environmental Protection Agency; 2000

[64] Wilson MA. Ecosystem Services at Superfund Redevelopment Sites. Prepared for U.S. EPA, Office of Solid Waste and Emergency Response, Policy Analysis and Regulatory Management Staff; 2004

[65] Nowak DJ, Crane DE. Carbon storage and sequestration by urban trees in the USA. Environmental Pollution. 2002;**116**:381-389

Chapter 13

Nanophytoremediation: An Overview of Novel and Sustainable Biological Advancement

Silpi Sarkar, Manoj Kumar Enamala, Murthy Chavali, G.V.S. Subbaroy Sarma, Mannam Krishna Murthy, Abudukeremu Kadier, Ashokkumar Veeramuthu and K. Chandrasekhar

Abstract

Increased threat of metals simultaneous to the biota well-being and the environs is continually causing a major apprehension worldwide. The phytoremediation technique is highly advantageous involving the natural processes of plants viz., translocation, evapotranspiration, and bioaccumulation, thus degrading contaminants slowly. In particular, nanophytoremediation is a rapid green alternative as it reduces the ancillary impacts of the environment such as green gas emissions, waste generation, and natural resource consumption to the present scenario as there is a great potential of nanoparticles from plants which can be synthesized. Nanophytoremediation is a current methodology for remediation of pollutants, contaminants by using synthesized nanoparticles from plants. In this, the use of different strategies enhances the selective uptake capabilities of plants. The metal elements in excess are affecting the physiological processes in plants; thus, it is necessary to apply nanophytoremediation technology through transgenic plants. In this review paper, we focused on plant species, which can be used as metal tolerant, hyperaccumulators. Due to the insurmountable pressure of a sustainable cleaner environment, bioremediation can be concurrent with nanoparticles for efficient and effective sustainable measures.

Keywords: nanoparticles, phytoremediation technologies, hyperaccumulators, bioelements, contaminants, transgenic plants

1. Introduction

Plants are autotrophic in nature, thus are self-sufficient in the utilization of sunshine and CO_2 as energy and carbon sources. The vegetation mostly depends on its roots for water, nutrients, and minerals from groundwater and soil. The maintenance of the greener environment is mostly integrated with plants. Further, the sustainability of these plants depends on the environment, which is contaminated mostly from anthropogenic activities and pollution. In contrast, plants also absorb

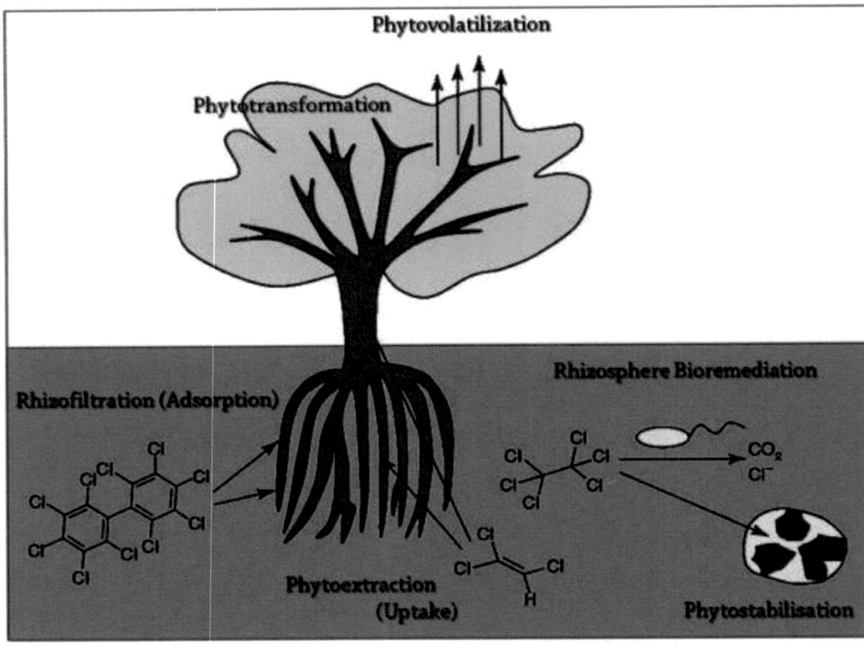

Figure 1.
Illustration of physiological processes occurring in plants during phytoremediation.

Treatment	Mechanism	Medium
Phytodegradation	Degradation of plant uptake organics	Surface and groundwater
Rhizofiltration	Roots can uptake metals	Surface waters and water pumped through troughs
Bioremediation supported by plants	Enhanced microbial degradation in the rhizosphere	Soils and groundwaters within the rhizosphere
Phytoextraction	Metal uptake and the presence of metal concentration directly via plant tissue with the subsequent exclusion of plants for biomass degradation.	Soils
Phytostabilization	Root exudes which causes metal precipitation, thus decreases the bioavailability	Soils, groundwaters, and tailings in a mine
Phytovolatilization	Evapo transpires Se, Hg, and volatile organics	Soils and groundwaters
Phytomining	Inorganic substance extraction from mine ore	Soil
Removal of organics	Volatile organics are left out through the plant	Air
Rhizosecretion	Molecular farming methodology, which secretes natural products and recombinant proteins from roots.	Soil
Vegetative caps	Rainwater is evapotranspiration, preventing contaminant leaching from a waste disposal site	Soil

Table 1.
Technologies related to phytoremediation.

diverse compounds that are toxic in nature, thus can be considered as an efficient detoxification mechanism for the removal of contaminants. Thus, from this viewpoint, plants are employed effectively in the treatment of contaminants viz., organic contaminants, polyaromatic hydrocarbons, which are potentially viable in contaminant detoxification. Previously, the traditional remediation of metal-contaminated soil includes on-site management and subsequent disposal of wastes to another landfill site. However, this makes the site hazardous with additional risks of migration of contamination. There are various clean-up techniques for soils that can be categorized as physical, chemical, and biological. There are reports of the chemical and physical processes, which have limitations viz., great price, labor intensive, variations in properties of soil, and disturbance of the native soil microflora, whereas chemical techniques increase secondary pollution problems with large volumetric sludge which increases the cost. The biological remediation processes consist of bioventing, bioleaching, bioremediation, bioreactors, bioaugmentation, biostimulation, and land forming. In this context, the phytoremediation technology has been in existence in par with other remediation technologies as a novel natural ecological, biological remediation process.

Phytoremediation created from Greek prefix "*phyto*" means plant and Latin suffix "*remedium*" means remedy or restore. Phytoremediation is a versatile technology to treat polluted soils, pollutants, deposits, and groundwater, in a profitable as well as environmental welcoming the usage of plants [1], thus can be referred to as natural green biotechnology **Figure 1** denotes the different phytoremediation technologies. Phytoremediation technology is suitable against several types of contaminants [2] in the atmosphere in a variety of media, as mentioned in **Table 1**.

Nanoparticles	Plant
Silicon-Germanium (Si-Ge) nanoparticles	Freshwater diatom *Stauroneis* sp.
Au and Ag nanoparticles	*Pelargonium graveolens*, *Hibiscus rosasinensis*, *Citrus sinensis*, *Diospyros kaki (Persimmon)*, *Emblica officinalis*, *Phyllanthium*, *Mushroom extract*, *Coriandrum sativum*
Ag nanoparticles	*Elettaria cardamom*, *Parthenium hysterophorus*, *Euphorbia hirta*, *Ocimum* sp., *Nerium indicum*, *Brassica juncea*, *Azadirachta indica*, *Pongamia pinnata*, *Clerodendrum inerme*, *Opuntia ficus-indica*, *Gliricidia sepium*, *Desmodium triflorum*, *Carica papaya*, *Coriandrum sativum*, *Peargoneum graveolens*, *Avicennia marnia*, *Aloe vera extract*, *Capsicum annum*, *Rhizophora mucronata*, *Ceriops tagal*, *Rumex hymenosepalus*, *Pterocarpus santalinus*, *Sonchus asper*
Au nanoparticles	*Terminalia catappa*, Banana peel, *Mucuna pruriens*, *Medicago sativa*, *Allium cepa* L., *Camellia sinensis* L., *Chenopodium album* L., *Justicia gendarussa* L., *Macrotyloma uniflorum* (Lam) Verde, *Azadirachta indica* A. Juss, *Magnolia kobus and Diospyros kaki*, *Cinnamomum zeylanicum*, *Mentha piperita* L., *Mirabilis jalapa* L., *Syzygiuma romaticum*, *Terminalia catappa* L., and *Amaranthus spinosus*
Ag, Ni, Co, Zn and Cu nanoparticles	*Brassica juncea*, *Medicago sativa*, and *Helianthus annuus*
Platinum nanoparticles	*Diospyros kaki* and *Ocimum sanctum* L.,
Palladium nanoparticles	*Cinnamomum zeylanicum* Blume, *Cinnamomum camphora* L., *Gardenia jasminoides* Ellis., *Soybean (Glycine max)* L.,
Lead nanoparticles Indium oxide nanoparticles Gold/Silver bimetallic nanoparticles	*Vitis vinifera* L. and *Jatropha curcas* L. *Aloe vera* (*Aloe barbadensis* Miller), *Azadirachta indica* (Neem)

Table 2.
Numerous nanoparticles synthesized from the plants.

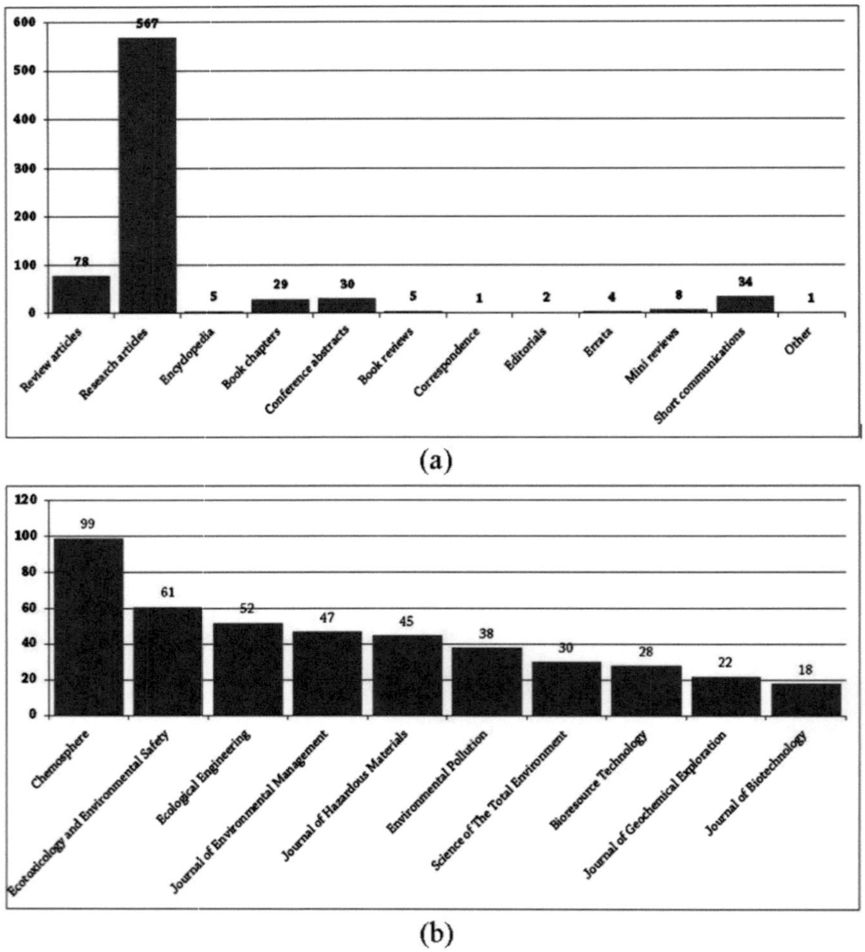

Figure 2.
Publication trends for phytoremediation as per the ScienceDirect database—year-wise publications, (a) category wise and (b) journal wise.

Phytoremediation technique has its own limitations:

a. Slow remediation time

b. Plant waste after phytoremediation

It is seen previously that plants [3] have a tendency to produce nanoparticles under appropriate conditions, as mentioned in **Table 2**. The deployment of contained contaminants remains equally *in situ* and *ex situ*. One of the newer techniques of *in situ* remediation, nanotechnology has been in focus with the usage of nanomaterials in various laboratory investigations and field applications, mostly in North America and Europe. But in India, nanophytoremediation is not practiced. Although nanophytoremediation can be an economically viable process, proper utilization can be ecologically useful.

Several studies report the usage of nanoparticles to have an affirmative effect on plants. Mixed TiO_2 (nano) and SiO_2 (nano) were presented into soybean (*Glycine max*) increasing activity of nitrate reductases, which sped the plant propagation by increasing the water absorption and fertilizer utilization (Lu et al., 2001).

Similarly, it was found by studies that carbon dots (CDs) promote growth in mung bean at 0–1.0 mg/mL concentration (Li et al., 2016). This result supports that nanoderivatives like carbon dots can absorb and utilize nutrients that induce a physiological response. Although there are studies on nanoparticles that can cause toxicity, it has not been yet elucidated for most nanoparticles. It is vividly important to study nanoparticles and their effect on plant growth mechanisms to prevent the ecological risk of nanoparticles and to promote sustainable development of nanotechnology in the near future, particularly in the Indian context. Thus, the different integrated approaches to producing nanoparticles and apply nanoderivatives eliminating the metal impurities from soil and water; thus, a flawless, in-depth study of nanoparticles is required, which can be applied. Nanophytoremediation study is based as an alternative remediation advanced technology in addition to the phytoremediation, the current scenario of reducing the contaminants in a safer way.

1.1 Publications

Publications wise not many were found in the literature databases; for example, probing ScienceDirect database, it has found none on nanophytoremediation. Since the year 1995 to date, 2018, the number of publications found to be 764. Of which highest published were found to be research articles (567) followed by review articles (78), short communications (34), and rest others.

Among journal trends, the highest number was found to be in journal: Chemosphere (99) followed by Ecotoxicology and Environmental Safety (61), Ecological Engineering (52), the lowest number published was in Journal of Biotechnology (18) over the years 1995–2018. Publication trends for phytoremediation, as observed from the ScienceDirect Database year-wise publications: (a) category wise and (b) journal wise were shown in **Figure 2**. Nanophytotechnological remediation was published in the *J. of Environ. Protec.* (JEP) (2016, http://dx.doi.org/10.4236/jep.2016.75066).

2. Phytoremediation classification

Phytoremediation technologies are classified in general into:

a. **Phytoextraction:** Metal concentration reduction in the soil through plants that can accumulate metals in the shoots.

b. **Phytostabilization:** Immobilize the utilization of soil metals via adsorption onto roots; rhizosphere precipitation.

c. **Phytostimulation:** The process where root releases certain compounds enhancing the microbial activity in the rhizosphere of the plant. It is a type of rhizosphere phytoremediation which is used as an inexpensive approach to remove soil organic pollutants.

d. **Phytovolatilization:** A technique, where the soil contaminants are cleaned up by plants and discharge them as atmospheric volatiles through transpiration.

e. **Phytotransformation/phytodegradation:** Breaking down of organic contaminants seized through plants via

i. *Plant metabolic processes or*

ii. *The outcome of metabolites, such as enzymes, produced by the plant*

f. **Phytoresaturation:** Re-vegetation of the drylands by plants can prevent the spread of pollutants into the environment [4].

Type of nanoparticles	Biochemical agents	Size/morphology	Environmental applications
Stabilized bimetallic Fe/Pd nanoparticles	Starch	14.1 nm distinct, well dispersed	Degradation of chlorinated hydrocarbons in water
Fe_3O_4	Na-Alginate	27.20 nm spherical	Urea decomposition
Fe_3O_4-Polymer Composite	Agar (reducing and stabilizing agent)	50–200 nm spherical, 24 nm diameter and hexagonal	Magnetic storage media
Nano-shell (Fe, Cu)	Ascorbic acid (antioxidant)	<100 nm cubic	Functions in catalysis, biosensors, energy storage problems, nanodevices
nZVI	Ascorbic acid (Vit-C)	20–75 nm, spherical	Cd removal
Superparamagnetic Iron oxide (coatings and functionalization)	Ascorbic acid (Vit-C)	5–30 nm (hydrodynamic size)	Contrast enhancement agent for MRI applications
Fe_3O_4 (MNPs)	L-Lysine (A. Acid)	17.50 nm and spherical crystalline	Biosensors, drug delivery
nZVI	L-Lysine (A. Acid) L-Glutamic Acid L-Glutamine L-Arginine and L-Cysteine	—	Low molecular, biocompatible
FeNPs	Hemoglobin and myoglobin	2–5 nm aggregates, crystalline	Bioconjugated nanoparticles for biological applications
Fe_3O_4	D-glucose gluconic Acid	12.5 nm roughly spherical, crystalline	Drug delivery, cell transplantation
Fe_3O_4	Glucose & Glyconic acid	4–16 nm crystalline	Removal of waste in the biomedical field
Carbon capsulated Iron NPs	Wood-derived sugar	100–150 nm nanospheres, 10–25 nm diameter of iron core	Acts as catalysts in the conversion of wood-derived syngas to liquid hydrocarbons
Iron oxide	Tannic acid	<10 nm	Utilization of biomass causes the reduction of metal ions
Fe core-shell structure	Chitosan-gallic acid	11 nm cubic	Increased thermal stability of drug gallic acid, anticancer activity was higher for HT29 and MCF7 cell lines

Table 3.
Synthesis of iron nanoparticles/derivatives.

An overview of metal contaminants in several phytoremediation processes is provided in **Table 3**. In the case of contaminated water, the following processes in phytoremediation technologies are utilized as:

a. **Rhizofiltration:** Roots were used to remove aqueous toxic metals, mainly the heavy metals like, lead (Pb) and radioactive elements [5]. The plants are employed as filters in wetlands or as a hydroponic setup [6]. Wetlands are often widely considered as sinks for pollutants, and there are countless instances where the wetlands plants are considered to remove contaminants [7] used which include metals viz., Se, perchlorate, cyanide, nitrate, and phosphate [8].

b. **Hydraulic control:** It is a process in which bulk amount of water is absorbed by the wildly growing plants preventing the increase of pollutants into the unpolluted surrounding zones [4].

The phytoremediation methods chosen depend upon:

i. Specifically high growth rates in the polluted sites

ii. Huge surface area proportionately in contact with the water body

iii. High translocation potential [9]

These factors say both the bioconcentration factor (BCF) and translocation potential (TP) are related to plants' sensitivity for phytoremediation.

In Brake fern (*Pteris vittata*), the best phytoremediation process is established as it consists of a high root to shoot metal transduction; thus, it is observed that the BCF value is greater than one. Out of the several phytoremediation technologies, phytoextraction is the most effective, which depends upon hyperaccumulation of metals into the whole plants. For phytoextraction, a heavy metal tolerant plant that grows rapidly with high biomass yield per hectare also should possess a prolific root system. When the cultivation is over by the season's end plants are harvested, dehydrated and the enriched mass with contaminants is dumped or sent into the smelter. To be active phytoextraction, the dehydrated biomass, ash extracted from the aboveground parts of a phytoremediator crop, consists of a greater concentration of the pollutants than the contaminated soil [10]. The biomass rich product exudes as the secondary metabolic waste, which requires further treatment. The phytoextraction process can be natural and induced. The energy can be recovered from biomass burn or pyrolysis; thus, phytoextraction can be used as a cost-effective technology by giving biomass yields. *Salix* and *Populus* species are also used for phytoremediation technology.

3. Bioelements and their effects on pollution

Pollution is an undesirable change observed, which is deteriorating our raw materials, especially land and water. An overall representation of the contamination process, which can cause microorganisms to pollute soil and surface water, is shown in (**Figure 3**). At normal concentration, soil comprises bioelements, particularly metals. These bioelements serve as micronutrients and macronutrients for the soil. They can be classified as light metals (Mg and Al) metalloids (As and Se)m and heavy metals viz., Cd, Hg, Pb, Cr, Ag, and Sn. Light metals have a greater significance to health and environment [11], whereas substantial metals are the bioelements

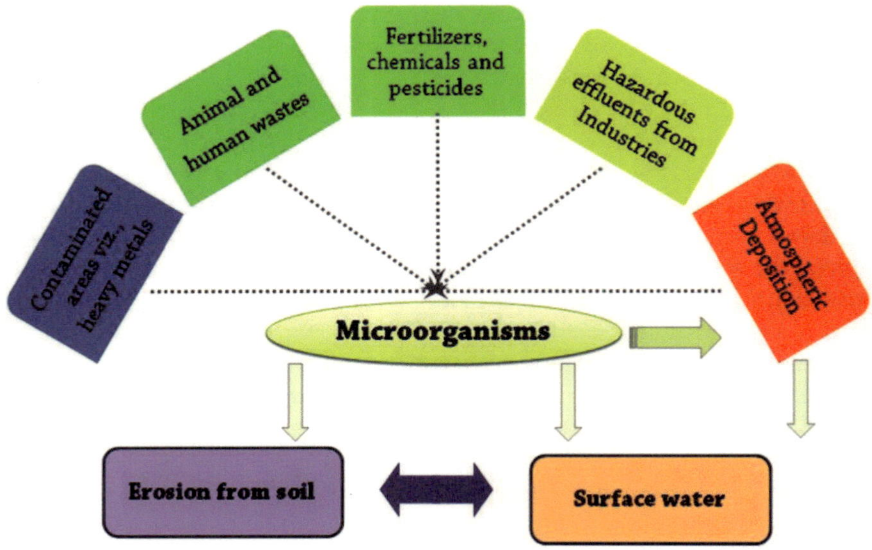

Figure 3.
An overall representation of the contamination process—that can cause microorganisms to pollute soil and surface water.

(At. No., Z > 20) with a density > 5.0 g/cc and have definite metal properties such as conductivity, ductility, ligand specificity, cationic stability. Beneficial heavy metals include elements such as Cu, Cr, Zn, Mn, Fe, Co, and Ni, which are essential in smaller amounts in metabolism but may be lethal in higher concentrations. Geogenic and anthropogenic contaminations by heavy metal is shown and can cause microorganisms [12] to affect the normal molecular process as shown in (**Figure 4**). Heavy metals sieve through the soil and are terminated into the soil by geogenic and anthropogenic processes [13].

Geogenic contamination can be exemplified by extensive arsenic contamination, as seen in the ground waters of Indian state of West Bengal and Bangladesh [14]. The other contamination source includes anthropogenic activities like generating huge amounts of effluents, which is a constant threat to environmental pollution. Fertilizers incorporate phosphate compounds containing Cd, which are being used in horticulture, agriculture as well as in animal industries as a trace element nutrient. Cd, Hg, and Pb metals attack the activity of the enzyme, which contains the –SH group which initiates chronic diseases. These heavy metals/metalloids and organics form a grave danger to animals (including humans) and plants. Heavy metal pollution on land and water shows a severe impact on the ecosystem. In Western Europe, a large mass of approximately 14,00,000 sites affected as the reports of [15], out of which 3,00,000 are contaminated, but the projected number in Europe could be greater, as the problem was progressively occurring in the Central and East European countries. In the United States, around 600,000 contaminated brownfields with heavy metals requiring reclamation [16]. Land pollution has been a great challenge in the Asian continent as seen in China, where one-sixth of arable land is with heavy metal pollution, and over 45% has been ruined either due to erosion or desertification. This becomes the consequence because of human-dominated ecological problems viz., urban ecology and agricultural ecology [17]. Thus, it is vital to eliminate these pollutants from the contaminated sites in which phytoremediation is one of the processes that include complexation, accumulation, volatilization, and degradation of pollutants both of organic and inorganic origins.

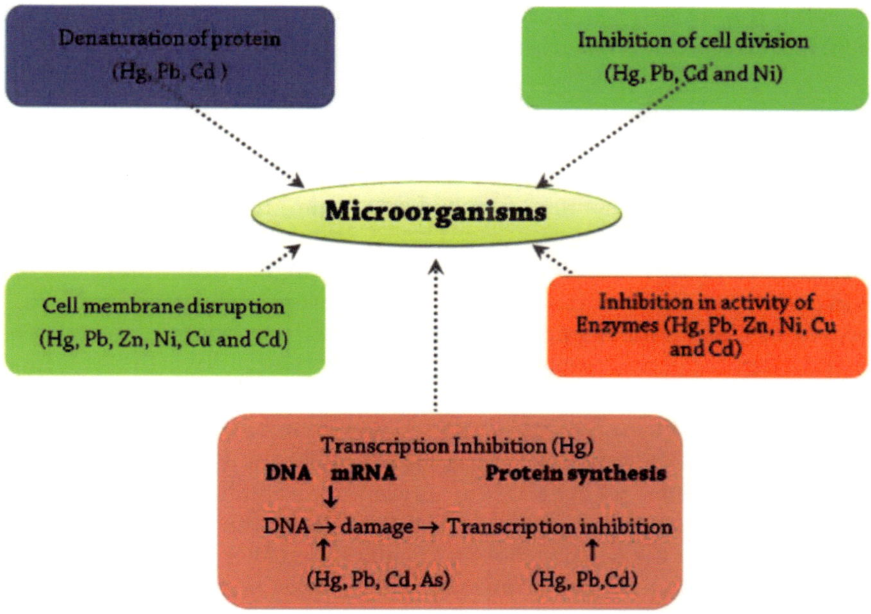

Figure 4.
Geogenic and anthropogenic contaminations by heavy metal is shown and can cause microorganisms to affect the normal molecular process.

4. Biosynthesis of nanoparticles from plants

Nanoparticles are aggregates between 1 and 100 nm; this particular size that alters the physicochemical properties equated to other material. A variety of nanoparticles are produced by bacteria, fungi, and plants [18], which have wider applications in several sectors. Plants are more appropriate than bacteria or fungi toward the synthesis of NPs, as less incubation time is required for metal ion reduction. The procedures such as plant tissue culture (PTC) and downstream processing techniques make more promising in synthesizing metal and oxide NPs at a larger scale. The documentation of hyperaccumulator exclusive genes and their succeeding transfer to the other species of transgenic plants can improve phytoremediation capacity. The plant's remediation volume shall be greatly enhanced by genetic manipulation and other viable plant-based transforming techniques. In plants, it is seen to have an inherent ability to lessen metals through their specific metabolic pathways [19]. Stampoulis et al. [20] have examined the impact of ZnO, Cu, Si, and Ag NPs on the root elongation, seed germination, and biomass production of *Cucurbita pepo* grown as hydroponics. Accordingly, experimental findings suggested that root length is reduced by 77% when seeds are exposed to Cu nanoparticles and 64% when exposed to bulk Cu powder when equated to the untreated controls.

Plant biomass was reduced by 75% when exposed to Ag NPs. Shekhawat and Arya [21] used *Brassica juncea* seedlings to produce Ag NPs *in vitro*. There are reports from of synthesized gold nanoparticles by *Terminalia catappa* leaf extract in an aqueous medium [22]. The authors [4, 23] examined metal ions Ag^+ and Au^{3+} to Ag^0 and Au^0 NPs in *Brassica juncea* for the reduction sites. Nevertheless, Ag NPs in plants are mostly modeled as Ag not only forms NPs in plants but it also exhibits higher catalytic properties as it consists of high electrochemical reduction potential and several additional useful properties. Although the research on the production of nanoparticles is in a nascent stage in plants, more qualitative work is required to realize the physiological, biochemical, and molecular mechanistic process relative to nanoparticles.

4.1 Nano-iron and its derivatives

Reactive nanoscale iron product (RNIP) and nanoscale zero-valent iron (NZVI) are mostly the elementary forms of iron (nano) technology [24]. Nano zero-valent iron because of its nano-size (1–100 nm) enables high-level remedial adaptability. NZVI, a product of nanotechnology, is used to treat a range of impurities in perilous wastewater (see **Table 3**) and represents the synthesis of iron nanoparticles [25]. As for example, NZVI was tested in the removal of As(III) seen in groundwater. NZVI can be used in permeable reactive barriers (PRBs) form to intercept plumes on the subsurface and remediate them. The sustained zero-valent iron nanoparticle "*ferragels*" swiftly dispersed and immobilize Cr(VI) and Pb(II) from aqueous solutions, reducing the Cr(VI) to Cr(III) and Pb(II) to Pb(0) while oxidizing Fe to goethite (α-FeOOH) [26]. Anionic hydrophilic carbon (Fe/C) and poly (acrylic acid)-supported (Fe/PAA); Fe(0) NPs were further considered as a sensitive material for the dehalogenation of chlorinated HCs in soils and ground waters [27]. Nickel-iron NPs in the ratio 1:3 were employed in the dehalogenation of trichloroethylene (TCE) [28].

4.2 Single-enzymed nanoparticles

Enzymes serve as effective biocatalysts in bioremediation. Nevertheless, less stability as a result of diminutive catalytic lifetimes of enzymes limits their effectiveness being inexpensive due to oxidation. The usage of nanotechnology provides a novel method where the enzymes are stabilized in the form of single enzyme nanoparticles (SENs). Enzymes can be devoted to the magnetic iron NPs increasing stability, longevity, and reusability. The enzyme separation from the magnetic iron NPs is usually done by the use of a magnetic field. The two different catabolic enzymes—trypsin and peroxide subjected to unvarying core-shell magnetic nanoparticles (MNPs). SEN requires the involvement of modification of enzyme surface, vinyl polymer growth from the enzyme surface. There are immobilized enzymes in biopolymers and carbon nanotubes, which can add as environmental biosensors.

4.3 Exopolysaccharides

Exopolysaccharides (EPSs) are polymers of the polysaccharide of high molecular weight, secreted by microorganisms. EPSs are sustainable as it has good adsorption capacity and environmental friendly. Therefore, the usage of EPS for bioremediation in the metallic and dye-based environmental pollution attracted researchers in the past years. Polysaccharides are very rich in –OH groups using them as a stabilizer for the production of metal NPs, an environment friendly alternate for the chemical-reducing method [29].

EPSs are used as a reducing agent and stabilizer. They are further used for the synthesis of metal NPs viz., lentinan, carboxymethylated chitosan, glucan, carboxymethyl cellulose, and carboxylic curdlan [30]. Apart from exopolysaccharides, the Au and Ag nanoparticles also consist of good dispersible capability and uniformity. EPS produced from *A. fumigatus*, [31] *Lyngbya putealis*, *Lactobacillus plantarum* [32], and *Bacillus firmus* [33] removed heavy metals viz., Cu^{2+}, Pb^{2+}, Cr^{4+}, Cd^{2+}, and Zn^{2+} within the adsorption capability of 50–1120 mg/g. EPS-605 obtained from newly identified *L. plantarum*-605 was obtained from a Chinese fermented food, Fuyuan pickles. When EPS-605 was self-assembled in H_2O, monodispersed nanoparticles were detected that are useful for bioremediation and record heavy metal and dye adsorption.

4.4 Dendrimers

Dendrimers are multivalent, globular, highly branched, and monodispersed molecules with synthetic elasticity. Dendrimers have proper architecture and controlled composition, which consist of three components and have an extensive assortment of applications ranging from catalysis, electronics to drug release. With unique structural characteristics viz., nanoscopic size, spheroidal surface, vast interior with exhilarating properties which consists of low viscosity, extraordinary solubility, and reactivity. Dendrimers' first dendrimers were synthesized by Fritz Vogtle in 1978 [34] consists of three constituents—a vital core, internal branch cells or radiated symmetry, and terminal branch cell or marginal group. The void spaces in dendrimers interact with nanoparticles, which enhances the catalytic activity. The dendrimer nanocomposites were also set for treatment of water and dye removal from industrial waters to enhance the reactivity by creating more surface area with a reduced amount of toxicity. PAMAM dendrimers using group of hydroxyl-terminated (G4-OH) poly (amidoamine) also acts as templates in the production of Cu NPs formed by coordination of Cu ions with dendrimer interior amines and subsequent reduction forming dendrimer-encapsulated Cu NPs (Cu-DEN).

Cowpea mosaic virus (CPMV), a plant virus, is adequate to endorse the templated mineralization of metal and metal oxide. CMV particles used for templated fabrication of metallic NPs by an electron less deposition metallization process. In the virus capsid, Pd ions are electrostatically bound to the virus capsid and upon reduction acts as a nucleation site to deposit metal ions from solution. Further, dendrimer-modified and plain magnetite nanoparticles (MNPs) have been widely studied in environmental decontamination. Dendrimers can enhance drug targeting efficacy mainly to be used in drug delivery systems [34].

4.5 Nanocrystals and carbon nanotubes

Nanomaterial-based applications in the field of environment are in multiples that provide both large and portable scale and also clean up impurities that are present in our environment. Carbon-based nanomaterials viz., nanocrystals and carbon nanotubes (CNT) have wider applications as antimicrobial agents, environmental sensors, biosensors, sorbents, depth filters, renewable energy technologies, high flux membranes, and in pollution prevention [35]. CNTs are both single walled (SWCNT) or multiwalled (MWCNT); functionalized hybrids were evaluated for the elimination of Et-C_6H_6 from aqueous solution and remediating pollution to avert diseases from ethylbenzene (Et-C_6H_6) viz., cyclodextrins (CD). Nickel ions from water were remediated using MWCNT-based materials [36]. CNT-based polymeric materials incorporating nanomaterials, Calixarenes, and Thiacalixarenes were synthesized to remove both organic (p-NO_2-C_6H_5OH) and inorganic contaminants (Cd^{2+}, Pb^{2+}) from water bodies [37]. CNTs immobilized by calcium alginate (CNTs/CA) materials investigated the Cu removal efficiency (69.9% at pH 2.1) via equilibrium studies [37]. Magnetic-MWCNT nanocomposites reported eradicating cationic dyes in aqueous solutions [38].

4.6 Engineered polymeric nanoparticles application in bioremediation for removal of hydrophobic contaminants

Hydrophobic contaminants, say, polycyclic aromatic hydrocarbons (PAHs), are globally persistent in the atmosphere. PAHs are hydrophobic, strongly sorbed to the soil; thus, sorption limits the bioavailability of these pollutants on the surface. Sequestration in nonaqueous phase liquids (NAPLs) shrinks the mobility and bioavailability of hydrophobic contaminants [39]. Though surfactant micelles have shown an increased rate of PAHs and hydrocarbon solubilization in contrast also causes biodegradation.

Synthesis of nonionic amphiphilic polyurethane (APU) NPs from a mixture of polyethylene glycol (PEG) altered polyurethane acrylate (PMUA), and polyurethane acrylate precursor chains solubilize PAHs from the contaminated soil. Unlike surfactant micelles, PMUA NPs are cross-linked, so not easily breakable when it comes in contact with soil interacting with liposomes of microorganisms but have excellent properties to improve desorption and the agility of phenanthrene (PHEN) in aquifer sand [40].

4.7 Polymeric nanoparticles used in soil remediation

Research based on nanoparticles usage in soils and groundwater remediation processes increased greatly with promising results. Using nanotechnologies, polluted soils remediation becomes an emerging area with an enormous impending to advance the performance over traditional remediation technologies in a large way [41, 42]. Effective application for soil contaminants contexts, predominantly, for heavy metals, other inorganic and organic contaminants, and emerging contaminants, such as pharmaceutical, cosmetic, personal care products.

Polynuclear aromatic hydrocarbons (PAHs) that absorb intensely to soil are very challenging to eliminate. In such cases, amphiphilic polyurethane (APU) nanoparticles are used in soil remediation which is polluted with PAHs. Desired properties of APU particles can be achieved by engineering, and experimental results have shown that these designed particles make sure hydrophobic interior regions that confer a high affinity for PHEN and hydrophilic surfaces that encourage soil particle mobility. APU NPs (17–97 nm) are prepared of polyurethane acrylate (PA) and ionomer (UAA) or PEG, modified urethane acrylate (PMUA) precursor chains which are emulsified and cross-linked in water. APU particles are stable, independent to their concentration in the aqueous phase, and have interiors regions exhibiting hydrophobic property enhances PAH desorption. APU particles contrived to give the anticipated properties. APU particles affinity toward pollutants like PHEN is precisely managed by varying hydrophobic segment size required for the chain propagation. Mobility of soil APU suspensions is controlled by the charge density or the size of the water-soluble chains [40].

4.8 Biogenic uraninite nanoparticles

There is evidence of the widespread prevalence of uranium in India's groundwater. A variety of sources and studies have indicated the link between exposures to uranium in drinking waters which causes chronic kidney diseases. Although the main source is geogenic but still anthropogenic factors play their part in the decline in groundwater table and nitrate pollution promote uranium mobilization. The term *Uraninite* defines compositionally complex, nonstoichiometric, cation-substituted forms of UO_2, which are found in nature. Biogenic uraninite being nanoscale biogeological material is significant due to usage in bioremediation strategies. Uraninite is utmost preferred product in situ stimulated subsurface uranium U(VI) and has its solubilization much lesser compared to other uranium species.

Uraninite nanoparticles have its properties viz., solubility and dissolution kinetics, which are crucial for microbial bioremediation which mitigates subsurface uranium contamination through uranium reduction. Uraninite exhibits structural chemistry, thus derives its properties from its open fluorite structure. Biogenic biocontrol properties, etc. The detrimental effects of PHCs are significantly reduced by the synergistic action of plant microbes' association.

8.2 Evaluating phytoremediation as a potential remediation technology

According to DOE [62], previous remediation processes posed with risk in human health or ecological imbalance, while remedial process like phytoremediation must offer advantages of reduction of health risk or cost-effectiveness over excavation and landfilling of polluted material as compared to the traditional approach.

8.3 Benefits of phytoremediation

Various benefits of phytoremediation have been explained or established:

- Phytoremediation can be more eco-friendly than other technologies.
- Phytoremediation may be suitable for cost savings more than 50% over traditional technologies [63].
- Phytoremediation offers the restoration of ecosystems, habitat to animals, biodiversity conservation, and reduces anthropogenic activities [62–64].
- Phytoremediation provides better environment for sustainable life [64].
- Erosion caused by wind or water may be significantly reduced by vegetation [64].
- Trees plantation decreases energy consumption and provides shade to buildings [65].
- Sequesters carbon and facilitates as carbon sink.

9. Conclusion and recommendations

Due to their hydrophobic nature, PHCs are pollutants with higher priority as they are difficult for degradation. Retention of these pollutants poses threat to biodiversity and environmental health. The reclamation of environment from PHCs' contamination is a global problem. Bio-phytoremediation proved to be an economic and alternative approach as compared to the physico-chemical process. Many factors that influence the remediation process, which include nutrition, physical conditions, microbial diversity, contaminant bioavailability, etc., can play an important role in the bio-phytoremediation of PHCs. Due to plant and microbe synergy, both possess enzymes, root exudates, etc. for better interaction of plant microbes with the contaminant. Therefore, phytoremediation with the involvement of microorganisms can be considered a key process of PHCs' remediation. The increase in our understanding of the bio-phytoremediation and the mechanisms by which petroleum hydrocarbons biodegradation occur will prove helpful for predicting the environmental fate of these compounds and for developing practical PHCs' bioremediation strategies in the future. It is crucial to continue in developing a technology which is cost-effective, feasible, and can remediate PHCs and other environment contaminants. Further study could be conducted for scaling up this technology or approach.

uraninite forms by reduction of U(VI) to U(IV) considered as the first stage. After the reduction process, the second step formation requires the precipitation of the mineral. In situ U(VI) reduction has been observed and reported at a large number of contaminated U.S. Department of Energy (DoE) nuclear legacy sites and has shown potential results. The success in uranium bioremediation should be maintained strictly in anaerobic conditions. The surface chemistry of nanoparticulate uraninite is important for the construction of geochemical models of uranium behavior, which follows the bioremediation. This may be challenging for research in nano-bio geosciences in the future [43].

5. Soil trace element biomonitoring plants

Soil contamination manifested by trace elements, organic, and inorganic compounds is an extensive problem occurring worldwide. Common techniques in soil remediation include waste disposals, incinerations, leaching of soil thermal desorption, and vapor abstraction, but all these types of actions may be responsible for secondary pollution, which ultimately affects soil properties. Plants are the major factors to keep our environment clean and green by remediation of soil and water. The soil organic and inorganic contaminants are removed by phytoremediation. Ryegrass, oat plant, tall fescue, sunflower, and green gram grow in diverse contaminated conditions useful for phytoremediation. Certain plants known as hyperaccumulators are good in phytoremediation in particularly toward heavy metal removal. Some hyperaccumulator families represent their metal content [44].

Table 4 defines the hyperaccumulator plants of various families, which are used to accumulate specific metals at different concentrations. Phytoextraction seems to be a feasible alternate to the traditionally conventional practice used in the decontamination of soils with heavy metals [45]. In phytoextraction, methodology plants absorb pollutants from soil. Metals that are deposited as ions in the plant's roots, stems, leaves, and inflorescences are burnt to recover metals, and the subsequent biomass is removed to dispose of safely. The build-up of heavy metals is connected to the total concentration of the metals and suggestively segregated as macro nutrients and micronutrients and soil acidity.

5.1 Vascular plants

Water pollution is dangerous, and one of the ecological risk factors suggests the need to cultivate water plants that absorb trace elements. Usually, there is a quick dilution of the contaminants in water; thus, investigating the plant tissues provides combined evidence about the quality and components of water and the method of phytoremediation [46]. The various nanomaterials that can be synthesized through several methods have been represented in **Table 5**. Further, it is observed that species viz., duckweed (*Lemna gibba*), water spinach (*Ipomoea aquatica*), and fern (*Azolla pinnata*) are prominent to phytoremediate metals [47]. like boron, chromium, and manganese, respectively [48–50]. Aquatic macrophytes such as water hyacinths are used extensively in phytoremediation of water contaminated with dyes [51]. Hasan et al. [52] stated the efficacy of water hyacinth in sorption of Zn(II) and Cd(II) from the water. The species from Lemnaceae family, eliminate dyes such as acid blue (azo dye, AB92) undergoes a transformation to form dissimilar transitional compounds [53]. Aquatic plants viz., *Azolla pinnata* (water-fern) and *Hydrilla verticillata* (water-thyme) are used for elimination of fly ash and uranium, respectively [54, 55]. *Micranthemum umbrosum* observed [56] removal of As and Cd by phytofilteration method. *Oenothera picensis* plant was quite extensively

Metals	Plant species	Accumulated metal concentration (mg/kg)
Cadmium		
Thlaspi caerulescens	Brassicaceae	2130
Zinc		
Thlaspi caerulescens	Brassicaceae	43,710
Thlaspi rotundifolium	Brassicaceae	18,500
Dichapetalum gelonioides	Brassicaceae	30,000
Nickel		
Thlaspi Sps.	Brassicaceae	2000-2031,000
Allyssium Sps.	Brassicaceae	1280–29,400
Berkheya codii	Asteraceae	11,600
Pentacalia Sps.	Asteraceae	16,600
Psychotria coronata	Rubiaceae	25,540
Copper		
Ipomoea alpina	Convolvulaceae	12,300
Lead		
Minuartia verna	Caryophyllaceae	20,000
Agrostis tenuis	Poaceae	13,490
Vetiveria zizanioides	Cyperaceae	>1500
Cobalt		
Crotalaria cobalticola	Fabaceae	30,100
Haumaniastrum robertii	Lamiaceae	10,232

Table 4.
Hyperaccumulator plants for varied metals.

Nanomaterials	The methodology used in the synthesis	Examples
Nanoparticles biosynthesis from metals (NPs)	Photochemical	Cu, Au, CoNi, CdTe, CdSe, ZnS, Rh, Pt, Ir, Pd, Co, Ag, Au, Cu, Fe & Ni
	Biochemical	
	Electrochemical	
	Thermochemical	
Nanomaterials from carbon	Arc-discharge	Cylindrical nanotubes (SWNT, MWNT) Fullerenes
	Chemical vapor deposition	
	Laser ablation	
Nanomaterials from polymers	Electrochemical Polymerization	Nanowires of PPy, PANI, Poly (3–4 ethylene dioxy thiophane, PAMAM, dendrimers
Metal oxide Nanoparticles	Hydrothermal	$BaCO_3$, $BaSO_4$, TiO_2,
	Reverse micelles solvo-thermal	ZnO, Fe_2O_3, Fe_3O_4, MgO
	Sol-gel method	
	Electrochemical deposition	
Bionanomaterials	Biological	Plasmids, nanoparticles from protein viruses

Table 5.
Synthesis of diverse nanomaterials.

considered toward phytoextraction of copper [57]. Algae such as charaphytes viz., *Chara aculeolata* and *Nitella opaca* were used to remove Pb, Cd, and Zn [58].

Cystoseira indica (brown algae) after its chemical treatment become greatly effective against chromium. Metal uptake is seen in algae species such as *Spirulina* used for chemisorptions of metals with few heavy metals like chromium and copper [59]. *Ranunculus peltatus*, *Ranunculus trichophyllus*, *Lemna minor*, *Azolla caroliniana* viz., serve as an arsenic indicator [60]. *Ulothrix cylindricum* (green algae) has biosorption capacity of 65.6 mg/g, forming an inexpensive method for biosorption of As(III) [61]. Aquatic macrophytes grow quickly, and due to their high biomass production, the greater capacity in accumulating heavy metals widely used for wastewater treatment compared to soil-grown plants.

A macrophyte grows in or near the water body and is emergent, submerged or floating. Aquatic plants have adjusted to living in aquatic environments (hydrophytes or macrophytes) to differentiate from algae and other microphytes. Water hyacinth (*Eichhornia crassipes*), Sensitive Plant (*Neptunia aquatica*), Lucky 4-Leaf Clover (*Marsilea mutica*) water lettuce (*Pistia stratiotes*), Moneywort (*Bacopa monnieri*), Mosaic Flower (*Ludwigia sedioides*), Water poppy (*Hydrocleys nymphoides*), and duckweed (*Lemna minor*) are a few of the aquatic macrophytes widely intended for heavy metal phytoremediation [62]. *Pistia stratiotes* have relatively high growth rate thus ideally chosen in phytoremediation study as it is proposed to accumulate As [63]. Water lettuce is observed to be a probable plant for phytoremediation for manganese contaminated waters [62]. In the elimination of Pb, Cd, Cr from the water, *Lemna minor*, a native of Europe, North America, Asia, and Africa is naturalized for its advantage to grow in several climatic conditions and also a potential accumulator of Cd to remediate the aquatic environment. *Eichhornia crassipes* was used for the tertiary treatment of wastewater phytoremediation as it has broader leaves and fibrous root system which assists in the absorption of heavy metals [64]. There has been experimentation on water hyacinth (*Eichhornia crassipes*), two algal species (*Chlorodesmis sp.* and *Cladophora sp.*) found in As-contaminated water bodies are used to determine the arsenic tolerance capability. Cladophora species are found to be appropriate for co-treatment of sewage and As-contaminated brine in algal ponds. *Typha latifolia* and *Eichhornia crassipes* are freshwater plants used to clean up the effluents that usually contain high concentrations of Co, Cd, and As. *Eleocharis acicularis* commonly known as dwarf hair grass and needle spike rush acts as hyperaccumulators as it uptakes several metals Fe, Pb, Mn, Cr, and Zn from drainages and mines [65, 66]. *Myriophyllum aquaticum* consists of enzymes that play a vital part in the transformation of organic compound contamination and is effective in the phytoremediation of an aquatic environment [9]. *Ludwigia palustris* (marsh seedbox; creeping primrose) and *Mentha aquatica* (water mint) effectively remove Cu, Fe, Hg, and Zn. Among the freshwater vascular plants, the most efficacious plants are *E. crassipes* and *L. minor.*

6. Hyperaccumulator plants for different metals

Bioconcentration factor and factor of translocation are multiplied to get the phytoextraction efficiency. It is observed that accumulated metal concentration in soil modifies its biological properties. Different plant species vary with regard to uptake of heavy metal. The hyperaccumulation of heavy metals mainly rest on several factors viz., plant species, soil circumstances (pH, temperature, humidity, soil organic content, and cation capacity), and types of heavy metals. The uptake of metals is determined by the metal type and metal chemical speciation and habitat characteristics of the plant [67]. Hence, the plant selection became significant for the remediation of the containment location. The accumulation efficacy of heavy metals in any plant species is calculated via a bioconcentration factor [68]. The willow plant consists of the highest biomass, thus identified itself as an appropriate plant for soil remediation [69]. In a prior experiment, plant species of Brassicaceae family, such as *Brassica juncea* L., *Brassica napus* L., and *Brassica rapa* L. are able to accumulate Zn and Cd moderately. In *Brassica juncea*, the nuts showed the bioaccumulation ability toward Cu [70]. *Pistia stratiotes* L. (water lettuce) has the potential to remove Cd from surface water [71]. Canola (*Brassica napus* L.) is very effective with respect to Cu, Cd, Pb, and Zn in comparison to *B. juncea* L. (Indian mustard). Application of Ethylene diamine tetra acetic acid (EDTA) increases heavy metal availability, thus making the plant uptake showing the prominence of organic chelates in increasing metal solubility/availability, thus applicable to enhancing the efficiency of phytoremediation technique.

Table 6 represents the advantages and limitations of phytoremediation technologies. In Brassicaceae family, plants are used for biofumigation. *Helianthus annuus* (Sunflower) has the capability for soil remediation contaminated by Pb. Soybean plants characteristically synthesize homophytochelatins alternative to phytochelatins when heavy metals are exposed. For the soybean seeds and young seedlings, Cr metal is found to be extremely toxic at higher concentrations [72]. Crops are affected as it is seen that soil contamination by heavy metals causes a considerable loss in seed production of soybean canopies [73]. Agricultural soils accumulate toxic metals in edible portions of crops which grow in contaminated soils that described in crops viz., rice, soybean, maize, and vegetables.

Advantage	Limitation
Phytoextraction	
Plant with high biomass within lesser time should be successful to remove contaminants from soil.	1. Hyperaccumulators exhibit slow growth and less bioproductivity due to shallow root systems 2. Biomass/phytomass must be disposed of cautiously
Phytostabilization	
Cost-effective and less disruptive which enhances the ecosystem restoration/re-vegetation.	1. The requirement of extensive fertilization/soil modification. Proper maintenance is required to prevent leaching
Phytovolatilization	
Contaminants/pollutants are transformed into less toxic forms, for example, volatilization of mercury(Hg) by conversion to the elemental form in transgenic *Arabidopsis* and yellow poplars which contains bacterial mercuric reductase (*merA*)	1. Contaminants/hazardous metabolites might accumulate in vegetation viz., fruits/lumber 2. Low levels of metabolites can be found in plant tissues
Phytofiltration/rhizofiltration	
In situ (pond floating rafts) or *ex-situ* (tank system); aquatic Absorption and adsorption play an important role	1. Constant pH monitoring of the medium is required for optimizing the uptake of metals 2. Influent chemical speciation and all the species interactions are to be understood 3. Intensive maintenance is needed 4. Large root surface area is usually required

Table 6.
Advantages and limitations of phytoremediation.

7. Effect of metals on the physiological process

Generally, metals play a significant part in the metabolic pathways in plants during the growth and development in appropriate amounts but lethal in excess. Soil gets contaminated due to several activities such as mining, disposal of solid wastes, automobile exhausts, and engineering activities. Therefore, there is a possibility of augmented uptake of metals by food crops, which cause human health risks, thus affecting food quality and safety. Metals viz., iron (Fe), molybdenum (Mo), copper (Cu), cobalt (Co), manganese (Mn), and zinc (Zn) are crucial for plant growth, categorized as essential micronutrients. The nonessential metals found as pollutants comprise mercury (Hg), chromium (Cr), selenium (Se), uranium (U), nickel (Ni), cadmium (Cd), arsenic (As), lead (Pb), vanadium (V), and wolfram (W). Prior published reports by [74] provided information on the impact of metal on the seed of crops and medicinal plants regarding biochemical and molecular implications, which provide an important role in seed germination. It has been noted that metals applied exogenously in the range of micromolar to milimolar concentrations could affect seed variability. Seeds from metal tolerant plants and hyperaccumulators possess higher threshold toxicity than the seeds of nontolerant plants. Nonetheless, data on their effects on *in situ* seed germination are in the nascent stage, which is required to be investigated. Cd and Cu inhibit water uptake, obligatory for seed germination. One can overcome seed dormancy with metal treatment, although the actual mechanism of action yet to be understood. But the process of deposition and toxicity of metals are unknown in developing seeds, to embryos and cotyledons.

Similarly, few experiments have focused on the detoxification of metals by phytochelatins (PC) and metallothioneins (MT). Similarly, Shanker et al. [75] have studied extensively about the chromium toxicity in plants which predominantly hinge on valence states of chromium ions. Cr has toxic effects on plant development which includes modifications in the germination process, development of roots, leaves, and stems which ultimately affects entire dry mass production and yield. Chromium too has harmful effects on the plant's physiological processes such as photosynthesis, water channeling, and mineral nutrition. Shukla et al. [76] inspected the effects of cadmium in wheat (*Triticum aestivum* L.) plant. Gupta and Gupta [77] reported in their publication that nutrient toxicities in crops due to manganese and boron are more compared with other nutrients. The foremost toxicity symptoms in crops include burning, chlorosis, and yellowing of leaves. The toxicity of metals is influenced by metal concentration, the composition of minerals, and organics in the soil, pH, redox potential, and the existence of other metals in the soil. Metal toxicity is also affected by the association to mineral constituents of the polluted sites. Since, there is a lack of basic understanding of metal behavior for a precise condition a precise protective method toward metal additions to soils is warranted [78].

In addition, the requirement to know the proper metal toxicity in food products and their nutritional intake in evaluating their risk to human well-being is more. However, the problem of metal toxicity persists due to contamination of the environment, which worsens intensively due to negative human activities. Hyperaccumulators grow on metalliferous soils; leaves possess toxic metal accumulation compared with other plant species. Studies aimed regarding these hyperaccumulators to understand their physiological role and molecular mechanisms, and thus, these plants can be used as a tool in removing metals from natural metal-rich soils (ores) and contaminated areas. Metal tolerant species *Hordeum vulgare*, *Brassica juncea*, *Triticum aestivum*, *Brassica napus*, and *Helianthus annuus* accumulates toxic metals in high concentrations in their shoot system.

8. Transgenic plants usage in phytoremediation

Transgenic plants with wide geographic distribution are used owing to their enhanced tolerance and phytoextraction potential. Transgenic plants are fast growing and seem to possess high biomass, much-elongated roots, and greener leaves than unmodified plants. Herbivores are repulsive to transgenic plants, thus making it greatly an encouraging candidate in phytoremediation efforts [79].

Transgenic plants, when grown in Cu-contaminated soil, and leaves contain two to –three times more Cu compared to other plants [80]. *Arabidopsis thaliana* also possess greater Cu accumulation as reported by overexpression of a pea MTgene [81]. PsMTA from *Pisum sativum*, when overexpressed in *A. thaliana*, accumulated eight times more Cu in roots [82]. *Nicotiana glauca* (shrub tobacco) has a high tolerance toward Pb and Cd when grown in a metal-contaminated soil; the transgenic plants accumulated higher Pb concentrations in the shoot system (50% more) and in the root system (85% more).

An attempt was made toward transferring and expression of genes from bacteria, yeast, animals, or other plants and improvised for potentially high yield. One of the encouraging advances in transgenic technology is the use of multiple genes (cytochrome P450s, GSH, GT, etc.) for thorough degradation of xenobiotics within the plant system that was involved in metabolism, uptake, and transport of specific pollutants in transgenic plants [1, 83, 84]. A published review focused on the development of transgenic plants for remediation of 2,4,6-trinitrotoluene, hexahydro-1,3,5-trinitro-1,3,5-triazine, and glycerol trinitrate [85] by introducing and expressing bacterial nitro-reductases and cytochrome p450s.

As hyperaccumulators have a high metal tolerant trait, probable detoxification capacity is maximum thus efficiently used in phytoremediation. But there is an alternative to hyperaccumulators due to sluggish growth and condensed biomass production; hence, it requires numerous years for sanitization of contaminated sites. Thus, to facilitate faster decontamination, the remedial property can be extensively improvised by genetic manipulation, plant tissue culture, imbursement of transgenic approaches viz., genes, traits can be manipulated and thus the production of transgenic plants, mainly industrialized for remediating heavy metal contaminated soil sites. Examples include *Nicotiana tabaccum* expressing a yeast metallothionein gene for higher cadmium tolerance or *Arabidopsis thaliana* overexpressing a mercuric ion reductase gene for higher mercury tolerance [86]. Dhankher et al. [87] stated about arsenic sequestration which happens largely in vacuoles by complexation with glutathione (-GSH) and phytochelatins (PCs).

In another example, the arsenic fall was seen in the transgenic plant developed by using bacterial genes ArsC from *E. coli* with co-expression of γ-glutamylcysteine synthetase to provide sufficient -GSH for subsequent conjugation [88]. By the expression of bacterial genes merA gene encoding organo-mercurial lyase, transgenic plants show better resistance against the toxic effects of mercury [89]. When merB was expressed in endoplasmic reticulum, resistance was further improved. Therefore, findings on chloroplast are the primary target for mercury poisoning and are leading the ongoing research in chloroplast genome engineering. Further, the expression of bacterial genes atrazine chlorohydrolase (atzZ) and 1-aminocyclopropane-1-carboxylate deaminase has shown a promising result in the remediation of atrazine and alachlor [90]. Transgenic plants expressing these genes show significantly increased tolerance, uptake, and detoxification of targeted explosives. Expression of cytochrome p450 as in CYP2E1 in tobacco and poplar plants have not only increased TCE metabolism but also is metabolizing vinyl chloride, benzene, toluene, and chloroform [84]. Also, trace element detoxification systems have been implemented at the molecular level in yeast and bacteria. A vivid study and approaches by manipulation of molecular genetic techniques to regulate the discharge of metals as contaminants can be controlled through the use of the transgenic plant.

9. Metal homeostasis in plants

Metal homeostasis is defined as the metal uptake, trafficking, efflux, and sensing pathways, which allows organisms to maintain a narrow intracellular concentration range of essential transition metals. The molecular and genetic basis for these mechanisms will be vital in the development of plants that can be agents for phytoremediation of contaminated sites. One among the recurrent general mechanism requires metal homeostasis, chelation of the metal by a ligand, and subsequent compartmentalization of ligand-metal complex. Plants evolved a variety of mechanisms managing heavy metal stress, which include the synthesis of the sulfur-rich metal chelators, glutathione (GSH), phytochelatins (PCs), and metallothioneins (MTs) [91, 92]. Organic acids such as citrate and maleate which chelate extracellularly have significant tolerance to aluminum. Peptide ligands comprise metallothioneins (MTs) and small gene-encoded, Cys-rich polypeptides. GSH, abundantly the low-weight molecular SH-compound in plants, is synthesized through ATP-dependent enzymatic pathway. GSH protects plants from environmental and oxidative stresses, xenobiotics, and heavy metals. Glutathione acts as a precursor of phytochelatins (PCs) during excessive mental stress [93, 94]. The SH-peptide GSH (ç-Glu-Cys-Gly) and its variation homo-glutathione (h-GSH, ç-Glu-Cys-â-Ala) has a stimulus in the form and toxicity to heavy metals such as Cu, Cd, As, Hg, and Zn in different ways. Inventive measures of remediation technologies are of paramount importance; thus, plants can be an introduced as supplementary alternative renewable source and thus used in situ remediations.

9.1 Metallothioneins

Metallothioneins (MT) are cytoplasmic proteins [95], a family of small, vastly conserved, cysteine-rich metal-binding proteins (M.W. ~7000), that are rich in sulfhydryl groups (thiols, make them bind to a number of trace metals) that are significant small proteins that bind toward Zn and Cu homeostasis, small amounts of Fe, Hg and perhaps other heavy metals [96], safeguard against oxidative stress, and buffering against toxic heavy metals. MTs were recognized firstly as Cd-binding proteins in mammalian tissues. Comparably, proteins are recognized in large numbers of animal species [97]. Cysteine-rich proteins are known for their high affinity toward cations Cd, Cu, Zn, etc. and also known for deliberating heavy-metal tolerance and accumulation in yeast and plants.

To mention,

a. Enhanced Cd tolerance is a result of overexpression of MT genes in tobacco and oilseeds.

b. A 16-fold greater Cd tolerance was observed by MT yeast gene (CUP 1) overexpression in cauliflower.

c. The yeast metallothionein (CUP1) encourages Cu uptake in tobacco—seven times more in older leaves than fresh leaves, during Cu stress.

d. Likewise, high accumulation of Cu was found in *Arabidopsis thaliana* by overexpression of a pea MT gene.

9.2 Phytochelatins

Phytochelatins (PC) are oligomers of glutathione [98] produced by the enzyme phytochelatin synthase from GSH, seen in plants, fungi, nematodes, and all the algal groups including cyanobacteria. Phytochelatins are central for heavy metal detoxification and act as chelators [99], Cysteine-rich metal-chelating (post-translationally synthesized) peptides which suggestively show heavy-metal tolerance in plants and fungi by chelation and thus decrease their unrestricted availability. It is projected that PCs are the functionally alike MTs [100].

PCs are not reported in animal species, which supports that MTs performs normal functions well in animals, as a contribution by PCs in plants. Heavy-metal toxicity in plants is seen in diverse ways; these include chelation, exclusion, compartmentalization of the metal ions, immobilization, and the expression of more stress response mechanisms in general such as ethylene and other stress proteins [11].

To mention,

a. In the *Agrobacterium*-mediated transformation, the induction and overexpression of phytochelatin synthase (PCS1) in *Nicotiana glauca* bring about high concentrations of Pb and Cd.

b. Accumulation of high Pb concentrations in aerial parts and roots were also observed in transgenic plants.

c. Longer roots, greener higher leaves than unmodified plants were seen in transgenic seedlings.

d. Overexpression of an Arabidopsis PC synthase (AtPCS1) in transgenic which increases PC synthesis thus accumulating and tolerating metals.

As PCs are found in tissues of the plants and cell cultures upon open to trace levels of crucial metals and the level of PCs were seen in cell cultures is correlated with the medium by reduction of metal ions. These remarks are inferred to designate the role of PCs in the crucial metal ion metabolism homeostasis [94, 101].

10. Conclusion

Among several regions of the world, cultivation of plants is significant in the maintenance of the ecosystem. Environmental contamination occurs due to geogenic and anthropogenic activities as discussed in the review paper. Although a few metals are true bio elements at normal concentration, they can cause a potentially hazardous impact on excessive usage causing environmental contamination. There are a variety of measured steps taken through the different aspects of phytoremediation to curb the menace of contaminants and pollution, but there is always a step of further progress which can be implemented in this scenario.

Plants are naturally found to synthesize nanoparticles. Nanophytoremediation is an innovative and encouraging technology which has gathered a wider reception due to its current area of research in plants. As in the review paper, there are several plant families which act in the biosynthesis of nanoparticles. It is significant to study on metal nanoparticles formation, types of nanoparticles, and derivatives of these nanoparticles, and their action on the physiological process will further eliminate the bioaccumulation of toxic nanoparticles in the plants. Numerous countries globally use plants as a primary source of energy for food; fodder; thus, toxicity and contamination of metals in crops and medical plants may have a huge impact. In our review paper, we have made a significant effort to understand the phytoremediation processes, in general, the nanoparticles occurrence, the need to biomonitor the trace elements in the environment, the physiological effects of the bioelements, transgenic plants which can be used effectively in nanophytoremediation. Thus, in conclusion, nanophytoremediation can be a complementary biological clean-up technique, thus maintaining the sustainability of the environment.

Conflicts of interest

The authors declare no conflict of interest.

Author Contributions

Silpi Sarkar, Manoj Kumar Enamala, and Murthy Chavali wrote the chapter; Mannam Krishnamurthy contributed to the scope of the manuscript; Enamala Manoj Kumar planned the review of the literature and reorganized the chapter; verification was done by Subbaroy Sarma and Murthy Chavali critically reviewed the manuscript. All the authors contributed to this book chapter.

Financial and Ethical disclosures

This work was not funded by any organization; the authors have done on their own.

Author details

Silpi Sarkar[1], Manoj Kumar Enamala[2], Murthy Chavali[3,4,5*],
G.V.S. Subbaroy Sarma[6], Mannam Krishna Murthy[7], Abudukeremu Kadier[8],
Ashokkumar Veeramuthu[9] and K. Chandrasekhar[10]

1 School of Biotechnology, Vignan Foundation for Science, Technology and Research (VFSTR) University, Guntur, Andhra Pradesh, India

2 Bioserve Biotechnologies Private Limited Unit, Hyderabad, Telangana, India

3 Department of Chemistry (PG Studies), Shree Velagapudi Rama Krishna Memorial College, Guntur, Andhra Pradesh, India

4 PG Department of Chemistry, Dharma Appa Rao College, Krishna District, Andhra Pradesh, India

5 NTRC-MCETRC and Aarshanano Composites Technologies Pvt. Ltd., Guntur District, Andhra Pradesh, India

6 Department of Basic Sciences and Humanities, Vignan Lara Institute of Technology and Science, Guntur, Andhra Pradesh, India

7 Varsity Education Management Limited, Hyderabad, Telangana, India

8 Department of Chemical and Process Engineering, Faculty of Engineering and Built Environment, National University of Malaysia (UKM), Selangor, Malaysia

9 Department of Chemical Technology, Chulalongkorn University, Bangkok, Thailand

10 Green Processing, Bioremediation and Alternative Energies (GPBAE) Research Group, Faculty of Environment and Labour Safety, Ton Duc Thang University, Ho Chi Minh City, Vietnam

*Address all correspondence to: chavalim@gmail.com; chavalim@outlook.com

IntechOpen

References

[1] Van Aken B. Transgenic plants for phytoremediation: Helping nature to clean up the environmental pollution. Trends in Biotechnology. 2008;**26**(5):225-227. DOI: 10.1016/j.tibtech.2008.02.001

[2] Warrier RR. Phytoremediation for environmental clean-up. Forestry Bulletin. 2012;**12**(2):1-7. Available from: https://pdfs.semanticscholar.org/a855/2eb54d818442b99072f5c39676278d9feeff.pdf

[3] Sinha R, Valani D, Sinha SS, Herat S. Bioremediation of contaminated sites: A low-cost Nature's biotechnology for environmental clean-up by versatile microbes, plants and earthworms. In: Faerber T, Herzog J, editors. Solid Waste Management and Environmental Remediation. NY, USA: Nova Science Publisher; 2010. pp. 1-73

[4] Masarovicova E, Králova K. Plant-heavy metal interaction: Phytoremediation, biofortification and nanoparticles. In: Monatanaro G, Dichio B, editors. Advances in Selected Plant Physiology Aspects. Croatia: InTech; 2012. pp. 75-102

[5] Gajewska E, Skłodowska M, Słaba M, Mazur J. Effect of nickel on antioxidative enzyme activities, proline and chlorophyll contents in wheat shoots. Biologia Plantarum. 2006;**50**(4):653-659. DOI: 10.1007/s10535-006-0102-5

[6] Lytle CM, Lytle PW, Yang N, Qian JH, Hansen D, Zayed A, et al. Reduction of Cr(VI) to Cr(III) by wetland plants: Potential for in situ heavy metal detoxification. Environmental Science & Technology. 1998;**32**(20):3087-3093. DOI: 10.1021/es980089x

[7] Weis JS, Weis P. Metal uptake, transport and release by wetland plants: Implications for phytoremediation and restoration. Environment International. 2004;**30**(5):685-700. DOI: 10.1016/j.envint.2003.11.002

[8] Nzenguang VA, McCutcheon SC. Phytoremediation of perchlorate. In: McCutcheon SC, Schnoor JL, editors. Phytoremediation: Transformation and Control of Contaminants. New Jersey: John Wiley and Sons, Inc.; 2003. pp. 863-885

[9] Susarla S, Medina VF, McCutcheon SC. Phytoremediation: An ecological solution to organic chemical contamination. Ecological Engineering. 2002;**18**(2):647-658. DOI: 10.1016/S0925-8574(02)00026-5

[10] Krämer U. Metal hyperaccumulation in plants. Annual Review of Plant Biology. 2010;**6**(1):517-534. DOI: 10.1146/annurev-arplant-042809-112156

[11] Sanità Di Toppi L, Gabbrielli R. Response to cadmium in higher plants. Environmental and Experimental Botany. 1999;**42**(2):105-130. DOI: 10.1016/S0098-8472(98)00058-6

[12] Gupta A, Joia J, Sood A, Sood R, Sidhu C, Kaur G. Microbes as potential tool for remediation of heavy metals: A review. Journal of Microbial and Biochemical Technology. 2016;**8**:364-372

[13] Mandal A, Purakayastha T, Ramana S, Neenu S, Bhaduri D, Chakraborty K, et al. Status on phytoremediation of heavy metals in India- a review. International Journal of Stress Management. 2014;**5**(4):553-560. DOI: 10.5958/0976-4038.2014.00609.5

[14] Mahimairaja S, Bolan NS, Adriano DC, Robinson B. Arsenic contamination and its risk management in complex environmental settings.

Advances in Agronomy. 2005;**86**:1-82. DOI: 10.1016/S0065-2113(05)86001-8

[15] McGrath SP, Zhao FJ, Lombi E. Plant and rhizosphere processes involved in phytoremediation of metal-contaminated soils. Plant and Soil. 2001;**232**(1):207-214

[16] Dotaniya ML, Thakur JK, Meena VD, Jajoria DK, Rathor G. Chromium pollution: A threat to environment-a review. Agricultural Reviews. 2014;**35**(2):153-157. DOI: 10.5958/0976-0741.2014.00094.4

[17] Wu J, Overton C. Asian ecology: Pressing problems and research challenges. Bulletin of Ecological Society of America. 2002;**83**(3):189-194

[18] Yadav KK, Singh JK, Gupta N, Kumar V. A review of nanobioremediation technologies for environmental clean-up: A novel biological approach. Journal of Materials and Environmental Science. 2017;**8**(2):740-757

[19] Handy RD, Owen R, Valsami-Jones E. The ecotoxicology of nanoparticles and nanomaterials: Current status, knowledge gaps, challenges, and future needs. Ecotoxicology. 2008;**17**(5):315-325. DOI: 10.1007/s10646-008-0206-0

[20] Stampoulis D, Sinha SK, White JC. Assay-dependent phytotoxicity of nanoparticles to plants. Environmental Science & Technology. 2009;**43**(24):9473-9479

[21] Shekhawat GS, Arya V. Biological synthesis of Ag nanoparticles through in vitro cultures of *Brassica juncea* C. zern. Advances in Materials Research. 2009;**67**:295-299. DOI: 10.4028/www.scientific.net/AMR.67.295

[22] Ankamwar B. Biosynthesis of gold nanoparticles (green-gold) using leaf extract of *Terminalia catappa*. E-Journal of Chemistry. 2010;**7**(4):1334-1339. DOI: 10.1155/2010/745120

[23] Beattie IR, Haverkamp RG. Silver and gold nanoparticles in plants: Sites for the reduction of the metal. Metallomics. 2011;**3**(6):628-632. DOI: 10.1039/c1mt00044f

[24] Watlington K. Emerging nanotechnologies for site remediation and wastewater treatment. Report prepared for National Network of Environmental Management (NNEM) studies the grantee under a fellowship from the U.S. Environmental Protection Agency (US-EPA). 2005

[25] Saif S, Tahir A, Chen Y. Green synthesis of iron nanoparticles and their environmental applications and implications. Nanomaterials. 2016;**6**(11):1-26. DOI: 10.3390/nano6110209

[26] Ponder SM, Darab JG, Mallouk TE. Remediation of Cr (VI) and Pb (II) aqueous solutions using supported, nanoscale zero-valent iron remediation of Cr (VI) and Pb (II) aqueous solutions using supported, nanoscale zero-valent iron. Environmental Science & Technology. 2000;**34**(12):2564-2569. DOI: 10.1021/es9911420

[27] Schrick B, Hydutsky BW, Blough JL, Mallouk TE. Delivery vehicles for zerovalent metal nanoparticles in soil and groundwater. Chemistry of Materials. 2004;**16**(11):2187-2193. DOI: 10.1021/cm0218108

[28] Schrick B, Blough JL, Jones AD, Mallouk TE. Hydrodechlorination of trichloroethylene to hydrocarbons using bimetallic nickel-iron nanoparticles. Chemistry of Materials. 2002;**14**(12):5140-5147

[29] Li C, Zhou L, Yang H, Lv R, Tian P, Li X, et al. Self-assembled exopolysaccharide nanoparticles for bioremediation and green synthesis

of noble metal nanoparticles. ACS Applied Materials & Interfaces. 2017;**9**(27):22808-22818. DOI: 10.1021/acsami.7b02908

[30] Tan J, Liu R, Wang W, Liu W, Tian Y, Wu M, et al. Controllable aggregation and reversible pH sensitivity of AuNPs regulated by carboxymethyl cellulose. Langmuir. 2010;**26**(3):2093-2098. DOI: 10.1021/la902593e

[31] Yin Y, Hu Y, Xiong F. Sorption of Cu (II) and Cd(II) by extracellular polymeric substances (EPS) from *Aspergillus fumigatus*. International Biodeterioration and Biodegradation. 2011;**65**(7):1012-1018. DOI: 10.1016/j.ibiod.2011.08.001

[32] Feng M, Chen X, Li C, Nurgul R, Dong M. Isolation and identification of an exopolysaccharide-producing lactic acid bacterium strain from Chinese Paocai and biosorption of Pb(II) by its exopolysaccharide. Journal of Food Science. 2012;**77**(6):T111-T117. DOI: 10.1111/j.1750-3841.2012.02734.x

[33] Salehizadeh H, Shojaosadati SA. Removal of metal ions from aqueous solution by polysaccharide produced from *Bacillus firmus*. Water Research. 2003;**37**(17):4231-4235. DOI: 10.1016/S0043-1354(03)00418-4

[34] Baig T, Nayak J, Dwivedi V, Singh A, Tripathi PK. A review about dendrimers: Synthesis, types, characterization and applications. International Journal of Advances in Pharmacy, Biology and Chemistry. 2015;**4**(1):44-59

[35] Mauter MSC, Elimlech M. Environmental applications of carbon-based Nanomaterials. Environmental Science & Technology. 2008;**42**(16):5843-5859. DOI: 10.1021/es8006904

[36] Kandah MI, Meunier J-L. Removal of nickel ions from water by multi-walled carbon nanotubes. Journal of Hazardous Materials. 2007;**146**(1-2):283-288

[37] Li Y, Liu F, Xia B, Du Q, Zhang P, Wang D, et al. Removal of copper from aqueous solution by carbon nanotube/calcium alginate composites. Journal of Hazardous Materials. 2010;**177**(1-3):876-880. DOI: 10.1016/j.jhazmat.2009.12.114

[38] Gong J-L, Wang B, Zeng G-M, et al. Removal of cationic dyes from aqueous solution using magnetic multi-wall carbon nanotube nanocomposite as adsorbent. Journal of Hazardous Materials. 2009;**164**(2-3):1517-1522

[39] Mackay DM, Cherry JA. Groundwater contamination: Pump-and-treat remediation. Environmental Science & Technology. 1989;**23**(6):630-636. DOI: 10.1021/es00064a001

[40] Tungittiplakorn W. Engineered polymeric nanoparticles for bioremediation of hydrophobic contaminants. Environmental Science & Technology. 2005;**39**(5):1354-1358

[41] Liao C, Xu W, Lu G, Liang X, Guo C, Yang C, et al. Accumulation of hydrocarbons by maize (*Zea mays* L.) in remediation of soils contaminated with crude oil. International Journal of Phytoremediation. 2015;**17**(7):693-700. DOI: 10.1080/15226514.2014.964840

[42] Mojiri A. The potential of corn (*Zea mays*) for phytoremediation of soil contaminated with cadmium and Lead. Journal of Biological and Environmental Sciences. 2011;**5**(13):17-22

[43] Bargar JR, Bernier-Latmani R, Giammar DE, Tebo BM. Biogenic uraninite nanoparticles and their importance for uranium remediation. Elements. 2008;**4**(6):407-412. DOI: 10.2113/gselements.4.6.407

[44] Cherian S, Oliveira MM. Transgenic plants in phytoremediation: Recent

advances and new possibilities. Environmental Science & Technology. 2005;**39**(24):9377-9390. DOI: 10.1021/es051134l

[45] Čechmánková J, Vácha R, Skála J, Havelková M. Heavy metals phytoextraction from heavily and moderately contaminated soil by field crops grown in monoculture and crop rotation. Soil and Water Research. 2011;**6**(3):120-130

[46] Baldantoni D, Maisto G, Bartoli G, Alfani A. Analyses of three native aquatic plant species to assess spatial gradients of lake trace element contamination. Aquatic Botany. 2005;**83**(1):48-60. DOI: 10.1016/j.aquabot.2005.05.006

[47] Rizwan M, Singh M, Mitra CK, Morve RK. Ecofriendly application of nanomaterials: Nanobioremediation. Journal of Nanoparticle Research. 2014;**2014**:431787 8, 1-7. DOI: 10.1155/2014/431787

[48] Bharti S, Banerjee TK. Phytoremediation of the coal mine effluent. Ecotoxicology and Environmental Safety. 2012;**81**:36-42. DOI: 10.1016/j.ecoenv.2012.04.009

[49] Chen JC, Wang KS, Chen H, Lu CY, Huang LC, Li HC, et al. Phytoremediation of Cr(III) by *Ipomonea aquatica* (water spinach) from water in the presence of EDTA and chloride: Effects of Cr speciation. Bioresource Technology. 2010;**101**(9):3033-3039. DOI: 10.1016/j.biortech.2009.12.041

[50] Marín CMDC, Oron G. Boron removal by the duckweed *Lemna gibba*: A potential method for the remediation of boron-polluted waters. Water Research. 2007;**41**(20):4579-4584. DOI: 10.1016/j.watres.2007.06.051

[51] El-Khaiary MI. Kinetics and mechanism of adsorption of methylene blue from aqueous solution by nitric-acid treated water-hyacinth. Journal of Hazardous Materials. 2007;**147**(1-2):28-36. DOI: 10.1016/j.jhazmat.2006.12.058

[52] Hasan SH, Talat M, Rai S. Sorption of cadmium and zinc from aqueous solutions by water hyacinth (*Eichhornia crassipes*). Bioresource Technology. 2007;**98**(4):918-928. DOI: 10.1016/j.biortech.2006.02.042

[53] Khataee AR, Movafeghi A, Torbati S, Salehi Lisar SY, Zarei M. Phytoremediation potential of duckweed (*Lemna minor* L.) in the degradation of C.I. acid blue 92: Artificial neural network modelling. Ecotoxicology and Environmental Safety. 2012;**80**(1):291-298. DOI: 10.1016/j.ecoenv.2012.03.021

[54] Pandey VC. Phytoremediation of heavy metals from fly ash pond by *Azolla caroliniana*. Ecotoxicology and Environmental Safety. 2012;**82**:8-12. DOI: 10.1016/j.ecoenv.2012.05.002

[55] Srivastava S, Bhainsa KC, D'Souza SF. Bioresource technology investigation of uranium accumulation potential and biochemical responses of an aquatic weed *Hydrilla verticillata* (L.f.) Royle. Bioresource Technology. 2010;**101**(8):2573-2579. DOI: 10.1016/j.biortech.2009.10.054

[56] Islam MS, Saito T, Kurasaki M. Phytofiltration of arsenic and cadmium by using an aquatic plant, *Micranthemum umbrosum*: Phytotoxicity, uptake kinetics, and mechanism. Ecotoxicology and Environmental Safety. 2015;**112**:193-200. DOI: 10.1016/j.ecoenv.2014.11.006

[57] González I, Neaman A, Cortés A, Rubio P. Effect of compost and biodegradable chelate addition on phytoextraction of copper by Oenothera picensis grown in Cu-contaminated acid soils. Chemosphere. 2014;**95**:111-115. DOI: 10.1016/j.chemosphere.2013.08.046

[58] Sooksawat N, Meetam M, Kruatrachue M, Pokethitiyook P, Nathalang K. Phytoremediation potential of charophytes: Bioaccumulation and toxicity studies of cadmium, lead and zinc. Journal of Environmental Sciences. 2013;**25**(3):596-604. DOI: 10.1016/S1001-0742(12)60036-9

[59] Siva Kiran RR, Madhu GM, Satyanarayana SV, Bindiya P. Bioaccumulation of cadmium in blue-green algae Spirulina (Arthrospira) Indica. Journal of Bioremediation & Biodegradation. 2012;**3**(3):1-4. DOI: 10.4172/2155-6199.1000141

[60] Favas PJC, Pratas J, Prasad MNV. Accumulation of arsenic by aquatic plants in large-scale field conditions: Opportunities for phytoremediation and bioindication. Science of the Total Environment. 2012;**433**:390-397. DOI: 10.1016/j.scitotenv.2012.06.091

[61] Tuzen M, Sari A, Mendil D, Uluozlu OD, Soylak M, Dogan M. Characterization of the biosorption process of As(III) on green algae *Ulothrix cylindricum*. Journal of Hazardous Materials. 2009;**165**(1-3):566-572. DOI: 10.1016/j.jhazmat.2008.10.020

[62] Hua J, Zhang C, Yin Y, Chen R, Wang X. Phytoremediation potential of three aquatic macrophytes in manganese-contaminated water. Water and Environment Journal. 2012;**26**(3):335-342. DOI: 10.1111/j.1747-6593.2011.00293.x

[63] Maine MA, Duarte MV, Suñé NL. Cadmium uptake by floating macrophytes. Water Research. 2001;**35**(11):2629-2634. DOI: 10.1016/S0043-1354(00)00557-1

[64] Mishra VK, Upadhyay AR, Pandey SK, Tripathi BD. Concentrations of heavy metals and aquatic macrophytes of Govind Ballabh Pant Sagar an anthropogenic lake affected by coal mining effluent. Environmental Monitoring and Assessment. 2008;**141**(1-3):49-58. DOI: 10.1007/s10661-007-9877-x

[65] Ha NTH, Sakakibara M, Sano S. Accumulation of indium and other heavy metals by *Eleocharis acicularis* an option for phytoremediation and phytomining. Bioresource Technology. 2011;**102**(3):2228-2234. DOI: 10.1016/j.biortech.2010.10.014

[66] Sakakibara M. Phytoremediation of toxic elements-polluted water and soils by aquatic macrophyte *Eleocharis acicularis*. AIP Conference Proceedings. 2016;**1744**:1-6. DOI: 10.1063/1.4953512

[67] Sarma H. Metal hyperaccumulation in plants: A review focusing on phytoremediation technology. Journal of Environmental Science and Technology. 2011;**4**(2):118-138. DOI: 10.3923/jest.2011.118.138

[68] Ladislas S, El-Mufleh A, Gérente C, Chazarenc F, Andrès Y, Béchet B. Potential of aquatic macrophytes as bioindicators of heavy metal pollution in urban stormwater runoff. Water, Air, and Soil Pollution. 2012;**223**(2):877-888. DOI: 10.1007/s11270-011-0909-3

[69] Landberg T, Greger M. Differences in uptake and tolerance to heavy metals in Salix from unpolluted and polluted areas. Applied Geochemistry. 1996;**11**(1):175-180. DOI: 10.1016/0883-2927(95)00082-8

[70] Ariyakanon N, Winaipanich B. Phytoremediation of copper contaminated soil by *Brassica juncea* (L.) Czern and *Bidens alba* (L.) DC. var. radiata. Journal of Scientific Research, Chulalongkorn University. 2006;**31**(1):49-56

[71] Das S, Goswami S, Talukdar AD. A study on cadmium phytoremediation potential of water lettuce, *Pistia*

stratiotes L. Bulletin of Environmental Contamination and Toxicology. 2014;**92**(2):169-174. DOI: 10.1007/s00128-013-1152-y

[72] Amin H, Arain BA, Amin F, Surhio MA. Analysis of growth response and tolerance index of *Glycine max* (L.) Merr . under hexavalent chromium stress. Advances in Life Sciences. 2014;**1**(4):231-241

[73] Imtiyaz S, Agnihotri RK, Ganie SA, Sharma R. Biochemical response of glycine max(L.) Merr to cobalt and lead stress. Journal of Stress Physiology and Biochemistry. 2014;**10**(3):259-272

[74] Kranner I, Colville L. Metals and seeds: Biochemical and molecular implications and their significance for seed germination. Environmental and Experimental Botany. 2011;**72**(1):93-105. DOI: 10.1016/j.envexpbot.2010.05.005

[75] Shanker AK, Cervantes C, Loza-Tavera H, Avudainayagam S. Chromiumtoxicityinplants.Environment International. 2005;**31**(5):739-753. DOI: 10.1016/j.envint.2005.02.003

[76] Shukla UC, Singh J, Joshi PC, Kakkar P. Effect of bioaccumulation of cadmium on biomass productivity, essential trace elements, chlorophyll biosynthesis, and macromolecules of wheat seedlings. Biological Trace Element Research. 2003;**92**(3):257-273. DOI: 10.1385/BTER:92:3:257

[77] Gupta UC, Gupta SC. Trace element toxicity relationships to crop production and livestock and human health: Implications for management. Communications in Soil Science and Plant Analysis. 1998;**29**(11-14):1491-1522. DOI: 10.1080/00103629809370045

[78] McBride MB, Sauve S, Hendershot WH. Solubility control of Cu, Zn, Cd and Pb in contaminated soils. European Journal of Soil Science. 1997;**48**(2):337-346

[79] Gisbert C, Ros R, Haro AD, Walker DJ, Bernal MP, Serrano R, et al. A plant genetically modified that accumulates Pb is especially promising for phytoremediation. Biochemical and Biophysical Research Communications. 2003;**303**:440-445

[80] Thomas JC, Davies EC, Malick FK, Endreszl C, Williams CR, Abbas M, et al. Yeast metallothionein in transgenic tobacco promotes copper uptake from contaminated soils. Biotechnology Progress. 2003;**19**(2):273-280

[81] Pan A, Yang M, Tie F, Li L, Chen Z, Ru B. Expression of mouse metallothionein-I gene confers cadmium resistance in transgenic tobacco plants. Plant Molecular Biology. 1994;**24**:341-351

[82] Evans KM, Gatehouse JA, Lindsay WP, Shi J, Tommey AM, Robinson NJ. Expression of pea metallothionein-like gene PsMTA function. Plant Molecular Biology. 1992;**20**:1019-1028

[83] Abhilash PC, Jamil S, Singh N. Transgenic plants for enhanced biodegradation and phytoremediation of organic xenobiotics. Biotechnology Advances. 2009;**27**(4):474-488. DOI: 10.1016/j.biotechadv.2009.04.002

[84] Doty SL. Enhancing phytoremediation through the use of transgenics and endophytes. The New Phytologist. 2008;**179**(2):318-333. DOI: 10.1111/j.1469-8137.2008.02446.x

[85] Van Aken B. Transgenic plants for enhanced phytoremediation of toxic explosives. Current Opinion in Biotechnology. 2009;**20**(2):231-236. DOI: 10.1016/j.copbio.2009.01.011

[86] Rugh CL, Senecoff JF, Meagher RB, Merkle SA. Development

of transgenic yellow poplar for mercury phytoremediation. Nature Biotechnology. 1998;**16**(10):925-928. DOI: 10.1038/nbt1098-925

[87] Dhankher OP, Li Y, Rosen BP, Shi J, Salt D, Senecoff JF, et al. Engineering tolerance and hyperaccumulation of arsenic in plants by combining arsenate reductase and γ-glutamylcysteine synthetase expression. Nature Biotechnology. 2002;**20**(11):1140-1145. DOI: 10.1038/nbt747

[88] Assunção AGL, Da CostaMartins P, De Folter S, Vooijs R, Schat H, Aarts MGM. Elevated expression of metal transporter genes in three accessions of the metal hyperaccumulator *Thlaspi caerulescens*. Plant, Cell and Environment. 2001;**24**:217-226. DOI: 10.1046/j.1365-3040.2001.00666.x

[89] Bizily SP, Rugh CL, Meagher RB. Phytodetoxification of hazardous organomercurials by genetically engineered plants. Nature Biotechnology. 2000;**18**:213-217. DOI: 10.1038/72678

[90] Wang X, Wu N, Guo J, Chu X, Tian J, Yao B, et al. Phytodegradation of organophosphorus compounds by transgenic plants expressing a bacterial organophosphorus hydrolase. Biochemical and Biophysical Research Communications. 2008;**365**(3):453-458. DOI: 10.1016/j.bbrc.2007.10.193

[91] Gasic K, Korban SS. Transgenic Indian mustard (*Brassica juncea*) plants expressing an *Arabidopsis* phytochelatin synthase (AtPCS1) exhibit enhanced As and Cd tolerance. Plant Molecular Biology. 2007;**64**(4):361-369. DOI: 10.1007/s11103-007-9158-7

[92] Hall JL. Cellular mechanisms for heavy metal detoxification and tolerance. Journal of Experimental Botany. 2002;**53**(366):1-11. DOI: 10.1093/jxb/53.366.1

[93] Cobbett C, Goldsbrough P. Phytochelatins and metallothioneins: Roles in heavy metal detoxification and homeostasis. Annual Review of Plant Biology. 2002;**53**(1):159-182. DOI: 10.1146/annurev.arplant.53.100301.135154

[94] Rauser WE. Phytochelatins and related peptides (structure, biosynthesis and function). Plant Physiology. 1995;**109**(4):1141-1149. DOI: 10.1104/pp.109.4.1141

[95] Sigel H, Sigel A, editors. Metallothioneins and Related Chelators (Metal Ions in Life Sciences). Vol. 5. Cambridge, England: Royal Society of Chemistry; 2009. ISBN: 1-84755-899-2

[96] Margoshes M, Vallee BL. A cadmium protein from equine kidney cortex. Journal of the American Chemical Society. 1957;**79**(17):4813-4814. DOI: 10.1021/ja01574a064

[97] Kagi JHR, Schaffer A. Biochemistry of metallothionein. The Biochemist. 1998;**27**(23):8509-8515

[98] Ha S-B, Smith AP, Howden R, Dietrich WM, Bugg S, O'Connell MJ, et al. Phytochelatin synthase genes from Arabidopsis and the yeast *Schizosaccharomyces pombe*. The Plant Cell. 1999;**11**(6):1153-1164. DOI: 10.1105/tpc.11.6.1153.PMC 144235. PMID: 10368185. Retrieved: January 13, 2014

[99] Olena KV, Elizabeth AB, James TW, Philip AR. A new pathway for heavy metal detoxification in animals: Phytochelatin synthase is required for cadmium tolerance in *Caenorhabditis elegans*. The Journal of Biological Chemistry. 2001;**276**(24):20817-20820. DOI: 10.1074/jbc.C100152200.PMID 11313333

[100] Grill E, Winnacker E-L, Zenk MH. Phytochelatins, a class of heavy-metal-binding peptides from

plants, are functionally analogous to metallothioneins. Proceedings of the National Academy of Sciences. 1987;**84**(2):439-443. DOI: 10.1073/pnas.84.2.439

[101] Zenk M. Heavy metal detoxification in higher plants—A review. Gene. 1996;**179**(1):21-30. DOI: 10.1016/S0378-1119(96)00422-2

Chapter 14

Soil Management and Conservation: An Approach to Mitigate and Ameliorate Soil Contamination

Oluwatosin Ayobami Ogunsola, Odunayo David Adeniyi and Victoria Abimbola Adedokun

Abstract

The chapter mainstreamed Soil Management and Conservation approach as a potent remedy for Soil Contamination. Largely, microbial activities play significant role in maintaining balance within the ecosystem however changes in Land-use has a direct influence on soil biota, including the floral and fauna components. The introduction of contaminants, from varying sources such as agrochemicals, petrochemicals, landfills, sludge, effluents, etc., into the soil builds up the amount of heavy metals present in the deposits hence degrading the soil and polluting groundwater. Integrating soil management options to enhance biodiversity and strengthen microbial activities improve the soil ecology thus creating a buffer for neutralizing potential contaminants.

Keywords: degradation, land-use, ecology, biodiversity, soil conservation

1. Introduction

One of the central component of terrestrial ecosystem is soil. Loss in ecosystem is a representation of the degradation of soil. The soil plays a key role in the health of ecosystem, however, over-exploitation of these ecosystem by humans causes considerable degradation and migration of contaminants. The use of land for agriculture occupies 36.5% of the earth's land mass [1]. Though this human activities may be justified to provide greater benefit in other services termed development, but consistent degradation of this ecosystem and exposure of it to various contaminants is not in the best interest of the society and it is detrimental to the environment that sustains all life forms.

Soil conservation are various practices of farming operations and management strategies which are conducted with the purpose of controlling soil erosion by avoiding or minimizing soil particle detachment and movement of water or/and air. It also helps in preventing the loss of the top-most layer of the soil and fertility which could also be caused by soil contamination. Understanding the processes and factors that govern soil erosion is very important to implementing its control practice and will help to manage soil erosion thus leading to soil conservation. The mechanics involve fluid (wind/water) detachment or entrainment which is being accompanied by the transportation of soil particles and its subsequent deposition as soil sediments. Conservation approaches and management strategies that ensures these include crop rotation, cover cropping, planting windbreaks and conservation tillage, which have been harnessed for millennia. Soil conservation practices are said to be farming operations and soil management strategies carried out with the aim of achieving a goal which is to control soil erosion by preventing or reducing soil particle detachment and transport in air or water [2]. Soil conservation started with the aim to protect an ecology from agricultural production by making use of largely unproven technology that failed to adapt with the natural requirements of the land. The evolving land degradation trend could only be understood by determining if the causes were as a result of natural occurrences or by unwise use [3].

In Europe, Common Agricultural Policy (CAP) is put in place in a bid to target the application of best management practices such as winter cover crops, reduced tillage, plant residues and grass margins in order to address conservation [4]. This traditional approaches which enhanced the productivity, environmental benefits and profits are based on procedures of no-tillage, and the broader concepts of agricultural conservation and land management sustainability. These concepts are one and not divided, but part of a continuous land management practices which range from detailed soil management practices such as zero-tillage, to the enhnaced concepts, principles and objectives of agricultural conservation and land management for sustainability.

2. Soil conservation methods

2.1 Cover cropping and mulching

This method is effective in reducing migration of top soil by leaving a cover over the soil in a bid to reduce soil displacement which is associated with the impacts from raindrops on the soil particles. Cover crops and mulching also reduces the amount of runoff and its velocity over the soil. Mulching, which is the application of organic materials over exposed soil to confer a form of covering to it over a period before decomposing. Straw can be used as mulch but hay is proven to be the best and it is important to ensure that it is harvested before the weeds mature. These crops are necessary to control erosion especially when the main crops planted do not give sufficient residue for more conventional residue management-based erosion control [5]. Where precipitation is adequate, cover crops like peas can help protect against wind erosion and also add nitrogen to the soil. The nitrogen released from the roots of these legumes are energy source for microbial metabolic activities hence such live mulch or cover crop give rise to an active microbial community in the rhizosphere soil.

2.2 Crop rotation

Crop rotation is an indigenous and practical way for managing agro-ecosystem biodiversity by enhancing soil health, minimizing pests and disease outbreaks [6]. This method enables farmers to improve the structure of the soil, increase the soil organic matter and rooting depth. This happens when secondary crops are grown in order to enhance soil health. As a result of the extensive shattering of soil aggregrates during seedbed preparation and harvesting, root crops are particularly destructive to the soil structure. Therefore, it is advised that root crops should be grown once in every three years. Corn can be grown in the following year with two years of silage followed in succession by three or more years of forage. Leguminous crops (such as pea and chickpea) during crop rotation helps in modifying soil functional microbial communities. In the rotation, cover cropping or mulching, and zero tillage should be incorporated too. Crop rotations can provide better opportunities for the growth of some soil functional microorganisms. This brings about rich biodiversity within the soil ecosystem as both the shallow feeding crops and deep rooted crops activates varying species of microorganismsper time thus creating a build up of microbes exhibiting varying characteristics to colonize the soil. Thus, different crops can produce various residues and root exudates to boost soil microbial diversity and activity, and increase soil microbial biomass as well as enhance C and N cycling [7, 8].

2.3 Conservation tillage

This method is aimed atpreserving soil aggregates, organic matter and crop residues [9, 10]. Conservation tillage include changes in making use of less destructive tillage implements (for instance, instead of using mouldboard plow, use chisel plow), minimum tillage (that is, one turn instead of two), leaving crop residue on the soil surface to prevent erosion. Plowing and tilling land for the preparation of the seed bed are basis of the traditional agricultural practices. However, these practices have been proven to be highly destructive to the soil with 24% of global agricultural land degraded as a result of this [11]. New approach which is centered on conserving and improving soil is gradually replacing soil tillage. The soil is typically inverted to a depth of less than 20 cm using mouldboard plow during conventional tillage system, however, in conservation tillage system, the soil is not disturbed or disturbed to a lesser degree [12]. This conservation method has shown to improve soil structure, reduce soil erosion, improve drainage and water holding capacity of the soil, increase soil organic matter and also increase microbial and earthworm activity [13].

2.4 Ridges, terraces and contours

The ridges are made across wind and they consist of tall listed seed beds that are being formed over the entire field or as trap strips which is in a position perpendicular to the direction of the prevailing wind. The formation of an earthen embankment along a common elevation contour gives an elevated terrace structure that can directly reduce wind erosion by potential reduction in wind speed and interception of soil particles. Indirect wind erosion control benefits of terraces and the related contour tillage and cropping practices expand overall crop grain and residue productivity by controlling runoff for increased water storage in the soil [14]. The underlying layer of soil becomes relatively less disturbed by the action of erosion hence making room for an increased microbial population within the micro-climate.

2.5 Strip cropping/planting windbreaks

This is another method of conserving the soil and for controlling wind erosion. A windbreak serves as a barrier with the purpose of deflecting the flow of air and reducing leeward wind speed [15]. However, the availability of irrigation makes this conservation method useful in a difficult environment. The crops may be cultured in strips perpendicular to the prevailing wind where field orientation is not restricted as a means to reduce the near surface wind speed [16]. This practice is broadly accommodating of various width of crop strips depending on the crop tolerance to eroding soil or potential to trap soil grains. The interplay between erosivity and erodibility potential of soil determines the gradient of detachment experienced within varying soil types. This confers significance on the efficacy of windbreaks/strip crops to band soil particles together thereby curtailing dislodgement.

2.6 Residue management

This is the most preferred method for controlling wind erosion for most crops and climates [17]. It is made up of several tillage practices that maintain residue from a previously harvested crop as a surface cover to prevent soil erosion. Residue management also maintains mulches which may be standing or flat to intercept soil grains by trapping their movement [18]. Leaving the residue of the previous crop on the surface of the soil is beneficial in that it improves soil water storage regardless of the runoff controlling contours, it helps to increase rain infiltration and reduce evaporation from the soil. The micro-climate here is well adapted for microbial activities as there exists a steady retrieval of energy from the decomposing biomass of residues thereby giving rise to mineralization of organic compounds and disintegration of complex molecules.

3. Effects of agriculture on environmental health

Soil health is the innate potential of a soil to function within ecosystem boundaries (either natural or managed), sustain plant productivity, maintain water and air quality, support human well-being, and provide habitats for biodiversity [19–21]. Agricultural intensification is placing huge pressure on the soil's potential to maintain its functions which is progressively leading to large-scale ecosystem degradation and loss of productivity in the long term [22–24]. Over a few decades, significant efforts have been made to enhance agricultural productivity through increased fertilization and pesticide application, improved irrigation, soil management regimes and crops, and massive land conversions [25]. However, there is a growing concern that the use of natural ecosystems for agricultural purposes has incurred substantial environmental costs, including desertification, increased emissions of greenhouse gasses, decreased organic matter in soils, loss of biodiversity, and alterations to biogeochemical and hydrological cycles [26, 27].

The quality of the soil, conversely, is an extrinsic feature of soils and changes with the desired usage of that soil by humans. This may be related to agricultural production and its capacity to support wildlife, watershed production, or recreation outputs provision. Some of the environmental challenges that are related to agriculture are expressed as pollutants, climate change, soil degradation, and deforestation [28].

3.1 Climate change

Climate is described as general or average weather conditions of a certain region, including temperature, rainfall and wind, over a long period. Climate change has direct and indirect effect in speeding up or slowing down terrestrial microbial community composition and their functional activities. Climatic change alters the relative population of microorganisms and their functions within soil communities since soil community members differ in their physiology, temperature sensitivity, and growth rates [29–34]. The direct effects of climatic change on microbial population, composition and function have been reviewed extensively [35–39]. Temperature and water are essential environmental factors for microbial growth. Increased temperature alters microbial community structures and processes such as respiration, fermentation and methanogenesis are also accelerated. This directly affects enzyme activity and microbial physiological property. Both agriculture and climate change are interrelated processes, of which they both take place on a global scale. Climate change impacts microbial community structure and activities both directly, through alteration of the soil chemical and physical environment, and indirectly through changes in land use. Environmental changes such as global warming are directly altering microbial soil respiration rates because soil microorganisms, and the processes they mediate, are temperature sensitive. The role of the prevailing changing climate, visibly expressed with elevated temperature, in microbial metabolism has been accorded considerable attention of recent [40–43]. This stresses the effects of climatic changes on soil microorganisms which are essential components in the ecosystem since they play a key role in maintaining soil health through ecological intensification.

3.2 Deforestation

Deforestation is a major driver of climate change and cause of the loss of habitat for millions of species. The soil is the basis for agriculture, natural plant communities and natural climate regulation, with 75% organic carbon stored in terrestrial habitat [44–46]. Vegetation has extensive contribution in sustaining ecosystem services of both surface and subsurface soil. Deforestation exacerbates climate change in that trees are completely or selectively removed to create farmland. Land use changes have several undesirable consequences, with significant effect on radical losses in soil fertility, soil carbon and nitrogen stocks have been recorded in the first 20–25 years after deforestation [47, 48].

3.3 Pollutants

Synthetic pesticides are the most common and widely use method of controlling pests in agriculture. A large number of agricultural chemicals (such as fertilizer, pesticides, etc.) are used and some become pollutants through their use, misuse or ignorance hence leaching through the soil to pollute the groundwater. Soil erosion has been instrumental in the horizontal and vertical movement of these pollutants (earlier bonded with soil particles but displaced) from agricultural fields to other places, especially water bodies (both surface and underground). Consequently, pollutants from agricultural fields do have large effect on the quality of water. Poorly managed animal feeding operations, overgrazing, heavy use of fertilizers, plowing, and improper, heavy use, or wrongly timed use of pesticides, causes pollution. These pollutants find their ways through the soil profile and across the gradient of slope hence affecting rivers, groundwater, wetlands, lakes, and estuaries [28] through continued deposition over a long period. In the same vein, untreated industrial pollutants discharged from the industries and factories have prevalent toxic concentration. Oftentimes, these wastes are discharged into the water body and affect aquatic cultures as well as flora and fauna life cycles. Usage of unsuitable contaminated water and the discharge of untreated industrial wastewater into water bodies form a main source of water pollution. Soil pollution occurs due to untreated disposal of industrial wastes (laden with high toxic contaminants) into soil. Wastes from industries have varying amount of toxic chemicals such that when deposited in soil, they cause the soil layer strength in the top soil to deteriorate, thus reducing fertility and microbial activity of the soil. In addition, the hazardous effect of these pollutants leads to ecological imbalances within the soil ecosystem.

3.4 Soil degradation

Soil degradation is the decrease in the quality of soil that can be as a result of many factors, most especially from agriculture. Soils hold the majority of the world's biodiversity, and healthy soils are essential for food production and adequate water supply [49]. Soil degradation shows expression in salting, waterlogging, compaction, pesticide contamination, decline in soil structure, loss of fertility, increase in soil acidity, alkalinity, salinity, and prevalence of erosion. Soil erosion is the wearing away of topsoil by water, wind, or farming activities [50]. At the same time, agriculture has been shown to contribute significantly to degradation, mainly through the continued dependence and improper use of inorganic fertilizers, synthetic pesticides, etc., which culminates in production and release of greenhouse gases such as carbon dioxide, methane, and nitrous oxide. Moreover, agriculture that practices conventional practices such as tillage, fertilization, and pesticide application also release ammonia, nitrate, phosphorus, and many other gases that pollute the air, water, and soil quality, as well as biodiversity. Agriculture also changes the land cover of the Earth, which can change its ability to absorb or reflect heat and light, hence contributing to radiative forcing. Soil degradation also has a large impact on biological degradation, which influence the microbial community of the soil negatively and alters nutrient cycling, pest and disease control, and chemical transformation properties of the soil.

4. Effects of microbial activities on soil contaminants

By 2050, it is projected that the world population will increase to 8.9 billion people and this will lead to higher demand for agricultural produce [51]. In the future, the high demand of food and shortage of new agricultural land development will require increasing crop yields making use of sustainable means. Improvement of soil conservation increases soil organic matter and reduces erosion in other to have a sustainable agricultural land management and improved soil health [52]. Assessment of soil is based on the quality of soil variables that guarantee crop production sustainability in agricultural lands [19, 53]. Soil biota components such as microbial community, activity, abundance, stability and diversity which are improved by soil conservation have been discussed in several studies to be important indicators of soil quality [19, 54]. The rhizosphere of the plant is the narrow zone of the soil that is closed to the root system and sustains the production of crops with agrochemical inputs level that is balance or minimized [55]. Rhizoremediation of organic pollutants [8] and organic compounds creates nutrient-rich environment that influence microbial communities and the degradation of organic contaminants [56]. Soil biota plays a great role in residues of plant mineralization to form plants nutrients which can be easily absorbed by the plants for their growth and development [57]. Also, soil biota increases the rate of decomposition by excreting different enzymes that support plants's nutrients kinetics in the soil [58]. Microorganisms in the soil especially bacteria and fungi, transforms N between organic and inorganic forms which improves plant minerals uptake [59]. Microbial communities support the fundamental processes that provide productivity and stability of agroecosystems [60].

Soil conservation activities such as cover crops and minimum tillage as earlier mentioned can favorably improve soil health by increasing the number of soil organisms that break down organic matter, and in the process, release nutrients for the plant uptake. This soil organism breaks organic soil contaminants and several factors can interfere with the soil–microbe–plant complex hence influencing its functionality. Soil type [61], organic carbon level [60], temperature and moisture [62], oxygen level [63], electrical conductivity, calcium level and pH [64] are all factors that can change the composition and functionality of soil microbial communities. Of the soil macrofauna, earthworms are a major component and are very important in the soil fertility dynamics as their burrowing activities helps in improving the soil aeration and infiltration of water into the soil. The population of earthworm is influenced by soil conservation. [65, 66] discussed how minimum tillage which is part of soil conservation affects the population of earthworm. The increase of earthworms could encourage biological-remediation of contaminated soil known as vermiremediation [67]; soils contaminated with metallic contaminants [68] and organic pollutants and some chlorinated compounds inclusive [69]. The earthworms makes holes through the soil, mix the soil, affects its structure, and alters its nutritional profile and fungal and bacterial communities [70].

Fungi are chemoorganotrophic organism that are present everywhere and plays fundamental roles in geological and ecological processes [71, 72]. They can transform a large varieties of organic substrates, in addition with natural polymers not only lignin, cellulose, starch and chitin, but also other anthropogenic products such as explosives, pesticides and other xenobiotics [73, 74]. Mycoremediation, that is, the use of fungi to remove soil contaminant, has emerged as one of the most promising and cost-effective soil remediation techniques [75–79]. Bacterial genera, namely, *Gordonia*, *Brevibacterium*, *Aeromicrobium*, *Dietzia*, *Burkholderia*, and *Mycobacterium*, Fungal genera, namely, *Amorphoteca*, *Neosartorya*, *Talaromyces*, and *Graphium* as well as terrestrial fungi, namely, *Aspergillus*, *Cephalosporium*, and *Penicillium* and yeast genera, namely, *Candida*, *Yarrowia*, and *Pichia* which were isolated from soil that has been contaminated by petroleum proved to be organisms that has the potential for degrading hydrocarbon while yeast species, namely, *Candida lipolytica*, *Rhodotorula mucilaginosa*, *Geotrichum* spp., *and Trichosporon mucoides* isolated from water that has been contaminated were discovered to degrade petroleum compounds [80–82]. When soil microorganism is improved by soil conservation, mycoremediation will be facilitated in order to remove soil contaminant. For instance, fungi is a potential approach for specific site Arsenic bioremediation [78, 79]. This adaptation of fungi towards soil that has been contaminated could be the high surface area to volume ratio and their various detoxification of metal mechanisms [83].

The physical and chemical properties of the soil significantly influence the soil fungal community structure and this is determined by agricultural practices [84, 85]. Increase in fungal biomass and bacterial is termed as changes in soil microbial communities and it has been observed in zero tillage than in conventional tillage practices [86]. Various land management practices has been examined to increase fungal biomass in the soil. Total fungal hyphal biomass and fungal propagules were discovered to be more in soil collected from organically managed agricultural systems [87–89]. The density of fungi in soil were found to be affected by crop rotation, animal grazing and soil tillage [90–98].

5. Conclusion

The type of land management practices in agroecosystems as an impacts on the structure of microbial community and function through a variety of different mechanisms. Land-use changes also impact on soil microbial community structure through alterations in carbon availability and quality, pH and nutrient availability. Since the ratio of fungal population to bacterial population are commonly measured as indicators of microbial community structure, and the relative proportions of fungi are increased by no-till practices, crop rotations, and use of cover crops, thus biological mechanisms are regulating carbon and nitrogen exchanges between the land, water and atmosphere. This reveals the importance of soil management and conservation approach in enhancing microbial activity for soil ecological intensification as well as buffering the soil to neutralize contaminants. Albeit, microbial ecology to assess terrestrial carbon cycle plays a crucial role in maintaining balance within the ecosystem.

Conflict of interest

There is no conflict of interest.

Author details

Oluwatosin Ayobami Ogunsola[1]*, Odunayo David Adeniyi[2] and Victoria Abimbola Adedokun[3]

1 Department of Crop, Soil and Pest Management, Federal University of Technology Akure, Nigeria

2 Department of Earth and Environmental Sciences, University of Pavia, Italy

3 Department of Environmental Health Sciences, University of Ibadan, Nigeria

*Address all correspondence to: oluwatosinogunsola8@gmail.com

References

[1] FAOSTAT. (2008). Fact sheet. *FAO, Rome, Italy*.

[2] Dabney, S.M., Shields, F.D., Binger, R.L., Kuhnle, R.A., Rigby, J.R. 2012. Watershed management for erosion and sedimentation control case study. Godwin Creek, Panola County, MS. In: Lal R., Stewart, B.A. (Eds.) Advances in Soil Science: Soil Water and Agronomic Productivity. Boca Raton, FL: Taylor and Francis. Pp; 539-568.

[3] Tanner, T. (Ed.), (2012). Aldo Leopold: The Man and His Legacy. Ankeny, IA Soil and Water Conservation society.

[4] Panagos, Panos; Borrelli, Pasquale; Meusburger, Katrin; Alewell, Christine; Lugato, Emanuele; Montanarella, Luca. (2015). Estimating the soil erosion cover-management factor at the European scale. *Land Use Policy, 48*, 38-50. doi:doi:10.1016/j.landusepol.2015.05.021

[5] Keeling, W., Segarra, E., Abernathy, J.R 1989. Evaluation of conservation tillage cropping systems for cotton on the Texas High Plains. *Journal of Production Agriculture*. 2; 269-273.

[6] Barbieri, P., S. Pellerin, V. Seufert, T. Nesme (2019). Changes in crop rotations would impact food production in an organically farmed world*Nat. Sustain*., 2, pp. 378-385.

[7] Gurr,G.M., Z. Lu, X. Zheng, H. Xu, P. Zhu, G. Chen, X. Yao, J. Cheng, Z. Zhu, J.L. Catindig, S. Villareal, H. VanChien, L.Q. Cuong, C. Channoo, N. Chengwattana, L.P. Lan, L.H. Hai, J. Chaiwong, H.I. Nicol, D.J. Perovic, S.D. Wratten, K.L. Heong (2016). Multi-country evidence that crop diversification promotes ecological intensification of agriculture *Nature Plants*, 2.

[8] Liu, S.H., G.M. Z.-Y.-H. (2017). Bioremediation mechanisms of combined pollution of PAHs and heavy metals by bacteria and fungi: a mini review. *Bioresources Technology, 224*, 25-33.

[9] Lal, R., D. R. (2007). Evolution of the plow over 10,000 years and the rationale for no-till farming. *Soil Tillage Resources, 93*, 1-12.

[10] Skaalsveen, K., J. I. (2019). The effect of no-till farming on the soil functions of water purification and retention in north-western Europe: a literature review. *Soil Tillage Resources, 189*, 98-109.

[11] Bai, Z.G., D.L. Olson, Schaepman, M.E. (2008). Report 2008/01. *Global assessment of land degradation and improvement., ISRIC-World Soil Information Report, Wageningen*.

[12] Morris, N., Miller, P., Orson, J., & Froud-Williams, R. (2010). The adoption of non-inversion tillage systems in the United Kingdom and the agronomic impact on soil, crops and the environment—a review. *Soil Tillage Resources*, 1-15.

[13] Abdollahi, L., & Munkholm, L. (2014). Tillage system and cover crop effects on soil quality: I. Chemical, mechanical, and biological properties. *Soil Water Management Conservation*, 9.

[14] Duncan, D., Burns, K. (2012). The Dust Bowl: An Illustrated History. San Francisco, FL: Chronicle Books LLC.

[15] Brandle, J.R., Hodges, L., Zhou, X.H. (2004). Windbreaks in North American agricultural systems. *Agroforestry Systems* 61, 65-78.

[16] Woodruff, N.P., Lyles, L., Siddoway, F.H, Fryrear, D.W., 1972. How to Control Wind Erosion Washington DC: Government Print Office US Department of Agriculture Bulletin. 354.

[17] Fryrear, D.W., Skidmore, E.L. 1985. Methods for controlling wind erosion.

In: Follett, R.F., Stewart B.A (Eds.), Soil Erosion and Crop Productivity. Madison, WI: *American Society of Agronomy, Crop Science Society of America, Soil Science Society of America*, pp. 443-457.

[18] Miner, G.L., Hansen, N.C., Inman, D., Sherrod, L.A., Peterson, G.A. 2013. Constraints of no-till dryland agroecosystems as bioenergy production systems. *Agronomy Journal*; 105, 364-376.

[19] Doran, J., & Zeiss, M. (2000). Soil health and sustainability: Managing the biotic component of soil quality. *Applied Soil Ecology, 15*, 3-11. doi: 10.1016/S0929-1393(00)00067-6

[20] Doran, J.W. (2002). Soil health and global sustainability: translating science into practice. *Agric. Ecosyst. Environ.* 88, 119-127. doi: 10.1016/S0167-8809(01)00246-8

[21] Gugino, B.K., Idowu, O.J., Schindelbeck, R.R., van Es, H.M., Wolfe, D.W., Moebius, B.N. (2009). *Cornell Soil Health Assessment Training Manual*, 2nd Edn. Geneva: Cornell University.

[22] Tilman, D., Fargione, J., Wolff, B., D'Antonio, C., Dobson, A., Howarth, R. (2001). Forecasting agriculturally driven global environmental change. *Science* 292, 281-284. doi: 10.1126/science.1057544

[23] Foley, J.A., DeFries, R., Asner, G.P., Barford, C., Bonan, G., Carpenter, S.R. (2005). Global consequences of land use. *Science* 309, 570-574. doi: 10.1126/science.1111772

[24] Vitousek, P.M., Naylor, R., Crews, T., David, M.B., Drinkwater, L.E., Holland, E. (2009). Nutrient imbalances in agricultural development. *Science* 324:1519. doi: 10.1126/science.1170261

[25] Tilman, D., Cassman, K.G., Matson, P.A., Naylor, R., and Polasky, S. (2002). Agricultural sustainability and intensive production practices. *Nature* 418, 671-677. doi: 10.1038/nature01014

[26] Balmford, A., Bennun, L., Brink, B.T., Cooper, D., Cote, I.M., Crane, P. (2005). The convention on biological diversity's 2010 target. *Science* 307, 212-213. doi: 10.1126/science.1106281

[27] Trivedi, P., Delgado-Baquerizo, M., Anderson, I.C, and Singh, B.K. (2016). Response of Soil Properties and Microbial Communities to Agriculture: Implications for Primary Productivity and Soil Health Indicators. *Front. Plant Sci.* 7:990. doi: 10.3389/fpls.2016.00990

[28] van der Warf, Hayo; Petit, Jean (2002). "Evaluation of the environmental impact of agriculture at the farm level: a comparison and analysis of 12 indicator-based methods". *Agriculture, Ecosystems and Environment.* **93** (1-3): 131-145. doi:10.1016/S0167-8809(01)00354-1.

[29] Castro, H.F., A.T. Classen, E.E. Austin, R.J. Norby, and C.W. Schadt. (2010). Soil microbial community responses to multiple experimental climate change drivers. *Applied and Environmental Microbiology* **76**: 999-1007.

[30] Gray, S.B., A.T. Classen, P. Kardol, Z. Yermakov, and R.M. Miller. (2011). Multiple climate change factors interact to alter soil microbial community structure in an old-field ecosystem. *Soil Science Society of America Journal* **75**: 2217-2226.

[31] Lennon, J.T., Z.T. Aanderud, B.K. Lehmkuhl, and D.R. Schoolmaster, Jr. (2012). Mapping the niche space of soil microorganisms using taxonomy and traits. *Ecology* **93**: 1867-1879.

[32] Briones, M.J.I., N.P. McNamara, J. Poskitt, S.E. Crow, and N.J. Ostle. (2014). Interactive biotic and abiotic regulators of soil carbon cycling: evidence from controlled climate experiments on peatland and boreal soils. *Global Change Biology* **20**: 2971-2982.

[33] Delgado-Baquerizo, M., F.T. Maestre, C. Escolar, A. Gallardo, V. Ochoa, B.

Gozalo, and A. Prado-Comesana. (2014). Direct and indirect impacts of climate change on microbial and biocrust communities alter the resistance of the N cycle in a semiarid grassland. *Journal of Ecology* **102**: 1592-1605.

[34] Whitaker, J., N. Ostle, A.T. Nottingham, A. Ccahuana, N. Salinas, R.D. Bardgett, P. Meir, and N. P. McNamara. (2014). Microbial community composition explains soil respiration responses to changing carbon inputs along an Andes-to-Amazon elevation gradient. *Journal of Ecology* **102**: 1058-1071.

[35] Blankinship, J.C., P.A. Niklaus, and B.A. Hungate. (2011). A meta-analysis of responses of soil biota to global change. *Oecologia* **165**: 553-565.

[36] Henry, H.A.L. (2012). Soil extracellular enzyme dynamics in a changing climate. *Soil Biology and Biochemistry* **47**: 53-59.

[37] Manzoni, S., J.P. Schimel, and A. Porporato. (2012). Responses of soil microbial communities to water stress: results from a meta-analysis. *Ecology* **93**: 930-938.

[38] A'Bear, A.D., T.H. Jones, and L. Boddy. (2014). Potential impacts of climate change on interactions among saprotrophic cord-forming fungal mycelia and grazing soil invertebrates. *Fungal Ecology* **10**: 34-43.

[39] Chen, S., J. Zou, Z. Hu, H. Chen, and Y. Lu. (2014). Global annual soil respiration in relation to climate, soil properties and vegetation characteristics: summary of available data. *Agricultural and Forest Meteorology* **198**: 335-346.

[40] Bradford, M.A. (2013). Thermal adaptation of decomposer communities in warming soils. *Frontiers in Microbiology*. doi: http://dx.doi.org/10.3389/fmicb.2013.00333

[41] Frey, S.D., J. Lee, J.M. Melillo, and J. Six. (2013). The temperature response of soil microbial efficiency and its feedback to climate. *Nature Climate Change* **3**: 395-398.

[42] Hagerty, S.B., K.J. van Groenigen, S.D. Allison, B.A. Hungate, E. Schwartz, G.W. Koch, R.K. Kolka, and P. Dijkstra. (2014). Accelerated microbial turnover but constant growth efficiency with warming in soil. *Nature Climate Change* **4**: 903-906.

[43] Karhu, K. (2014). Temperature sensitivity of soil respiration rates enhanced by microbial community response. *Nature* **513**: 81-84.

[44] Lal, R. (2004). Soil carbon sequestration impacts on global climate change and food security *Science*, 304, pp. 1623-1627.

[45] Lemenih, M. and Itanna, F. (2004). Soil carbon stocks and turnovers in various vegetation types and arable lands along an elevation gradient in southern Ethiopia *Geoderma*, 123, pp. 177-188

[46] Lemenih, M. (2004). Effects of Land use Changes on Soil Quality and Native Flora Degradation and Restoration in the Highlands of Ethiopia. Implication for Sustainable Land Management. Doctoral thesis Swedish University of Agricultural Sciences, Sweden.

[47] Tesfaye, M.A., F. Bravo, R. Ruiz-Peinado, V. Pando, A. Bravo-Oviedo (2016) Impact of changes in land use, species and elevation on soil organic carbon and total nitrogen in Ethiopian Central Highlands *Geoderma*, 261, pp. 70-79.

[48] Henok, K., S. Dondeyne, J. Poesen, A. Frankl, J. Nyssen (2017) Transition from Forest-based to Cereal-based Agricultural Systems: A Review of the Drivers of Land use Change and Degradation in Southwest Ethiopia

Land Degradation and Development, 28, pp. 431-449.

[49] Hemphill, D. (1993). "Agricultural Plastics as Solid Waste: What are the Options for Disposal?". *Hort Technology*. **3** (1): 70-73. Retrieved 23 April 2015.

[50] Kidd, G. (2000). "Pesticides and Plastic Mulch Threaten the Health of Maryland and Virginia East Shore Waters"(PDF). Pesticides and You. **19** (4): 22-23. Retrieved 23 April 2015.

[51] Lichtfouse, E., Navarrete, M., Debaeke, P., Souchere, V., Alberola, C., & Menassieu, J. (2009). Agronomy for sustainable. *A review Agron. Seustain. Dev., 29*, 1-6.

[52] Doran, J. (2002). Soil health and global sustainability: Translating science into practice. *Agric. Ecosyst. Environ., 88*, 119-127.

[53] Sahu, P., Singh, D., Prabha, R., Meena, K., & Abhilash, P. (2019). Connecting microbial capabilities with the soil and plant health: Options for agricultural sustainability. *Ecol. Indic., 105*, 601-612.

[54] Leskovar, D., Othman, Y., & Dong, X. (2016). Strip tillage improves soil biological activity, fruit yield and sugar content of triploid watermelon. *Soil Tillage Res., 163*, 266-273.

[55] Berendsen, R., Pieterse, C., & Bakker, P. (2012). The rhizosphere microbiome and plant health. *Trends Plant Science, 17*, 478-486.

[56] Kuiper, I., E. L. (2004). Rhizo-remediation: a beneficial plant-microbe interaction. *Mol. Plant Microbe Interact., 17*, 6-15.

[57] Meena, R., Bohra, J., Singh, S., Meena, V., Verma, J., Verma, S., & Sihag, S. (2016). Towards the prime response of manure to enhance nutrient use efficiency and soil sustainability a current need: A book Review.. *J. Clean. Prod.*, 1258-1260.

[58] Dotaniya, M., Meena, V., Basak, B., & Meena, R. (2016). Potassium uptake by crops as well as microorganisms. *n Potassium Solubilizing Microorganisms for Sustainable Agriculture; Springer*, 267-280.

[59] Van der Heijden, M., Bardgett, R., & van Straalen, N. (2008). The unseen majority: Soil microbes as drivers of plant diversity and productivity in terrestrial Ecosystems. *Ecol. Lett., 11*, 296-310.

[60] Singh, J. (2015). Plant–Microbe Interactions: A Viable Tool for Agricultural Sustainability Plant Microbes Symbiosis: Applied Facets. (N. E. Arora, Ed.) *Springer*, 384.

[61] Dai,M., C. Hamel, M.S.Arnaud, Y. He, C. Grant, N. Lupwayi, H. Janzen, S.S. Malhi, X. Yang, Z. Zhou (2012). Arbuscular mycorrhizal fungi assemblages in chernozem great groups revealed by massively parallel pyrosequencing *Can. J. Microbiol.*, 58, pp. 81-92.

[62] Wu, X., T. Ge, W. Wang, H. Yuan, C.E. Wegner, Z. Zhu, A.S. Whiteley, J. Wu (2015). Cropping systems modulate the rate and magnitude of soil microbial autotrophic CO_2 fixation in soil *Front. Microbiol.*, 6.

[63] Yang, C., C. Hamel, M.P. Schellenberg, J.C. Perez, R.L. Berbara (2010). Diversity and functionality of arbuscular mycorrhizal fungi in three plant communities in semiarid Grasslands National Park, Canada *Microb. Ecol.*, 59, pp. 724-733.

[64] Maarastawi, S.A., K. Frindte, M. Linnartz, C. Knief (2018). Crop rotation and straw application impact microbial communities in Italian and Philippine Soils and the rhizosphere of *Zea mays* *Front. Microbiol.*, 9.

[65] Bainard,L.D., P.L. Chagnon, B.J. CadeMenun, E.G. Lamb, K. LaForge, M. Schellenberg, C. Hamel (2017). Plant communities and soil properties mediate agricultural land use impacts on arbuscular mycorrhizal fungi in the Mixed Prairie ecoregion of the North American Great Plains *Agric. Ecosyst. Environ.,* 249, pp. 187-195.

[66] Rasmussen, K. (1999). Impact of ploughless soil tillage on yield and soil quality: A Scandinavian review. *Soil and Tillage Research, 53*, 3-14.

[67] Anderson, E. (1987). Corn root growth and distribution as influenced by tillage and nitrogen fertilization. *Agronomy Journal, 79*, 544-549.

[68] Sinha, R.K., G. B. (2008). Converting wasteland into wonderland by earthworms—a low-cost nature's technology for soil remediation: a case study of vermiremediation of PAHs contaminated soil. *Environmentalist, 28*, 466-475.

[69] Suthar, S. (2008). Metal remediation from partially composted distillery sludge using composting earthworm Eisenia fetida. *Journal of Environmenal Monitor, 10*, 1099-1106.

[70] Shi Z., J. L. (2020). Vermiremediation of organically contaminated soils: concepts, current status, and future perspectives. *Applied Soil Ecology, 147*, 103,377.

[71] Rodriguez-Campos J., L. D.-B.-R. (2014). Potential of earthworms to accelerate removal of organic contaminants from soil: a review. *Applied Soil Ecology, 79*, 10-25.

[72] Gadd, G.M. (2010). Metals, minerals and microbes: geomicrobiology and bioremediation. *Microbiology*, 609-643. doi:https://doi.org/10.1016/j.mycres.2006.12.001

[73] Gadd, G.M., R. Y. (2012). Geomycology: metals, actinides and biominerals. *Environmetal Microbiology Reports, 4*(3), 270-296. doi:https://doi.org/10.1111/j.1758-2229.2011.00283.x

[74] Gadd, G.M. (2013). Geomycology: fungi as agents of biogeochemical change. *PRAEGER REVIEW, 113B*(2), 139-153. Retrieved from https://www.jstor.org/stable/42912447

[75] Harms H, S. D. (2011). Untapped potential: exploiting fungi in bioremediation of hazardous chemicals. *Nature Reviews Microbiology, 9*, 177-192.

[76] Caporale, G., A., Sommella, A., Lorito, M., Lombardi, N., M.G.G., S., Azam,. .. Ruocco, M. (2014). Trichoderma spp. alleviate phytotoxicity in lettuce plants (Lactucasativa L.) irrigated with arsenic-contaminated water. *Journal of Plant Phsiology, 171*(15), 1378-1384. doi:https://doi.org/10.1016/j.jplph.2014.05.011

[77] Govarthanan, M., R. Mythili, T. Selvankumar, S. Kamala-Kannan, & H. Han. (2018). Myco-phytoremediation of arsenic-and lead-contaminated soils by *Helianthus annuus* and wood rot fungi, Trichoderma sp. isolated from decayed wood. *Ecotoxicology and Environmental Safety, 151*(30), 279-284. doi:https://doi.org/10.1016/j.ecoenv.2018.01.020

[78] Srivastava, P. K., Vaish, A., Dwivedi, S., Chakrabarty, D., Singh, N., & Tripathi, R. D. (2011). Biological removal of arsenic pollution by soil fungi. *Science of The Total Environment, 409*(12), 2430-2442. doi:https://doi.org/10.1016/j.scitotenv.2011.03.002

[79] Singh, M., P. S. (2015). Soil fungi for mycoremediation of arsenic pollution in agricultural soils. *Journal of Applied Microbiology, 119*, 1278-1290. doi:doi:10.1111/jam.12948

[80] Singh, R.K., R. T. (2020). Fungi as potential candidates for bioremediation. *Abatement of Environmental Pollutants, Elsevier*, 177-191.

[81] Chaillan, F., A. Le Flèche, E. Bury, Y.-H. Phantavong, P. Grimont, A. Saliot, and J. Oudot (2004). "Identification and biodegradation potential of tropical aerobic hydrocarbon-degrading microorganisms," *Research in Microbiology*, vol. 155, no. 7, pp. 587-595.

[82] Singh, H. (2006). *Mycoremediation: Fungal Bioremediation*, Wiley-Interscience, New York, NY, USA.

[83] Bogusławska-Was, E., and W. Dąbrowski (2001). "The seasonal variability of yeasts and yeast-like organisms in water and bottom sediment of the Szczecin Lagoon," *International Journal of Hygiene and Environmental Health*, vol. 203, no. 5-6, pp. 451-458.

[84] Kapoor, A., T. V. (1999). Removal of heavy metals using the fungus Aspergillus niger. *Bioresources Technology, 70*, 95-104.

[85] Wu, T., O. Chellemi, D., J. Martin, K., H. Graham, J., & N. Rosskopf, E. (2007). Discriminating the effects of agricultural land management practices on soil fungal communities. *Soil Biology and Biochemistry, 39*(5), 1139-1155. doi:https://doi.org/10.1016/j.soilbio.2006.11.024

[86] Jirout, J., Šimek, M., & Elhottová, D. (2011). Inputs of nitrogen and organic matter govern the composition of fungal communities in soil disturbed by overwintering cattle. *Soil Biology and Biochemistry, 43*(3), 647-656.

[87] Minoshima, H., L. J.-M. (2007, May 01). soil food webs and carbon dynamics in response to conservation tillage in California. *Soil Science Society of America Journal, 71*(3), 952-963. doi:https://doi.org/10.2136/sssaj2006.0174

[88] Sivapalan, A., W. M. (1993). Monitoring Populations of Soil Microorganisms during a Conversion from a Conventional to an Organic System of Vegetable Growing. *Biological Agriculture and Horticulture, 10*(1), 9-27. doi:https://doi.org/10.1080/01448765.1993.9754647

[89] Fließbach, A., P. M. (2000). Microbial biomass and size-density fractions differ between soils of organic and conventional agricultural systems. *Soil Biology & Biochemistry*, 757-768.

[90] Shannon, D., A. S. (2002). A comparative study of the microbiology of soils managed under organic and conventional regimes. *Soil Use & Management, 18*, 274-283.

[91] Wicklow, D. (1973). Microfungal populations in surface soils of manipulated prairie stands. *Ecology, 54*, 1302-1310.

[92] Martyniuk, S., & Wagner., G. (1978). Quantitative and qualitative examination of soil microflora associated with different management systems. *Soil Science, 125*, 343-350.

[93] Ploetz, R.C., D. M. (1985). Population dynamics of soilborne fungi in a field multicropped to rye and soybeans under reduced tillage in Florida. *Phytopathology*, 1447-1451.

[94] Beare, M., Parmelee, R., Hendrix, P., Cheng, W., Coleman, D., & Crossley, D. (1992). Microbial and Faunal interactions and effects on litter nitrogen and decomposition in agroecosystems. *Ecological Monograph, 62*, 569-591.

[95] Frey, S., Elliott, E., & Paustian, K. (1999). Bacterial and fungal abudance and biomass in conventional and no-tillage agroecosystems along two climatic gradients. *Soil Biology & Biochemistry, 31*, 573-585.

[96] Mazzola, M. (1999). Transformation of soil microbial community structure and Rhizoctonia-suppressive potential in response to apple roots. *Phytopathology, 89*, 920-927.

[97] Hedlund, K. (2002). Soil microbial community structure in relation to vegetation management on former agricultural land. *Soil Biology & Biochemistry, 34*, 1299-1307.

[98] Singh, S., & Rai, J. (2004). Soil microbial population and enzyme activity related to grazing pressure in alpine meadows of Nanda Devi Biosphere Reserve. *Journal of Environmental Biology, 25*, 103-107.

Chapter 15

Soil Contamination: A Menace to Life

Sonia Sethi and Payal Gupta

Abstract

The dire concern for soil contamination includes the safety of food, ecological environment, public's health and capacity of social sustainable development. Soil is polluted by heavy metals and pesticides which are far beyond pollution standards. The soil biodiversity and agricultural sustainability are adversely affected in long-term harmful effects by the prolonged intensive and indiscriminate use of agro-chemicals. It needs immediate attention for the whole world to curb continual deterioration of soil pollution and remediate contaminated soil as soon as possible to decrease harm on people's health and ecological environment. In fact, acceleration of related legislation, increased capital investment and technical development to remediate soil contamination and must achieve some progress. However, due to all sorts of the constraints, whether soil management system or technical capacity for decontamination is relatively outdated, so there remains a lot of work need to be done. Developing countries, including Brazil, India and so on, are also facing similar problems. Approaches to solve soil problems could benefit developing countries in process of industrialization and urbanization, so it's a very meaningful job to deep analyze and study the current situation and countermeasures soil pollution. In this Chapter, the overall situation of soil pollution is introduced, the concrete causes and hazards of soil contamination are discussed, and technologies and processes of soil remediation are suggested for improvement of the status of soil contamination and social sustainable capacity.

Keywords: soil quality, xenobiotics, soil microbiota, remediation, environment sustainability

1. Introduction

The fundamental to human life on Earth is Soil. In Natural environment soil forms the vital part and is as important as plants, animals, rocks, landforms, loch and rivers. The distribution of plant species are influenced by the soil and also it provides a habitat for a wide range of organisms. The flow of water and chemical substances between the earth and atmosphere is controlled by the soil and it acts as a source of all types of gases in the atmosphere. Natural processes are not only reflected by the soil but the human activities both at present and in past are also recorded [1].

The reduction in the productivity of soil is due to the presence of soil pollutants. The presence of any chemical substance or toxic chemicals (pollutants) in soil at a higher concentration than normal that adversely affects any non-targeted organism or pose a risk to human health and/or the ecosystem. Contaminants occurring

IntechOpen

naturally in soil even at low concentrations pose a risk. Direct assessment of contamination in soil cannot be achieved, which makes it a hidden danger [2]. According to the report of FAO, soil contamination concerns are growing in every region, they are not restricted to degradation but also, poisoning our food, air and water too.

Soil pollution is caused by fertilizers, pesticides, chemicals, organic manure, radioactive wastes, discarded food, plastics, clothes, carcasses, leather goods, paper, bottles, tins-cans etc. Industrial wastes contain chemicals like copper, iron, zinc, lead, mercury, cyanides, cadmium, aluminum, acids and alkalies etc. which reach the soil indirectly through air or directly with water. The basic composition of soils are getting altered and becoming toxic for plant growth due to continuous and improper use of herbicides, pesticides and fungicides to protect the crops from pests, fungi etc. [3].

Due to global warming, agricultural fertilizers and pesticides arable lands are turning to desert and becoming non-arable at ever-increasing rates, lessening the hope that we can feed our booming population. Food production should be increased by 40% and that to on the fertile soils that cover around 11% of whole surface of Land. But the problem is that there is very little new land that can be used for production because existing land id lost and degraded. Due to erosion, water logging and salination, annually 75 billion tons of soil (nearly 10 million hectares/25 million acres) of arable land is lost, as stated by UNFAO leading to degradation of soil [4].

Contamination of soil survives for many years with "memory". Analysis of the load by elements and substances of anthropogenic origin from the period of "industrial revolution" in the soils or sediments can still be done [5]. Inorganic contaminants form the natural part of soil and can be redistributed due to anthropogenic activities predominantly in the environment [6]. Soil contamination can result in many negative effects like reduction in the decrease of microbial activity leading to the humification process reduction, decrease in water retention in soil and increase of soil erosion vulnerability [7].

Exposure to these agents attributes towards epidemiological evidence which shows the increased incidence of a variety of human cancers, such as lymphoma, leukemia, and liver and breast cancers [8]. These effects are dependent upon the properties of soil which is responsible for the mobility of contaminants in food chains or transfer from root to shoot, bioavailability of toxic chemicals, and carcinogenicity and residence time of contaminants [9].

Around the world, intensification in agriculture, industrialization, mining and wars have left a bequest of contaminated soils [10]. Soil has been used as a sink for dumping solid and liquid wastes since urban expansion. It was reviewed that once concealed and out of sight, the contaminants would not cause any risk to health of human being or the environment and that they would disappear [11]. Assessing the adverse effects caused due to contamination and taking measures to meet the expectations of environmental standards according to current legal requirements comes under management.

Soil contamination is mainly caused due to industrial/commercial activities diffusing heavy metals, nuclear power plants, oil industry and military camps. Various industrial points are the anthropogenic sources of heavy metals, for example, present and former mining activities, foundries, smelters, and diffuse sources such as piping, constituents of products, combustion of by products, and traffic related to industrial and human activities [12].

On the basis of toxicity, bioaccumulation, mobility and environmental persistence, priority should be given to the pollutants, according to WHO [13]. Heavy metals are considered to be carcinogenic such as arsenic, cadmium, nickel, chromium,

DOI: http://dx.doi.org/10.5772/intechopen.94280

PAHs and dioxins, based on human and animal studies exposed to high levels [14]. Depending upon the exposure level and duration some of these substances produce toxic effects on animal organs like CNS, liver, heart, kidney, skin etc.

2. Key concepts in understanding soil contamination

2.1 Soil properties

Soil organic matter (SOM), Inorganic minerals, water and air comprise soil. Soil physical properties including texture, structure, and porosity, the fraction of pore space in a soil, are influenced by the composition and proportion of these components. The physical properties of soil in turn affect air and water movement in the soil, and thus the soil's ability to function.

2.2 Soil health

Physical, chemical and biological properties constitute a healthy soil which allows the soil to carry out important functions. Human health is linked to agricultural soil health, as poor soils with decreased nutritional value yield fewer crops. Erosion is limited by healthy soils, and helps improve air and water quality too [15]. Soil contamination affects the functions of soil in ecosystem. It is considered to be "functionally dead" once contamination exceeds threshold value. Pollution is sometimes irreversible which is caused by heavy metals and many organic contaminants.

2.3 Causes of soil pollution

Soil contamination is caused by two main causes (a) natural and (b) anthropogenic Natural causes include earthquakes, volcanic eruptions, tsunamis etc. while; anthropogenic causes include radioactive wastes, metals (trace and heavy metals) and chemicals. These hazardous substances persist for long duration in the environment during which they are absorbed by the atmosphere, stockpiled to higher concentration potentially toxic to organisms in the food chain.

2.3.1 Natural sources

Accumulation of chemicals and toxic compounds (e.g. Perchlorate) naturally in soil surface leads to soil contamination.

a. Volcanic Eruptions. Volcano produces hazardous substances in huge amount that can destroy the areas nearby and even living beings. The fertile top layer of the soil gets destroyed by the lava and ash particles released from the volcano which is called as Soil erosion.

b. Earthquakes. Due to the movement of tectonic plates below the surface of soil there is sudden release of energy in the surface of earth which is termed as Earthquakes. It results in the damage of the agricultural soil which becomes non-fertile.

c. Alterations in Rainfall Patterns. The composition of organic matter in the soil is sensitive to alterations in rainfall pattern which forms the soil structure framework; balance the nutrients, oxygen and water of soil. Alterations in rainfall results in acidification or alkalization of the soil [16].

d. Geographical Changes. Soil quality is affected by Changes in the geographical factors. The two recently developed techniques i.e. geographic information system (GIS) and remote sensing techniques (RS) are helpful in knowing the connection between soil erosion and geographical factors and also to obtain the information on soil quality and heterogeneity of soil surface as well as to investigate the extent of land degradation.

e. Tsunamis. Tsunami results in salting of agricultural land and drinking water. Different pollutants in large amount are carried away with the flooded water and get deposited resulting in water and land pollution [17].

2.3.2 Anthropogenic sources

Industrial, municipal, domestic and agricultural wastes are considered as major anthropogenic source of soil pollution [18]. Some of these wastes can be recycled into useful materials so all of them are not considered as contaminant. Solid wastes are the discarded materials which are of no use [19]. Depending on their source these wastes can be further classified as municipal waste, industrial wastes and hospital wastes.

a. Municipal Solid Wastes (MSW). Waste like kitchen wastes, livestock, poultry wastes, domestic waste, market wastes, slaughterhouse wastes, ceramic wastes, glass and metals waste comprises municipal wastes. Municipal wastes cause major environmental threat in developing countries due to improper waste disposal. Soil as well as ground water contamination results due to open disposal of municipal wastes.

b. Hospital Wastes. Wastes produced as a result of diagnosis, treatment and immunization of animals and human beings are considered as Hospital wastes. Improper disposal of these wastes and release of pollutants during incineration cause several health hazards on public health as well as the environment [20].

c. Industrial Wastes. Industrial wastes are high toxic in nature and they affect the chemical and physical nature of soils and fertility of soil. These chemicals get accumulated by the crops and pose serious health issues and environment related problems.

d. Agricultural Practices. In order to deal with the challenge of food and to enhance crop yield agrochemicals are used to protect the crops (from pathogens and insects). Agrochemicals can be classified as pesticides (herbicides, insecticides and fungicides), fertilizers, hormones and animal manure. These chemicals possess environmental and health risks at the value more than threshold value and they can persist for longer duration in environment and are non-biodegradable [21].

e. Radioactive Wastes. Byproducts of nuclear power plants and research stations contribute radioactive wastes. Unstable elements having atomic number > Bi-83 (Bismuth) are radioactive. Ionizing radiations are releases from these radioactive wastes which lead to several health issues [22]. Accumulation of radioactive material in soil not only affect the characteristics (Physical, chemical and biological), but also leads to magnification through food chain and affects living beings.

f. Chemical Wastes. Chemical wastes are classified as organic and inorganic which includes polynuclear aromatic hydrocarbons, petroleum hydrocarbons, various solvents and other heavy metals. These chemical wastes possess several health risks such as, allergies, cancers related to immune, reproductive and nervous system. Transformation of organic contaminants in soil occurs through different methods which include leaching, volatilization and biological transformation [23]. Furthermore, bioaccumulation in living organisms can result in various adverse effects at each trophic level of food chain.

g. Heavy metals. Heavy metals includes metalloids, lanthanides and actinides and are characterized by atomic number < 2 and atomic weight 22.98 to <40 [24]. Soil productivity, fertility and quality have been considerably affected by the heavy metal contamination in soil in the past decade. Heavy metals at high concentration are toxic for human beings.

3. Effects of soil pollution

3.1 Agricultural

- Reduction in soil fertility
- Reduction in ability of nitrogen fixation
- Increase in erosion of soil
- Nutrients loss
- Silt Deposition
- Crop yield Reduction
- Imbalance in soil fauna and flora

3.2 Industrial

- Underground water contamination
- Ecological imbalance
- Gas release causing health problems
- Release of radioactive rays
- Reduced vegetation

3.3 Environmental

- Unavailability of soil for food
- Low crop yield

- Soil Erosion due to lack of crops
- Change in makeup and microorganisms that live in soil

4. Techniques for controlling soil pollution

Soils are considered to have purification property which is due to their properties like physical, chemical and biological [25]. In order to prevent soil erosion, construction in sensitive area can be limited. In general we would also use less fertilizer and pesticides to adopt all the three R's: Reduce, Reuse, and Recycle for generation of less solid waste.

Measures to control soil pollution

- Minimal use of pesticides and chemical fertilizers
- Cropping techniques should be improved
- Wastes should be dumped properly
- Forest management
- Prevention of soil erosion
- Public awareness
- Recycling and Reuse of wastes
- Ban on Toxic chemicals

5. Interaction of pollutants with soil constituents

Pollutants interaction and behavior with soil depends upon different processes (physical, chemical, and biological) that occur in components of soil. Processes include

1. Detainment of pollutants on or within the soil medium
2. Transport, percolation and diffusion in soil medium
3. Chemical changes processes occurring within the soil medium

As soon as the pollutants enter into the soil, they undergo physical, physico-chemical, microbiological, and biochemical processes that help them to retain, reduce or get degraded [26].

5.1 Sorption of contaminants

A process by which a substance is accumulated within the phase of the boundary of phases physically or chemically is called sorption. It is of two types- chemical (as with ionic and hydrogen binding) and physical (as with van der Waals forces). Positively charged molecules participate in cation exchange while negatively charged molecules in anion exchange. The transition in cationic or anionic states of some ions or molecules in the soil is pH dependent which can control the mobility of contaminants [27].

5.2 Bioavailability, mobility and degradation of contaminants

Interactions between chemicals and organisms that determine the exposure is referred to as bioavailability. Major hurdle for applying bioremediation techniques for segregation of pollutants are sorption by strong bonding and slow release. Moreover, concepts related to soil screening and understanding risk level should also be reconsidered. According to Semple *et al.* (2004) [28] material is considered to be bioavailable which is "freely available" and is able to cross organism's cellular membrane from the medium where it inhabits.

6. Impacts of soil pollution on the food chain and ecosystem services

Quality and yields of crops are reducing due to the presence of soil contaminants which are affecting our food security. During the formation of research policies for use of poor natural resources in agriculture healthy crops should be the main motto. Key points that should be kept in mind during crops production are include monitoring of agricultural fields, tracking of wastewater units of sewage and industrial and inputs added on crops [29].

The top level predators have highest concentration of contaminants in bodies and experience the bad health effects in the food chain and also lost of apex predators occurs [30]. As a result, effect of pollution in food web scale increases. Response to contaminants may be sequential or remains inert may be serious with drastic change. The uptake and translocation of contaminants into above ground tissues are conditioned by genetic and physiological differences of plants as well as by the concentration of contaminants in the soil and the exposure time [31]. These changes results in abrupt degradation of ecosystem services which may not recover.

The exposure to environmental contaminants depends upon the routes, concentration, bioavailability, frequency and duration. It also depends upon the feeding behavior and habitat [32]. Difference in the fate of a contaminant within an organism and its toxicological effect is also seen among and within species [33]. This complexity leads to an impact on particular species and indirectly on the diversity. Indirect effects in food web is known as "Tropic cascade" in which disturbances in food chain is due to change in highest tropic level or change in resources [30]. Another well-known effect is "paradox of enrichment" where the increase of prey resources results in predators shift but the exposure to contaminants may inhibit the paradox of enrichment and drive them back to a fixed equilibrium [34].

Health implications ranging from minor to major fatalities including long term effects are observed due to contaminants in food chain [35]. Contaminants in food chain can adversely affect humans various systems [36].

7. Methods to assess soil contamination

Due to rapid development particularly urbanization and industrialization over the last century's contamination of the environment considerably has increased [37]. Due to this, assessment and control of soil contamination is an object of interest of researchers, scientists and authorities dealing with environmental protection. Data related with the spatial distribution of soil pollution are of great importance for the environmental protection, regional development, and spatial planning [38].

Soil assessment can be done through various approaches because of different types of soil contamination. This can be done both from a functional and structural perspective. The relationship between them is ecology which is receiving a lot of attention but assessments related to ecotoxoicology have not arisen. Part IIA of the Environmental Protection ACT (1990) [39] was put forward for assessing contamination related to land/soil which is causing significant harm to human health, water, livestock and ecological systems [40].

Data usage, accuracy and precision play an important role in the choice of monitoring methods of soil contamination. Biological, chemical and geophysical approaches are included in assessing methods. To measure specific contaminants using special instruments like MS, AAS or GC are included in chemical methods. In biological methods organisms or byproducts of biodegradation are used as indicators of soil contamination. And in Geophysical methods changes in physical properties of soil and contaminants are assessed.

Development in the methods of assessing soil contamination includes better extraction process for improved recovery and enhanced detection limits and also alternative methods development for soil contamination monitoring such as isotopic signatures or immunoassays. On site analysis, research based techniques and innovative methods that are cost effective, sensitive and easy to use should be developed for assessing organic, inorganic and radioactive contaminants in soil.

The overall success of analysis of soil contaminants depends upon the nature of soil matrix, association/interaction of pollutants with soil, forces including chemical and physical with which they bind to the soil particles Therefore, efficiency of assessment depends upon the procedure of extraction of contaminants from soil for analysis because soil contaminants migrate down with time and become less approachable [41].

Tool for evaluation of risk at contaminated sites are thought of as more appropriate and cost-effective and has the potential of focusing assessment and evaluating the contaminant. The risk assessment methods are the source to know the risk of undesired effects on ecosystems caused by various factors which are associated with human activities. Tools of assessing ecological effects includes: experiments related to ecotoxicity under controlled conditions, ex situ bioassays (simple laboratory assays) and mapping of community in field [42].

ERA allows the assessments of toxicants and their effects through changes in predation and competition. This could be achieved through the use of standardized terrestrial test procedures. Experiments related to ecotoxicity under controlled conditions have their own benefits of measuring direct toxicity of chemicals and their interpretation. In this context, Bioassays are one of the frequently used higher tier alternatives because of its advantage of assessing the toxicity in the soil. Contaminants in soil can be assessed using multispecies mesocosms or lysometers by evaluating intrinsic populations of the soil or by introducing species to system.

8. Soil remediation approaches

Selection of soil/sediment remediation approaches depend on various factors viz. type of soil, composition of soil, properties of soil, nature of contaminant etc. The properties of soil can be influenced by the addition of nutrients and chemicals for the growth of microbes [43]. These additions and approaches cause contamination and destruction of soil components yielding harmful products [44].

Economic solutions to cover the areas used for horticulture and agriculture are with low- or uncontaminated topsoil or with sandy and rocky soils [45].

Types of remediation methods

- In-situ: Contaminated soil is treated at the site where it is occurring [46]
- Ex-situ: These methods require the excavation of contaminated soil [47]

8.1 Containment technologies

Conventional civil engineering techniques are most frequently used approach for isolating contaminated media from surrounding environment. These approaches are typically convenient when the excavation process or the soil removal could lead to potential hazards. The advantages involved in using these technologies lie in, a) non requirement of soil excavation, b) low to moderate cost of treatment in spite of the requirement for long-term monitoring and maintenance of equipment.

To prevent migration of contaminants due to flow of groundwater, use of physical barriers are preferred which includes surface capping and subsurface barriers (vertical and horizontal). This result in limiting infiltration of surface water and reduce the migration of contaminated groundwater laterally or vertically [48].

8.2 Immobilization technologies

To prevent the migration of contaminants, addition of chemicals/reagents to soil to form insoluble low toxic matter is preferred. This technology covers a broad spectrum of inorganic contaminants however; it is a temporary solution because contaminants are still in soil. Therefore the immobilization technique should be applied only to surface soil [49]. Immobilization is achieved by working directly on the contaminants present in the soil and can be classified as solidification and vitrification.

8.3 Solidification and stabilization

Process that encapsulates or captures the contaminants within stabilized integrity and not involves any chemical interaction is called solidification. It involves reduction of contaminants hazard potential by their conversion in order to reduce their solubility, mobility, or toxicity. Use of cement, asphalt or phosphate, or alkalies that increase the pH helps in precipitation and immobilization of contaminants [50].

8.4 Vitrification

Vitrification involves the process of pyrolysis (1600–2000°C) and oxidation for melting and immobilization of contaminants. It can be applied for in and ex situ methods of remediation of inorganic substances, such as metals and radionuclides and organic compounds. Other process of vitrification involves heating by plasma, direct power, combustion, induction or microwave at a temperature of 1100–1400°C [51].

9. Technologies for treatment of contaminated soil

Treatment technologies can be classified into three main categories: (1) biological technologies (2) thermal technologies and (3) physicochemical technologies.

9.1 Biological technologies

Also called as Bioremediation, the use of microorganisms (mainly, bacteria and fungi) to clean up contaminated soils [52]. Microorganisms act on contaminants by following mechanism which includes biosorption, bioleaching, biomineralization, intracellular accumulation, and enzyme-catalyzed transformation [53]. Three main approaches are there for the bioremediation of contaminated areas [54] which includes

a. Natural attenuation

b. Biostimulation

c. Bioaugmentation and

d. Phytoremediation

Bioremediation is widely used to remediate organic contaminants which include:

- hydrocarbons
- halogenated organic solvents
- halogenated organic compounds
- non-chlorinated pesticides and herbicides
- nitrogen compounds
- metals (lead, mercury, chromium)
- radionuclides

Natural attenuation means remediation carried out by native population of microbes occurring in contaminated area. In natural attenuation process some factors affect the degradation process which includes: Native population of microbes and their metabolic capacity; soil physicochemical properties and chemical nature of contaminants. Some contaminants are efficiently degraded by natural attenuation process and some show null or low degradation especially aged contaminants [55].

In **Bioaugmentation,** inoculation of specific microbial strains which is having the ability to degrade the target contaminants is focused to stimulate the biodegradation. Consortium inoculation is more frequent as compared to individual strain inoculation as microorganisms in consortium show combined metabolic activities for remediation process. Also, selection of different strains for consortium should be based on their compatibility and ecological fitness in soil. To improve biodegradation efficiency of microorganisms, genetic modification for optimization of enzyme production and metabolic pathways relevant for degradation, has also been studied [56].

Various authors have been reported successful bioremediation of soil contaminated with hydrocarbon sources through bioaugmentation process. The efficiency of pollutant removal by selected microorganisms including five cultures of microbes and 3 bacterial strains- *Pseudomonas sp., Arthrobacter species and Mycobacterium* species was evaluated using hydrocarbon as sole carbon sources [57]. Bacterial consortium of *Bacillus cereus*, *Bacillus sphaericus, Bacillus fusiformis, Bacillus pumilus, Acinetobacter junii* and *Pseudomons* sp. results in degradation of diesel contaminated soil [58]. Ying et al., 2010 [59] studied PAH degradation using *Paracoccus* sp strain HPD-2 and observed 23.2% decrease in total PAH concentrations in soil after 28 days.

This process of bioaugmentation is not always an effective solution for soil contamination remediation as microbes from lab scale rarely grow and biodegrade contaminants as compared to indigenous microbes. Also, process of bioaugmentation is still not popular yet; mostly the use of microbes which are genetically engineered when added to soil may affect the ecology of the environment and also cause risk to environmental health if they persist for long time even after remediation.

And in case of **Biostimulation,** modification of the environmental conditions for the stimulation of biodegradation of target contaminants is focused. The efficiency of process can be enhanced by stimulation of degrading capacity of the indigenous microbial populations by providing them essential nutrients (organic and inorganic), available oxygen, moisture and temperature. Enrichment of soil by nutrients also called fertilization is a remediation approach in which fertilizers are added to contaminated environment for stimulation of indigenous microbial growth. Microorganism requires some key elements such as Carbon, Nitrogen, Oxygen and Phosphorous for their growth and activity, addition of fertilizers to soil fulfills their need and in turn enhance the process of biodegradation.

Studies have suggested that the carbon addition in form of pyruvate in soil stimulates the growth of microbes and increase the rate of PAH degradation [60]. Also, use of compost, organic wastes like banana skin or melon shell can enhance the rate of degradation process by mixing the compost with the contaminated soil [61]. In the treatment of organo pollutant contaminated sites mushroom compost and spent mushroom compost can be applied [62]. Efficiency of degradation by the addition of SMC was enhanced up to 82%, also it reduced the toxicity of PAH [63].

For cleaning hydrocarbons bioremediation is currently used commercially because of the capacity of microorganisms to biodegrade organic and inorganic contaminants (National Research Council 1993). Some microorganism can utilize molecular oxygen as electron acceptor and this process is called as Aerobic respiration. By utilizing oxygen, carbon is oxidized to carbon dioxide for energy generation and water molecule is formed.

Reaction: organic substrate (electron donor) + O_2 (electron acceptor) $\rightarrow$ biomass + CO_2 + H_2O + metabolites + energy.

While some microorganism cannot utilize molecular oxygen by process called anaerobic respiration in which metals such as Fe^{3+} and manganese Mn^{4+}, sulfate $SO_4{}^{2-}$, or even CO_2 can be used to accept electrons from contaminants being degraded [64]. Microorganisms involved in this remediation process include iron and manganese reducing bacteria, sulfur reducing bacteria and methanogenic bacteria. These microorganisms complete geochemical reactions such as bacterial corrosion, sulfur cycling, organic decomposition and methane production.

Reactions are as follows:

Iron reduction: organic substrate (electron donor) + Fe $(OH)_3$ (electron acceptor) + H^+ $\rightarrow$ biomass + CO_2 + Fe^{2+} + H_2O + energy.

Manganese reduction: organic substrate (electron donor) + MnO_2 (electron acceptor) + H^+ $\rightarrow$ biomass + CO_2 + Mn^{2+} + H_2O + energy.

Sulfanogenesis: organic substrate (electron donor) + SO_4^{2-} (electron acceptor) + H^+ $\rightarrow$ biomass + CO_2+ H_2O + H_2S + metabolites + energy.

Methanogenesis: organic substrate (electron donor) + CO_2 (electron acceptor) + H^+ → biomass + CO_2 + H_2O + CH_4 + metabolites + energy.

There are three ways by which contaminants can be demobilized by microorganisms:

1. Sorption of organic hydrocarbon molecules by biomass of microbes

2. Precipitation of metals (Oxidized/reduced) produced by microorganisms

3. Degradation of organic compounds bound to metals for its solubilization

Phytoremedation

Phytoremediation is a technique in which the plants are used to remediate environmental media. It is followed as a new approach for cleaning of contaminated soils and waters. It involves the interaction between plant roots and microorganisms associated with them for soil remediation. This technique is cost effective for the remediation of soil and groundwater contaminated with various types of wastes and also has less impact on environment as compared to other traditional remediation methods. All necessary nutrients are extracted from soil and water by the plants. Some plants have ability to store large amounts of contaminants, called hyperaccumulators, even though not required for plant functioning, while some can utilize these organic contaminants as a source for various physiological processes.

Plants act as filters or traps and break down the contaminants in soil or water. Process involves growing plants in contaminated area for a period required for the growth of plant to remove contaminants or facilitate immobilization or detoxification of the contaminants. Further plants can be harvested, processed and disposed off if required. This system focus on the symbiotic (Synergistic) relationships among microorganisms, plants, soil and water. Aerobic and anaerobic microorganisms both are present in the vicinity of plant roots supplied with both physical habitat and building blocks. Plant root and shoots provides colonisable surface area, organic exudates, leachates and oxygen to microbes for degradation of contaminants.

Mechanism of phytoremediation:

Contaminants are taken up by the roots of the plants which prevents the plants from toxicity. Root system provides large surface area that helps in absorption and accumulation of water and nutrients essential for growth [65]. Due to release of organic and inorganic root exudates at root-soil interface affects the number and activity of microorganisms, and in turn the availability of the contaminants through changes in chemical composition of soil. Process of phytoremediation is different for different environments and types of contaminants. Each of these processes has an effect on toxicity of contaminants and its mobility.

i. **Phytovoltalization:** In this process plants are used to transform the contaminants into volatile forms and transpire them in atmosphere with water vapor through leaves [66]. Diffusion of contaminant takes place from stem and travel through various parts before it reaches to leaves [67].

ii. **Phytoexraction:** Also called as phytoaccumulation and it refers to absorption, concentration, translocation and precipitation of contaminants from soil by the roots of plants. For this purpose hyperaccumulators are best for the removal of contaminants like Nickel, Zinc and copper. Phytoextraction process is cost effective and it removes the contaminant from soil permanently and up to 95% [68].

iii. **Rhizofilteration:** This process is used for the low concentration contaminant from groundwater, surface water and wastewater. It takes place by absorption of contaminants onto plant roots of in surrounding of root zone. Rhizofiltration is used for Chromium, Lead, Cadmium, Nickel and Zinc. Plants like sunflower, rye, tobacco, spinach, mustard and tobacco have been used for removal of contaminants significantly.

iv. **Phytostabilization:** Also called as in place activation and used for remediation of sludge, sediment and soil. This process immobilizes contaminants through adsorption, absorption and accumulation and reduces the mobility and bioavailability of the contaminant and prevents migration into the food chain. It can be used for treatment of lead, cadmium, chromium, copper, zinc and arsenic & also used to restore vegetation thereby decreasing migration and transport of contaminants.

v. **Phytodegradation:** Process involves degradation of complex to simple molecules and further incorporation into tissues of plant, also called as phytotransformation. During this process contaminants are absorbed and broken down.

vi. **Rhizodegradation:** It involves the breakdown of contaminants within the rhizosphere due to the secretion of plant exudates like sugars, amino acids, enzymes, and other compounds that can stimulate bacterial growth which carry out the degradation of contaminants. It has been investigated and found that variety of different chemicals like PAHs, Pesticides, PCBs, benzene and xylenes can be degraded by this process.

9.2 Physicochemical technologies

9.2.1 Stabilization and solidification

A process in which chemicals or reagents are mixed with contaminated soil in order to reduce toxicity and mobility of the contaminants is called stabilization. Stabilization involves trapping or binding the contaminants in soil and is permanent remedial solution. In stabilization process two types of chemicals can be used: Binder and Sorbent. Binder increases the strength of product and sorbent retains the contaminant. Examples include cement, pozzolans, pumice, ground blast furnace slag, lime, silicates etc. [69]. In solidification sufficient quantities of solidifying agents are added to contaminants for solidification. Mechanisms involved in stabilization/solidification are microencapsulation, absorption, adsorption, precipitation and detoxification. The process can be applied for PCBs, Oils, organic compounds and metals [70].

9.2.2 Soil flushing

Soil flushing is a type of in situ treatment technology in which an aqueous solution, that increases the mobility or solubility of contaminants adsorbed onto the soil matrix, is injected or infiltrated into the contaminated soil. The flushing solution consists of surfactants, cosolvents, acids, bases, oxidants, chelants, water or other solvents. Flushing is generally accompanied by other remediation technologies, namely activated carbon, biodegradation, and pump-and-treat [71].

9.2.3 Chemical reduction/oxidation

Hazardous contaminants are converted to less toxic compounds by reduction/oxidation (Redox) reactions. In this reaction, there is transfer of electrons from one to another i.e. reactant is oxidized another is reduced which results in the breaking of bonds. This process is a well established technology called as chemical redox which is used for drinking water and waste water disinfection.

9.2.4 Soil washing

Soil washing is an ex situ technique for removal of contaminants from the soil using a) physical separation and b) chemical leaching by aqueous solutions. In the initial steps of this process, the coarse particles are separated by homogenization based on the differences in their density. Since most organic and inorganic contaminants tend to bind to clay, silt, and inorganic particles, the washing processes separate the fine (small) clay and silt particles from the coarser. In the second step, the contaminants are selectively dissolved and then chemically converted or recovered. Based on the contaminant being treated, the additives and reagents that are added to water are decided [72].

9.2.5 Soil vapor extraction

Soil vapor extraction (SVE) is used to remediate unsaturated zone soil that uses application of vacuum on the soil to induce a controlled flow of air and helps in the removal of volatile and semi volatile organic contaminants. SVE is an in situ technology, although, in some cases, it can be used as an ex situ technology. In situ SVE, otherwise known as soil venting or vacuum extraction, vacuum is applied to the soil through the wells constructed near the source of contamination. Vacuum creates a negative pressure gradient, which in turn induces the controlled flow of air and remove the contaminants from the soil through an extraction well. Extracted vapor is treated before it is released into the atmosphere. The augmented airflow through the subsurface also stimulates the biodegradation of some of the contaminants, especially the less volatile substances. Advantageously, in situ SVE have greater depth of reach than other methods requiring removal of soil, the wells and the equipment are simple to install and maintain. On the other hand, ex situ SVE is a full-scale technology in which soil undergoes extraction and is placed over a grid of aboveground piping where it is subjected to vacuum in order to volatilize organic contaminants [73].

10. Thermal treatment

Treatments which involve destruction and remediation of the contaminants in soil by the use of heat including thermal destruction, thermal desorption, vitrification, and incineration [74]. Thermal treatment of contaminants results in volatilization of contaminants and removes them from the soil.

Thermal desorbers (100–300°C) are used for volatile and semi-volatile organic contaminants. The vapors formed are collected and treated in a gas treatment system. Technique is used for the removal of PCBs, pesticides, paint wastes, hydrocarbons, chlorinated solvents etc.

Vitrification is a process which involves melting and fusion of materials at temperature above 1200°C followed by rapid cooling. During this process, immobilization of nonvolatile metals within glass occurs and volatile materials are converted into vapors. Vitrification process is used to treat small quantities of contaminants and for radioactive wastes, asbestos containing waste and those that can not be treated by other technologies.

Thermal destruction involves destruction of contaminants by reduction, oxidation, hydrogenation and pyrolysis [75]. Catalytic oxidation involves the remediation of contaminants from industries by the use of electric heater, catalytic reaction, tube heat exchanger and scrubber. Pyrolysis is chemical process in which the wastes are heated in the absence of oxygen at temperatures 400 to 1200°C.

11. Advanced remediation technologies- zero valent iron nanoparticles and nanoremediation

Zero-valent iron is a reducing agent which dissolved in water in presence of oxygen is capable to oxidize organic pollutants. In the reaction, ZVI reacts with O_2 to produce H_2O_2which is reduced to water or can react with Fe^{2+}, also called as Fenton reaction, producing (hydroxyl radicals (·OH). This is able to degrade contaminants due to its oxidizing capability. Zero-valent iron nanoparticles (NZVI) are more effective remediators as compared to others [76] and also have great diversity in reacting towards different contaminants. Ability to oxidize to ferrous or ferric iron zero-valent iron provides electron for reducing other compounds making them less harmful [77]. NZVI helps in preserving characteristics of soil, enhance the process of remediation and improves the mobility and lowers the toxicity of the contaminants due to its nanometric size.

There are some studies which aims to develop new techniques in which the mixture or combination of techniques are applied for the remediation of contaminated soil like zero valent iron nanoparticles (nZVI) and compost. The application of nZVI for remediation of As and Cr showed a decrease in the concentration at contaminated industrial site and the addition of nanoparticles and compost results in decrease in aliphatic hydrocarbons upto 60% [78].

Nanotechnology use for the remediation of environment has received significant attention from community of scientists. Nanoparticles are effective against degradation of contaminants such as heavy metals [79], insecticides, dyes, organic halogenated hydrocarbons [80] and nitrates. Also, for the bioremediation of PAHs, SiO2 nanoparticles coated with a lipid derivative of choline have been used. Other nanomaterials like iron sulfide stabilized with carboxymethylcellulose is tested for immobilizing Hg in soils [81].

Nanoproducts can be applied in six main areas for remediation purpose which includes a) Photocatalytic degradation of organic pollutants [82], b) Propulsion area [83] c) biosensor technology [84] d) Water contaminant [85] e) Quick sensing of environmental stimuli [86].

Nanomaterials like nanocrystals and carbon nanotubes have provided a wide range of application to environment such as antimicrobial agents, sensors, pollution preventers etc. [87]. For example, hybrid carbon nanotubes (HCNTs), NM_s like single-walled carbon nanotubes (SWCNTs) and multiwalled carbon nanotubes (MWCNTs) have been used for the remediation of ethylbenzene.

Enzymes function as biocatalysts and are specific and effective, in bioremediation. However, enzymes have short catalytic lifetimes and less stability due to oxidation, which limit their use as an alternative to synthetic catalysts. Therefore, to increase the longevity, stability and reusability, attach them to magnetic iron NPs. This results in easy separation of the enzymes from reactants or products by applying a magnetic field. Two enzymes, peroxides and trypsin were attached to magnetic nanoparticles that increased the activity and longevity of enzymes making them more stable and efficient [88].

Hydrophobic organic contaminants get adsorb strongly to soils and are difficult to remove like polynuclear aromatic hydrocarbons (PAHs). A nanoparticle made of amphiphilic polyurethane (APU) has been synthesized for use in remediation of soil contaminated with PAHs. The particles are made of polyurethane acrylate anionomer (UAA) or poly(ethylene glycol), modified urethane acrylate (PMUA) precursor chains that can be cross-linked in water. The resulting particles have the ability to enhance PAH desorption and transport to the soil surface [89].

Uraninite can be used for bioremediation strategies of subsurface U (VI) contamination due to its small particle size, high dissolution rates and its molecular scale structure. Nanoparticles can be prepared from vegetation namely, *Gundelia tournefortii, Centaurea virgata, Reseda lutea, Scariola orientalis, Eleagnum angustifolia*, and *Noaea Mucronata*, have the ability to accumulate heavy metals. Based on the results, nanoparticles were prepared from the plants e.g. *N. mucronata* and were evaluated for the accumulation of heavy metals and found that the amount of heavy metals decreases during bioremediation process [90].

12. Sustainable remediation

Remediation of contaminated environment lays on the fact that application of any remediation technology should maximize the environmental benefit and minimize the impact through continous practices. With the focus on sustainable development as a key factor, Green remediation was introduced because it considers all effects and aspects of applying technology. Sustainable remediation reflects the perception that activities related to remediation can have wider, holistic approach with both positive and negative impacts on social, economic and environmental [91]. So this approach helps to contribute the solutions to short and long term problems generating through human health and ecosystem and facilitates risks assessment, benefits and future use for long term.

Therefore, some initiatives, efforts or progress have been observed towards disseminating sustainable approaches in the remediation and management of contaminated areas [92]. For this the Forum for sustainable remediation (SURF) was established in 2006, in collaboration between US remediation industry professionals, researchers and industries and focused on sustainable remediation. Now, other countries like UK, Brazil, China, Netherlands, New Zealand, Australia, Cananda etc. have joined and form an international forum. In addition to this SURF, the US environmental Protection Agency (USEPA), American Society for Testing and Materials (ASTM), Interstate Technology and Regulatory Council (ITRC) and Network for Industrially Contaminated Land in Europe (NICOLE), have developed focus on the application and evaluation of sustainable remediation.

Common issues like social equity, sustainability tripod, long term efficiency, democratic process and ecological integrity are generally addressed in the application of sustainability remediation. For this, analytical methods with the identification of indicators, metrics and tools are used for evaluation of the sustainable remediation. These methods help in decision making about the aspects of social, economic and environment for assessment and application of sustainable remediation project [93].

Although advances have occurred in area of sustainable remediation, problems and challenges related to adoption and acceptability of the sustainable remediation have been observed. There is no such method that can be used for standardization and assessment of degree of sustainability and the existing methods does not fulfill the criteria of sustainability assessment in remediation of contaminated areas [93]. In order to implement the sustainable remediation approach, a new way of thinking including social, economic and environmental variables should be considered as fundamental factor. Industries, government and academicians all should play key role in ensuring sustainable remediation approach for the incorporation of sustainability projects in developed and developing countries.

13. Conclusions

In this chapter, the main perspectives related to the development of technologies for soil remediation have been discussed. One of the approaches that are widely discussed is the use of green technologies as phytoremediation, biostimulation and biodegradation. Nanotechnology in the degradation of contaminants and the importance of sustainable remediation approaches are also currently under discussion.

Author details

Sonia Sethi and Payal Gupta*
Dr. B. Lal Institute of Biotechnology, Jaipur, India

*Address all correspondence to: soniakaura198@gmail.com

References

[1] Mishra R, Mohammad N Roychoudhury NVS. Soil pollution: Causes, effects and control. 2016; 3:1.

[2] Adriano DC, Bollag J, Frankenberger WT, Sims RC eds. Bioremediation of Contaminated Soils. Agronomy monograph 37: American; 1999.

[3] Miller RW, Gardiner DT. Soils in Our Environment. 8th edition. NJ: Prentice Hall; 2016.

[4] https://www.everythingconnects.org/soil-pollution.html

[5] Azoury S, Tronczynski J, Chiffoleau JF. Historical records of mercury, lead and polycyclic aromatic hydrocarbons depositions in a dated sediment core from the eastern Mediterranean. Environ Sci Technol. 2013; 47: 13; 7101–7109.

[6] Cachada A, Rocha-Santos T, Duarte AC. Soil Pollution. Chapter 1 - Soil and Pollution: An Introduction to the Main Issues. Academic Press; 2018, pp.1–28.

[7] Mühlbachova G, Sagova–Mareckova M, Omelka M. The influence of soil organic carbon on interactions between microbial parameters and metal concentrations at a long–term contaminated site. Sci Total Environ. 2015; 502:218–223.

[8] Badawi F, Cavalieri EL, Rogan EG. Effect of chlorinated hydrocarbons on expression of cytochrome P450 1A1, 1A2 and 1B1 and 2- and 4-hydroxylation of 17β-estradiol in female Sprague-Dawley rats. Carcinogenesis.2000; 21 (8); 1593–1599.

[9] FAO & ITPS. Status of the World's Soil Resources (SWSR) - Main Report. Rome, Italy, food and Agriculture Organization of the United Nations and Intergovernmental Technical Panel on Soils; 2015.

[10] EEA. Progress in management of contaminated site, European Environment Agency; 2014.

[11] Swartjes FA. Dealing with Contaminated Sites. Dordrecht, Springer Netherlands; 2011.

[12] Guvenç N, Alagha O, Tuncel G. Investigation of soil multi-element composition in Antalya, Turkey. Environment International. 2003: 29(5); 631–640.

[13] WHO European Centre for Environment and Health, Methods of Assessing Risk to Health from Exposure to Hazards Released from Waste Landfills, WHO Regional Office for Europe, Lodz, Poland, 2000.

[14] Johnson BL. Hazardous waste: human health effects. Toxicology and Industrial Health. 1997: 13(2–3); 121–143.

[15] Brevik EC. Soils and human health: an overview. In: Soils and Human Health (eds E.C. Brevik & L.C. Burgess), pp. 29–56. CRC Press, Boca Raton, FL; 2013.

[16] Wild A. Soils and the environment. Cambridge University Press, Cambridge; 1993.

[17] Moqsud MA, Omine K. Bioremediation of agricultural land damaged by tsunami. In: Rolando Chamy, Francisca Rosenkranz (ed) Biodegradation of hazardous and special products; 2013.

[18] Alloway BJ. Heavy metals in soils. Blackie Academic and Professional, an Imprint of Chapman and Hall, London; 1995.

[19] Singh RP, Singh P, Arouja ASF, Ibrahim MH, Sulaiman O. Management of urban solid waste: vermicomposting a sustainable option. Resourc Conser Recycl. 2011; 55:719–729.

[20] Batterman S. Findings on assessment of small-scale incinerators for health-care waste water, sanitation and health protection of the human environment. World Health Organization, Geneva; 2004.

[21] Navarro S, Vela N, Navarro G. Review. An overview on the environmental behaviour of pesticide residues in soils. Span J Agric Res. 2007; 5(3):357–375.

[22] Van der Perk M. Soil and water contamination. Taylor & Francis, London; 2006.

[23] Semple KT, Morriss AWJ, Paton GI. Bioavailability of hydrophobic organic contaminants in soils: fundamental concepts and techniques for analysis. Eur J Soil Sci. 2003; 54: 809–818.

[24] Afal A, Wiener SW. Metal toxicity. Medscape.org; 2014.

[25] Erfan-Manesh, Majid & Afyouni, Majid. Environment, Water, Soil and Air Pollution, Publications of Arkan Danesh; 2008.

[26] Blum WEH. Functions of Soil for Society and the Environment. Reviews in Environmental Science and Bio/ Technology. 2005; 4(3): 75–79.

[27] Navarro S, Vela N, Navarro G. Review. An overview on the environmental behaviour of pesticide residues in soils. Spanish Journal of Agricultural Research. 2007; 5(3): 357.

[28] Semple KT, Doick KJ, Jones KC, Burauel P, Craven A, Harms H. Defining bioavailability and bioaccessibility of contaminated soil and sediment is complicated. Environmental Science and Technology. 2004; 38: 228A – 231A.

[29] Dotaniya ML, Aparna K, Choudhary J, Dotaniya CK. Effect of Soil Pollution on Soil Microbial Diversity In book: Frontiers in Soil and Environmental Microbiology.2020; 255–272.

[30] Heath, Michael R, Douglas C, John H. Understanding patterns and processes in models of trophic cascades. In: Ecology Letters. 2014; 17(1):101–114.

[31] Rizwan M, Ali S, Adrees M, Ibrahim M, Tsang DCW, Zia-ur-Rehman M, Zahir ZA, Rinklebe J, Tack FMG, Ok YS. A critical review on effects, tolerance mechanisms and management of cadmium in vegetables. Chemosphere. 2017; 182: 90–105.

[32] Van Den Brink, Nico W, Dennis R, Wim J, Boerwinkel. Cadmium accumulation in small mammals: species traits, soil properties, and spatial habitat use. In: Environmental Science and Technology. 2011; 45(17):7497–7502.

[33] Fritsch, Clémentine, Richard P, Michaël C, Francis R, Patrick G, Nadia C, Annette DV, Renaud S. Responses of wild small mammals to a pollution gradient: host factors influence metal and metallothione in levels. In: Environmental Pollution 2010a; 158(3): 827–840.

[34] Prosnier, Loïc, Michel L, Florence DH. Modeling the direct and indirect effects of copper on phytoplankton–zooplankton interactions. In: Aquatic Toxicology. 2015; 162: 73–81.

[35] Khan S, Cao Q, Zheng Y, Huang Y, Zhu Y. Health risks of heavy metals in contaminated soils and food crops irrigated with wastewater in Beijing, China. Environ. Pollut. 2008; 152: 686–692.

[36] Androutsopoulos V, Hernandez A, Liesivuori J, Tsatsakis A. A mechanistic

overview of health associated effects of low levels of organochlorine and organophosphorous pesticides. Toxicology. 2013; 307: 89–94.

[37] Modrzewska B, Wyszkowski M. Trace metals content in soils along the state road 51 (north-eastern Poland). Environ Monit Assess. 2014; 186: 2589–2597.

[38] Osakwe SA. Chemical partitioning of iron, cadmium, nickel and chromium in contaminated soils of south-eastern Nigeria. Chem Speciation Bioavailability. 2013; 25: 71–78.

[39] Environmental Protection Act UK Public General Acts 1990

[40] Doick K, Hutchings T. Greenspace Establishment on Brownfield Land: the site selection and investigation process · Forestry Commission Information Note 91. Edinburgh: 2007.

[41] ENVIRONMENTAL MONITORING – Vol. II - Soil Contamination Monitoring - Aelion C.M.

[42] Bhattacharya A, Routh J, Jacks G. Environmental assessment of abandoned mine tailings in Adak, Vaster botten District (Northern Sweden). Applied Geochemistry. 2006; 21: 1760–1780.

[43] Gomes HI, Dias-Ferreira C, Ribeiro AB. Overview of in situ and ex situ remediation technologies for PCB-contaminated soils and sediments and obstacles for full-scale application. Science of the Total Environment. 2013; 445:237–60.

[44] Dermont G, Bergeron M, Mercier G, Richer-Laflèche M. Metal-contaminated soils: remediation practices and treatment technologies. Practice periodical of hazardous, toxic, and radioactive waste management. 2008;12(3):188–209.

[45] Yao Z, Li J, Xie H, Yu C. Review on remediation technologies of soil contaminated by heavy metals. Procedia Environmental Sciences. 2012;16:722–9.

[46] Karn B, Kuiken T, Otto M. Nanotechnology and in situ remediation: a review of the benefits and potential risks. Environmental health perspectives. 2009;117(12):1813.

[47] Gomes HI, Dias-Ferreira C, Ribeiro AB. Overview of in situ and ex situ remediation technologies for PCB-contaminated soils and sediments and obstacles for full-scale application. Science of the Total Environment. 2013; 445:237–60.

[48] Smith LA. Remedial Options for Metals-contaminated Sites. L.A. Smith et al. (eds.) CRC Lewis Publishers, Boca Raton, FL: 1995.

[49] Martin TA, Ruby MV. Review of in situ remediation technologies for lead, zinc and cadmium in soil. Remediation Journal. 2004; 14: 35–53.

[50] Sherwood LJ, Qualls RG. Stability of phosphorus within a wetland soil following ferric chloride treatment to control eutrophication. Environmental Science and Technology. 2001; 35(20): 4126–4131.

[51] Fingerman M, Nagabhushanam R. Bioremediation of aquatic and terrestrial ecosystems. Science Publishers Inc, Enfield; 2016.

[52] Lloyd JR. Bioremediation of metals: The application of microorganisms that make and break minerals. Microbiology Today. 2002; 29:67–69.

[53] Bento FM, Camargo FAO, Okeke BC, et al. Comparative bioremediation of soils contaminated with diesel oil by natural attenuation, biostimulation and bioaugmentation. Bioresource Technology. 2005; 96: 1049–1055.

[54] Lacalle RG, Gómez-Sagasti MT, Artetxe U. *Brassica napus* has a key role in the recovery of the health of soils contaminated with metals and diesel by rhizoremediation. Science of the Total Environment. 2018; 618: 347–356.

[55] Shen T, Pi Y, Bao M. Biodegradation of different petroleum hydrocarbons by free and immobilized microbial consortia. Environmental Science: Processes and Impacts. 2015; 17: 2022–2033.

[56] Fernández-Luqueño F, Valenzuela-Encinas C, Marsch R. Microbial communities to mitigate contamination of PAHs in soil - possibilities and challenges: a review. Environmental Science and Pollution Research. 2011; 18: 12–30.

[57] Bagherzadeh NA, Shojaosadati SA, Hashemi NS. Biodegradation of used engine oil using mixed and isolated cultures. International Journal of Environmental research.2008; 2 (4): 431–440.

[58] Bento FM. Camargo FOA, Okeke BC, Frankenberger WT. Comparative bioremediation of soils contaminated with diesel oil by natural attenuation, biostimulation and bioaugmentation. Bioresource technology. 2005; 96:1049–1055.

[59] Ying T, Yongming L, Mingming S, Zengjun L, Zhengo L, Peter C. Effect of bioaugmentation by *Paracoccus sp* strain HPD-2 on the soil microbial community and removal of polycyclic aromatic hydrocarbon aged contaminated soil. Bioresource technology. 2010; 101: 3437–3443.

[60] Ramadass K, Megharaj M, Venkateswarlu K, et al. Bioavailability of weathered hydrocarbons in engine oil-contaminated soil: Impact of bioaugmentation mediated by *Pseudomonas* spp. on bioremediation. Science of The Total Environment. 2018; 636: 968–974.

[61] Semple KT, Reid BJ, Fermor TR. Impact of composting strategies on the treatment of soils contaminated with organic pollutants: A review. Environmental Pollution. 2001; 112: 269–283.

[62] Trejo-Hernandez MR, Lopez-Munguia AR, Ramirez Q. Residua; compost of Agaricus bisporus as a source of crude laccase for enzymatic oxidation of phenolic compounds. Process Biochemistry. 2001; 36:635–639.

[63] Lau KL, Tsang YY, Chiu SW. Use of spent mushroom compost to bioremediate PAH-contaminated samples. Chemosphere. 2003; 52(9): 1539–1546.

[64] Hicks P. The use of oxygen release compound (ORC®) for enhanced bioremediation. Proc., the 1998 National Conference on Environmental Remediation Science and Technology, Greensboro, North Carolina, U.S.A. 1999;.63–79.

[65] Ouyang Y. Phytoremediation: Modeling plant uptake and contaminant transport in the soil plant-atmosphere continuum. Journal of Hydrology. 2002; 266:66–82.

[66] Raskin I, Ensley BD. Recent developments for in situ treatment of metal contaminated soils. In: Phytoremediation of Toxic Metals: Using Plants to Clean Up the Environment. John Wiley & Sons Inc., New York: 2000.

[67] Trap S, Kohler A, Larsen LC, Zambrano KC, Karlson U. Phytotoxicity of fresh and weathered diesel and gasoline to willow and poplar trees. J. Soil Sediments. 2005; 1: 71–76.

[68] Zhuang P, Ye ZH, Lan CY, Xie ZW, Hsu WS. Chemically assisted phytoextraction of heavy metal contaminated soils using three plant species. Plant Soil. 2005; 276: 153–162.

[69] Paria S, Yuet PK. Solidification stabilization of organic and inorganic contaminants using Portland cement: A literature review. Environmental Reviews. 2006;14(4):217–255.

[70] Sharma HD, Reddy KR. Geoenvironmental Engineering. John Wiley & Sons, Hoboken, N.J., U.S.A; 2004.

[71] Martel R, Gélinas PJ, Lefebvre R, Hébert A, Foy S, Saumure L, Roy A, Roy N. Laboratory and field soil washing experiments with surfactant solutions: NAPL recovery mechanisms. Emerging technogies in hazardous waste management

[72] Kuhlman MI, Greenfield TM. Simplified soil washing processes for a variety of soils. Journal of Hazardous Materials, 2006;.66(1–2):31–45.

[73] Murena F, Gioia F. Solvent extraction of chlorinated compounds from soils and hydrodechlorination of the extract phase. Journal of Hazardous Materials. 2009;162(2–3):661–667.

[74] Lee WJ, Shih SI, Chang CY, Lai YC, Wang LC, Chang-Chien GP. Thermal treatment of polychlorinated dibenzo-p-dioxins and dibenzofurans from contaminated soils. Journal of Hazardous Materials. 2008; 60(1):220–227.

[75] Renoldi F, Lietti L, Saponaro S, Bonomo L, Forzatti P. Thermal desorption of a PAH contaminated soil: a case study. Ecosystems and sustainable development IV, Vols. 1 & 2. Tiezzi E., Brebbia C.A. and Uso J.L., Editors. WIT Press, Southampton, U.K. 2003; 1123–1132.

[76] Tungittiplakorn W, Lion LW, Cohen C, Kim JY. Engineered polymeric nanoparticles for soil remediation. Environmental Science and Technology. 2004; 38(5):1605–1610.

[77] Mohsenzadeh F, Chehregani Rad A. Bioremediation of heavy metal pollution by nano-particles of noaea mucronata. International Journal of Bioscience, Biochemistry and Bioinformatics. 2012; 2:85–89.

[78] Galdames A, Mendoza A, Oeueta M, Garcia ISD, Sanchez M, Virto I, Vilas JL. Development of new remediation technologies for contaminated soils based on the application of ZVIN and bioremediation with compost. Resource-efficient technologies. 2017; 3 (2):166–176.

[79] Liang F, Fan J, Guo Y, Fan M, Wang J, Yang H. Reduction of nitrite by ultrasound-dispersed nanoscale zero-valent iron particles. Ind. Eng. Chem. Res. 2008; 47:8550.

[80] Shih YH, Tai YT. Reaction of decabrominated diphenyl ether by zerovalent iron nanoparticles. Chemosphere. 2010; 78: 1200.

[81] Gong Y, Liu Y, Xiong Z, Kaback D, Zhao D. Immobilization of mercury in field soil and sediment using carboxymethyl cellulose stabilized iron sulfide nanoparticles. Nanotechnology. 2012; 23:Article number 294007.

[82] Ali Tahir A, Ullah H, Sudhagar P, Asri MTM, Devadoss A, Sundaram S. The application of graphene and its derivatives to energy conversion, storage, and environmental and biosensing devices. Chem. Rec.2016; 16: 1591.

[83] Li J, Rozen I, Wang J. Rocket science at the nanoscale. ACS nano, 10, 5619, 2016c.

[84] Scognamiglio V, Antonacci A, Patrolecco L, Lambreva MD, Litescu SC, Ghuge SA, Rea G. Analytical tools monitoring endocrine disrupting chemicals. Trends Anal Chem. 2016; 80: 555.

[85] Elango G, Roopan SM. Efficacy of SnO_2 nanoparticles toward

photocatalytic degradation of methylene blue dye. J. Photochem. Photobiol. B. 2016; 155: 34.

[86] Begum R, Farooqi ZH, Khan SR. Poly (N-isopropylacrylamide-Acrylic acid) copolymer microgels for various applications: A Review. Int. J. Polym. Mater. Po. 2016; 65: 841.

[87] Mauter MS, Elimelech M. Environmental applications of carbon-based nanomaterials. Environmental Science and Technology.2008;42 (16): 5843–5859.

[88] Qiang Y, Sharma A, Paszczynski A, and Meyer D. Conjugates of magnetic nanoparticle-enzyme for bioremediation," in Proceedings of the 2007 NSTI Nanotechnology Conference and Trade Show. 2007;4:656–659.

[89] Tungittiplakorn W, Lion LW, Cohen C, Kim JY. Engineered polymeric nanoparticles for soil remediation. Environmental Science and Technology.2004; 38(5):1605–1610.

[90] Mohsenzadeh F, Chehregani Rad A. Bioremediation of heavy metal pollution by nano-particles of noaea mucronata. International Journal of Bioscience, Biochemistry and Bioinformatics.2014; 2:85–89.

[91] Rizzo E, Bardos P, Pizzol L, Critto A, Giubilato E, Marcomini A, Albano C, Darmendrail D, Döberl G, Harclerode M, Harries N, Nathanail P, Pachon C, Rodriguez A, Slenders H, Smith G. Comparison of international approaches to sustainable remediation. J Environ Manage.2016; 184:4–17.

[92] Barnett J. The meaning of environmental security, ecological politics and policy in the new security era. Zed Books, London. 2001.

[93] Reddy KR, Adams JA. Sustainable remediation of contaminated sites. Momentum Press, New York.2015.